Chapter 8
14
9
10

read chpters
4, 18, 16
9, 10, 11

SOIL SCIENCE
AND MANAGEMENT

SOIL SCIENCE AND
MANAGEMENT Second Edition

Edward J. Plaster

Consulting Editor for Agriculture: H. Edward Reiley

Delmar Publishers Inc.®

NOTICE TO THE READER

Cover photo by John Meyers

Delmar Staff
Senior Administrative Editor: Joan Gill
Project Editor: Carol Micheli
Production Supervisor: Larry Main
Art Supervisor: John Lent
Design Supervisor: Susan Mathews

For more information, address Delmar Publishers Inc.
3 Columbia Circle, Box 15-015
Albany, New York 12212-5015

Copyright © 1992 by Delmar Publishers Inc.

Printed in the United States of America
Published simultaneously in Canada
by Nelson Canada,
a division of The Thomson Corporation

10   9   8

Library of Congress Cataloging-in-Publication Data

Plaster, Edward J.
  Soil science and management / Edward J. Plaster.—2nd ed.
    p.   cm.
  Includes index.
  ISBN 0-8273-4050-8 (textbook)
  1. Soil science. 2. Soil management. I. Title.
S591.P513   1992   90-46969
631.4—dc20   CIP

Contents

Preface

The people of the world depend upon soil to produce food crops. Where soil is abundant and fertile, and where society has the means to improve fertility, good crops can be grown. Where fertile soil is scarce, or where misuse occurs, growing adequate crops is difficult.

Soil and water are limited natural resources. Their conservation is as important an issue as the improvement of soil fertility. These issues are discussed in detail in *Soil Science and Management*, Second Edition. The text presents not only the basic principles of soil science, but also covers those topics of soil and water conservation and land use that are so important to the intelligent use of these valuable resources.

The text was designed and written to achieve four major goals:

1. To acquaint the reader with the soil and water resources of the United States.
2. To present soil science theory insofar as it applies to soil use by the grower.
3. To show how soil is used by farmers and horticulturists.
4. To cover the basics of soil and water conservation.

In the years since the publication of the first edition, new issues have arisen that need our attention. The author has also used the text in his own classes, and recognized the need for some changes. This second edition, I believe, answers some of those needs.

A number of topics are covered in a bit more detail in the chapter on physical properties, including soil compaction, soil pans, and the Munsell system for soil color. The section on organic soils has been expanded for use in those states rich in such soils.

This new edition reflects the growing concern for water pollution and conservation. Chapter four, Water Conservation, is more detailed, and now discusses issues of water quality. However, other chapters also broach the subject, such as the topic of xeriscaping covered in chapter sixteen.

Low Input Sustainable Agriculture (LISA), another topic of growing interest, has been added to chapter fifteen. Topics related to LISA methods that were not in the first text, such as allelopathy, are included in their relevant chapters.

Recognizing that the horticulture content of many programs has increased, the chapter on horticulture soil use (chapter sixteen) has been reorganized and expanded, and new knowledge and methods included. It has been the author's experience that too many horticulturists understand the soil too little· and misuse it too much; this chapter may help.

Many growers have been affected by soil-conserving provisions of the 1985 Farm Act, so such topics as the Conservation Reserve and Sodbusting have been included in this second edition.

A new Appendix has been added that may be useful to those programs that cover soil judging. This appendix, on evaluating soils for various uses, can also be used in a regular lab exercise in agriculture, horticulture, or natural resource programs.

As in the first edition, chapters are arranged to enhance learning, with introductory comments and objectives, and chapter summaries. A list of important terms appears at the beginning of each chapter. Users of the first edition may have found many of the review questions cumbersome, so review questions have been simplified and made easier to follow and grade. The new chapter openings are visually pleasing and should better attract the attention of students. Instructors will also note a number of simple new "hands-on" activities have been added to most chapters.

FEATURES OF THE TEXT

- Presents the topics of soil chemistry in a thorough, clearly explained manner. Appendix material covering basic chemistry provides a background for those that have not studied chemistry.
- The text stresses soil use by growers and other soil users.
- The text provides applications of the principles of soil science for both agricultural and horticultural crops.
- A unit on urban soils discusses the special problems and treatments required for these special soils that are inhabited by most of the people of this country.
- Covers the principles and application of the Universal Soil Loss Equation (USLE).
- Soil and water conservation chapters stress the need to preserve these natural resources.

An instructor's guide accompanies the text. The guide contains a short precis of each chapter, followed by the answers to the review questions in the chapters. A set of additional questions with answers is also provided for each chapter.

AUTHOR'S ACKNOWLEDGMENTS

I would like to thank the Minnesota Office of the Soil Conservation Service which supplied many of the photographs in this text. A number of their staff aided in the preparation of this text, including Larry Nelson, David Breitbach, and Ray Genrich.

A number of individuals offered photographs for this text, including Doctors Elwin Stewart, Caroll Vance, J. Burton, and David McDonald. Howard Hobbs, of the Minnesota Geological Survey, offered a number of photographs for this second edition.

A number of photographs were obtained from the Journal of Soil and Water Conservation and from Agriculture Research magazine, issued by the USDA. A number of photos also first appeared in *Crops and Soils* magazine, formerly published by the American Society of Agronomy. Unfortunately, this useful magazine is no longer being published. I must thank those many individuals who offered the photographs to those publications.

chapter one

The Importance of Soil

No earthquake or hurricane; this building collapsed because of improper soil use. It was built on soil that slipped under the load. Before we really use soil properly, we must be fully aware of its importance. This chapter will underline how critical soil is to us, state what we use soil for, and present some basic soil concepts.

(Courtesy *Crops and Soils Magazine*, American Society of Agronomy)

OBJECTIVES

After studying this chapter, you should be able to:

- summarize the role of soil in recycling resources needed for plant growth
- describe the four ways plants use soil
- explain how soil is a three-phase system
- list and explain some agricultural and engineering uses of soil

TERMS TO KNOW

anchorage	photosynthesis	soil matrix
desertification	respiration	soil pores
hardpans	shrink-swell potential	soil solution
hydroponics	soil aeration	waterlogging
load-bearing capacity	soil air	

SOIL SHAPES HUMAN HISTORY

This seems a strong statement, but history provides ample proof of the role soils have played in our past. The Egyptian civilization of 4,000 years ago, for example, was made possible by rich sediments left by the yearly flooding of the Nile River. Historians suppose that the agriculture resulting from the fertile soil supported a large population and allowed enough leisure time for the Egyptians to engage in the activities that make a cvilization.

Recent United States history provides us another example of the importance of soils; this time, an example of soil misuse. The Dust Bowl of the 1930s, caused by drought, soil misuse, and widespread wind erosion, drove farmers out of several states. Photographs of this era create a striking image: the Okies (Oklahoma emigrants), driving down the road with their belongings tied to the back of the pickup truck, searching for a place to make a living.

In the future, soil will become even more crucial. The world population is doubling every 40 years; even now, at least 500,000,000 people have too little to eat. Although there is still much land left that could be farmed, most is marginal land that will be difficult to use. Some of it, like parts of the Amazon River Basin, cannot be prepared for farming without serious environmental side effects. At the same time, deserts continue to creep into croplands, a process called *desertification.* Additional acreage is lost to urban uses or other causes.

Human society is possible only because the earth's crust is dusted with a bit of soil we can grow food on. Let's look at that soil more closely.

SOIL IS A LIFE-SUPPORTING LAYER OF MATERIAL

Although we often take soil for granted, it is a very thin and often fragile layer of life-supporting material. The earth, as shown in figure 1-1, consists of a solid part (core, mantle, and crust) and the atmosphere surrounding it. Most of the crust is covered with sea, but, where the continents are, the crust is thicker. The continental crust, made of rock, is about 50 miles thick, and the atmosphere is about 170 miles deep. The soil forms a very thin interface between the two.

The atmosphere, crust, and soil interact to provide plants and animals with the resources they need. Living things need proper temperature, oxygen, water,

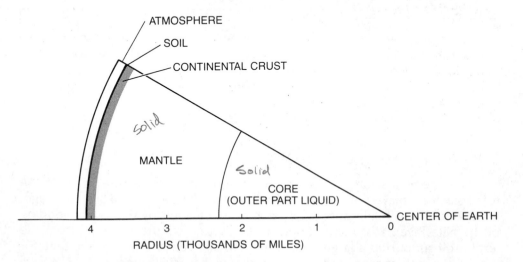

Figure 1-1. Cross section of a "slice" of the earth through a continent. The soil is a very thin layer between the atmosphere and the earth's crust. If the soil were drawn to scale in this figure, it would be 1/2,000,000 of an inch thick.

carbon (the basic element of all living bodies), and other nutrients. These factors are exchanged in the soil, as shown in figure 1-2, usually in cycles that allow elements to be recycled rather than lost. Although later chapters will discuss the cycles in greater detail, this chapter will preview them briefly.

Oxygen. Plant roots need oxygen to grow. Figure 1-2 shows that gases pass in and out of the soil to supply that oxygen for roots.

Temperature. Plants grow best in certain temperature ranges. For instance, most plant roots in temperate climates grow only when the soil temperature is above 40 to 50 degrees Fahrenheit. Seed germination also depends on soil temperature. Wheat seed, for example, germinates between 40°F and 50°F, while sorghum needs temperatures above 80°F. Soil temperature, and to some degree the air above the soil, is controlled by a heat-exchange mechanism.

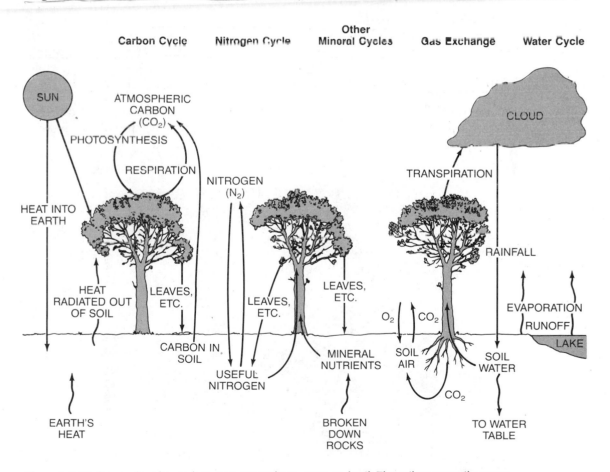

Figure 1-2. Cycling and exchange between atmosphere, crust, and soil. The soil temporarily stores resources needed for plant growth.

   Pedestrians standing on a tar road on a hot summer day sense how heat is exchanged as the road both gains and loses heat. Feet in contact with the pavement get hot, because the road is absorbing energy from the sun. Looking down the road, one sees heat waves rising from the tar, an effect of the road losing excess heat to the air. In the same way, soil maintains temperatures for growing plants.

Water.   Water seldom stays in one place long, always on its way to the next stage of the water cycle. Water evaporates from the land, lakes, and ocean and forms clouds in the atmosphere. Rain falls from the clouds, moistens the soil, and fills streams and lakes. Much of the water finally reaches the ocean. More evaporation begins the cycle again. Some water seeps deep into the ground where it is held as groundwater. When moisture falls on the soil, however, some water is temporarily stored for plant use (figure 1-3).

Carbon.   Plant leaves collect sunlight to use the sun's energy in the process known as *photosynthesis*, which involves converting atmospheric carbon (carbon dioxide) to "biological carbon" (simple sugars). Some of the carbon is recycled directly back to the atmosphere by plant and animal *respiration*, while other carbon is recycled by organic matter decay in the soil (figure 1-4).

Nutrients.   Plant *nutrients* (chemicals a plant needs to grow) also cycle through the soil. Two kinds of nutrient cycles are shown in figure 1-2: the nitrogen cycle and other mineral cycles.

Figure 1-3. Some moisture is captured in the soil for plant use. Here snow has been caught by corn stubble on a midwestern farm (Courtesy USDA-SCS)

The element most essential to all life is carbon, because carbon is at the heart of all organic chemicals. Plants obtain this element from the atmosphere by *photosynthesis*. In photosynthesis, the green pigment in plant leaves, *chlorophyll*, captures sunlight energy and uses it to fuel a series of reactions that combine the carbon from carbon dioxide and the hydrogen from water to produce simple sugars. Oxygen is released as a byproduct:

✳ **Photosynthesis**

$$6\ CO_2 + 12\ H_2O \xrightarrow[\text{chlorophyll}]{\text{light energy}} C_6H_{12}O_6 + 6\ O_2 + 6\ H_2O$$

$$\text{carbon dioxide + water} \xrightarrow[\text{in presence of chlorophyll}]{\text{using light energy}} \text{sugar + oxygen + water}$$

The sugars are converted to other compounds the plant needs. Some of these compounds are meant to store energy that can be reused later. The reaction that regains that energy is *respiration*. Respiration is the opposite reaction to photosynthesis: oxygen combines with sugars to produce carbon dioxide and water and energy is given off:

✳ **Respiration**

$$C_6H_{12}O_6 + 6O_2 \longrightarrow 6CO_2 + 6H_2O + \text{energy (691 Cal/mol)}$$
$$\text{sugar + oxygen} \longrightarrow \text{carbon dioxide + water + energy}$$

The energy given off by respiration can then be used to fuel other plant processes, for instance, the absorption of nutrients.

See appendix one if you need help understanding the chemistry in this figure.

Figure 1-4. Photosynthesis and respiration

Nitrogen comes entirely from the atmosphere, where it occurs as a gas that plants cannot use. Soil organisms change the gaseous nitrogen to forms that plants can use. Some nitrogen recycles as living creatures die and return nitrogen to the soil; some is carried deep into the ground by water; and some nitrogen returns to the air when other microbes change it back to its original form.

Other nutrients come from rocks in the earth's crust. These chemicals are released when rock is broken down by weather, plants, and other factors. These are continuously reused by plants, until some return deep into the ground by leaching or get washed into the ocean.

Soil Storage.   In the interchanges between the crust, soil, and atmosphere, soil temporarily stores resources for plant use. Water, for instance, which is always on its way to the atmosphere, the sea, or the water table, is held in the soil for plant use for a short time.

SOIL IS A MEDIUM FOR PLANT GROWTH

In the broad view, soil has an important function in recycling resources needed for plant growth. In the narrow view, an individual plant depends on soil to supply four needs: anchorage, water, nutrients, and oxygen for roots. Let's look at those four needs next.

Anchorage.   In a deep soil, where roots grow freely, plants are firmly supported, or anchored, so they can grow to reach for sunlight. When people grow plants in ways that deprive plants of soil support, artificial support is often required. Growers producing crops hydroponically (roots growing not in soil but in water with fertilizer dissolved in it) often support plants with a wire framework. Landscapers often stake or "guy" a newly planted tree until the tree is firmly rooted (figure 1-5).

Water.   Because roots are a plant's best water-absorbing body, soil supplies nearly all the water a plant uses. For each pound of dry matter produced by growth, different plants obtain between 200 and 1,000 pounds of water from the soil for photosynthesis, sap flow, and other uses. It is obvious that the water-holding capacity of a soil is important in its agricultural use.

Oxygen.   Except for a few microscopic organisms, all living creatures, including plants, need oxygen. Plants release oxygen during photosynthesis, but they consume it during respiration. The parts of a plant above ground, suspended in an atmosphere that is 21% oxygen (figure 1-6), have all the oxygen they need. Underground, plant roots and organisms that live in the soil use up the oxygen

Figure 1-5. Tree-guying. Because it was just planted, this tree is not firmly anchored by the soil. It has been guyed to keep it standing upright.

and give off carbon dioxide. As a result, *soil air* has less oxygen and more carbon dioxide than the atmosphere (figure 1-6).

In the absence of factors that limit it, the process known as *soil aeration* exchanges soil and atmospheric air to maintain adequate oxygen for plant roots. Aeration varies according to soil condition. A *waterlogged soil*, one that is completely soaked with water, is an example of a soil with poor aeration.

Nutrients. Of 16 nutrients usually considered to be needed by most plants, plants obtain 13 from the soil. Carbon, oxygen, and hydrogen come from air and water; the rest are stored in the soil. While plant leaves are able to absorb some nutrients, roots are specialized for the purpose. Root hairs absorb plant nutrients dissolved in soil water (called the *soil solution*) by an active process that moves nutrients into root cells. The energy that powers this process is produced by respiration in the roots.

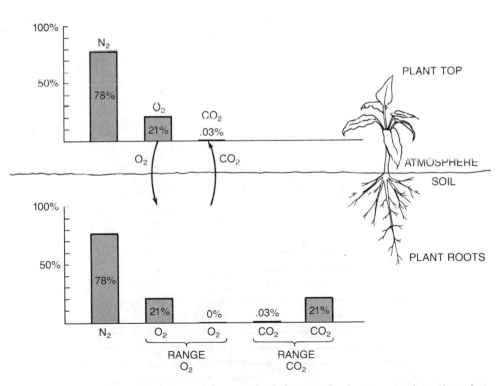

Figure 1-6. Soil air and aeration. Most of the gas in both the air and soil is nitrogen. Above the soil, air is about 21% oxygen. In the soil, respiration of living things tends to replace oxygen with carbon dioxide. Aeration is the process by which carbon dioxide and oxygen are exchanged. The amount of aeration depends on soil conditions, explaining why there is a range in the amounts of oxygen and carbon dioxide in the soil.

SOIL: A THREE-PHASE SYSTEM

How does soil fulfill the four functions described above? Any grower knows that soil is made of solid particles. Between these solid particles are *pore spaces* that contain gases and/or water. This arrangement of solid particles and pore spaces is called the *soil matrix*. The matrix is a three-phase system of solid, liquid, and gas. Most soil experts consider the ideal soil (figure 1-7) to be about 50% solid material (45% mineral particles and 5% organic matter), 25% water, and 25% gas.

The *actual* proportion of air and water varies according to conditions. When it rains, the pores fill with water, only to again empty as the soil dries. This allows air to reoccupy an increasingly greater percentage of pore space. Eventually, unless more rain falls, air will replace most of the water in the soil.

Root Growth.   As roots grow into the soil, they follow pore spaces between the solid particles (figure 1-8). The roots absorb water and nutrients from the soil solution. Water reaches roots in two ways: either water flows toward the root, or the root grows into moist soil. Water does not move far in the soil, so roots have to spread to contact as much soil as possible. Tree roots, for instance, typically cover

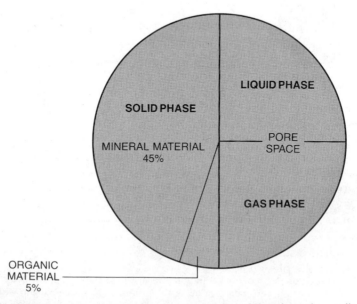

Figure 1-7. The soil system. The solid phase is mostly mineral particles with a small amount of organic matter. Between the solid particles are pores that contain air and/or water in varying proportions. Soil conditions are ideal for plant growth when pore space is half water, half air.

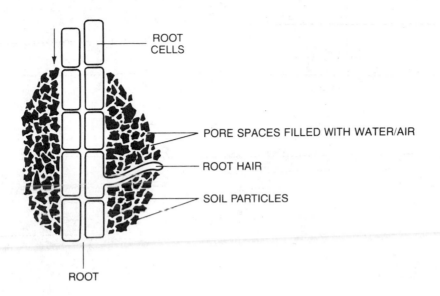

ROOT
CELLS

PORE SPACES FILLED WITH WATER/AIR

ROOT HAIR

SOIL PARTICLES

ROOT

Figure 1-8. Roots and root hairs grow between soil particles. By contacting as much soil as possible, roots absorb water, air, and nutrients. Conditions that obstruct root growth or that interfere with the right amounts of oxygen and water in the pores threaten the health of the plant.

an area 60% to 100% greater than the spread of the tree canopy (figure 1-9). One authority estimated that a mature oak tree has about one million live root tips. Another example is alfalfa roots, which grow to a depth of five or six feet and that may go much deeper in loose soils. However, for most plants in most soils, roots seldom grow below a five foot depth.

While roots try to grow wide or deep, they grow best and most thickly where air, water, or nutrients are present in optimal amounts. Oxygen levels especially determine where roots grow. Most roots—even those of large trees—occupy the upper 12 inches of soil, where the greatest amount of oxygen is found. An example of the importance of oxygen levels in setting root distribution is readily seen by removing the plastic mulch from an old shrub bed in in a landscape planting (figure 1-10). Landscapers commonly mulch shrub plantings with plastic and rock to keep weeds from growing. Since plastic is a barrier to air exchange, shrub roots grow almost entirely on the soil surface where some air is present.

Different soil types can affect how roots grow. For instance, some soils have hard layers under the surface, called *hardpans*, that do not allow roots to grow deeply. As a result, the plant has less soil to draw on for water and nutrients. The same effect results from bedrock or waterlogged soil near the surface.

To summarize, only where a soil has the proper proportion of solid, liquid, and gas can roots grow actively to obtain good anchorage and sufficient water and nutrients.

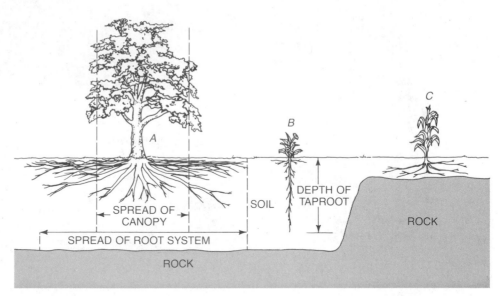

Figure 1-9. Plant roots spread far from the plant seeking water and nutrients. (*A*) Most roots spread laterally near the soil surface where the oxygen content is highest. (*B*) A few plants, like the taprooted alfalfa, send roots deep into soil. (*C*) Conditions that limit the spread of roots, like solid rock near the surface, also limit plant growth.

Figure 1-10. Shrub roots that grew under plastic mulch, now removed from the ground. The plastic acted as a barrier to oxygen, so the roots grew just below the plastic where most of the air was concentrated.

AGRICULTURAL USES OF SOIL

Human societies depend on soil to grow food, fiber, timber, and ornamental plants. In this text, we will call a use agricultural if the main point is to grow plants. Different agricultural uses require different soil management practices. Some categories of agricultural crops and their management needs are described in the following paragraphs.

Cropland. *Cropland* is land on which soil is worked and crops are planted, cared for, and harvested. Worldwide, the greatest acreage of cropland is devoted to annual crops—those planted and harvested within one growing season. Annual crops include agronomic products such as corn (figure 1-11) and soybeans, fiber plants such as cotton, and horticultural crops like most vegetables. Annuals require yearly soil preparation. This activity gives growers a chance each year to control weeds and to work fertilizer and organic matter into the soil. Because the soil surface is bare much of the time, growers must be careful to keep soil from washing away.

Perennial forages, such as alfalfa, are in the ground for a few years. They may be harvested for hay to feed animals, or they may be used for grazing. These crops cover the soil completely and so keep the soil from washing away. Because the soil is not worked each year, fertilization is different than for annual crops. Perennial crops also tend to build up and improve the soil.

Perennial horticultural crops include fruits, nuts, and nursery stock (figure 1-12). Crops stay in the ground for 3 to as many as 20 years. Many of the crops are clean-cultivated, or kept bare by tillage, to keep the ground bare and weed-free. The challenges to the grower of horticultural crops are to control weeds, reduce erosion, prevent soil compaction, and keep the level of organic matter stable.

Figure 1-11. Agronomic crops occupy most of the world's cropland. (USDA, Soil Conservation Service)

Figure 1-12. Young spruce trees. An important use of soil is to grow trees for nurseries, landscapes, farm shelterbelts and woodlots, and forests. (USDA, Soil Conservation Service)

Grazing Land.   Much land in the United States is used to graze animals like cattle and sheep. In the eastern half of the country, pasture is planted to perennial forage, so the above comments on forage apply here as well. In the western half of the country, which has a drier climate, most grazing is on rangeland (figure 1-13). Range consists largely of native grasses and shrubs, with some non-native grasses planted through the existing vegetation. Partly because of the size of much rangeland, it is usually loosely managed.

Forest.   Foresters probably disturb the soil the least, but soil management is still a concern. When trees are finally harvested after many years growth, logging equipment tears up the vegetative cover and compacts the soil. Increased erosion results, and the soil is a less desirable medium for growth of newly planted seedlings. Other concerns of forestry include choosing the best trees for each soil type and ensuring good conditions for the newly planted seedlings.

Other Uses.   Some crops, such as flowers, house plants, and some nursery stock, are grown in pots. Plants growing in the tiny root zone of a pot require great care. Growing media for containers are highly modified to improve their properties, or may be entirely soilless mixtures of peat, perlite, and other materials.
    Landscapers plant and maintain ornamental plants to beautify our surroundings. Their activities also conserve soil in urban areas. Soil knowledge is

Figure 1-13. Rangeland in Montana (Courtesy USDA-ARS)

needed to decide what plants to put where and how to manage the plantings. All too often, landscapers must plant in soils heavily disturbed by construction or in rocky fill with a thin layer of good soil on top. Landscapes stay in place for decades, during which time the soil can be subjected to injury from deicing salt and foot and equipment traffic.

NONAGRICULTURAL USES OF SOIL

Other human activities—in addition to growing plants—require soils. At its most basic, soil is a surface that people inhabit. Specific nonfarming soil uses include recreation and engineering projects such as building foundations and disposing of waste. Let's look at a few of these uses.

Recreation. Recreational uses of the soil surface have become important in America. Sit in an urban park and you will see children in the playground, softball teams on the field, and runners on jogging paths. Golf courses (figure 1-14), parks, and campgrounds are examples of large areas used for recreation. The design of recreational facilities is a specialized skill that requires knowledge of soil properties.

Sports playing fields are probably the most demanding of all soil uses. To grow good turf against the punishment of football cleats, soccer shoes, and field hockey sticks challenges even the best of managers.

Rather than native soil, the soils in the best playing fields are highly engineered mixes of loam, specific sizes of sand, and other ingredients. They may even include a plastic mesh, called a *geotextile*, to hold the soil together. The fields generally have several soil layers, are carefully graded and drained, and well maintained.

Soil managers dealing with playing fields must worry about sideways pressure from shoes tearing the soil surface, an action called *shear*. Playing fields are designed to have good *shear resistance*. The fields must be of a certain hardness, to provide a proper playing surface and reduce injuries. They must dry quickly after a rain, yet hold enough water to grow good turf. These and other considerations are plenty to occupy the field manager. He or she must know his/her soils!

Foundations. Before building a home, the builder or owner usually has the soil tested to a depth of several feet. People know that the structural soundness of a

Figure 1-14. The design, building, and maintenance of golf courses and parks, like farming, require some knowledge of soil.

building depends not only on the builder's skill but also on the soil under the house. Building foundations, for instance, will crack if the soil settles under the building. Even stricter requirements apply to soils for larger structures such as office buildings (figure 1-15). Civil engineers also need firm soils that settle little for the roadbeds of highways and foundations of bridges.

Examples of important engineering properties include the *shrink-swell potential* and the *load-bearing capacity.* Many soils swell when wet and shrink as they dry, cracking walls, destroying foundations, and breaking buried pipes. Soils high in clay or organic matter have low load-bearing capacity. Foundations of buildings constructed on such soils may shift and crack. Roads and other structures built on such soils may also have structural problems. In 1989 San Francisco shook to a major earthquake that brought down many buildings—most of them in an area of weak soils. These buildings were located on a loose "fill" soil that could not support structures when the earth began to shake.

Waste Disposal. Newspaper headlines about hazardous waste disposal focus attention on the difficulties of safely handling the wastes generated by modern society. Soil has been long used for waste disposal, sometimes with unfortunate results.

Figure 1-15. Foundation of a building under construction. The soil must have a good load-bearing capacity and low shrink-swell potential for the foundation to be sound.

Treatment of human sanitary waste often relies on soil, because it filters out some of the material, while microorganisms break down organic portions into less dangerous compounds. The common home drain field is an example.

One way for sewage treatment plants to handle their end products is to spread them on soil. Near many cities and towns, sludge remaining after sewage treatment is spread on farmland for disposal. Sewage sludge can be useful to farmers as a source of nutrients and/or organic matter, as long as possibly harmful materials in the sludge are taken into account. To avoid problems from sludge, its use is regulated by government agencies.

Sanitary or especially hazardous chemical or radioactive waste landfills require soils that will not allow hazardous materials to leach into the water table or run off into neighboring streams or lakes. The search for landfill sites often arouses conflict in a community. Many people feel landfills cannot be entirely safe, and even those that agree landfills are necessary do not want them nearby.

Building Materials.   Before long-distance shipping of building materials became practical, most often people built their homes of locally available materials, including soil. Early settlers in the Great Plains built huts out of sod, which is like a thick carpet made up of grass, its roots, and soil. Adobe, a sun-baked mixture of three parts sandy soil to one part clay soil, has been used as a building material for thousands of years, and continues to be used in the American Southwest. The walls of Jericho were adobe, and so is former President Ronald Reagan's California ranch home.

Modern applications of soils are being developed in the search for energy-efficient housing. Buildings can be built underground, into hillsides, or even with soil piled over them. These earth-sheltered buildings are warm in winter and cool in summer, lowering both heating and cooling costs. A few homes have been built of packed earthen walls, constructed by tamping earth into erected forms.

LAND USE IN THE UNITED STATES

Most of the land in the United States is used in one or more of the farming or nonfarming ways mentioned in the last two sections. Figure 1-16 shows how nonfederal land was used in 1982. Land use does not remain static, however. For example, in any given year some forest is cleared for cropland, while somewhere else cropland is changing to forest. In the last decade, there has been a major conversion of grazing land in Nebraska and neighboring states to cropland. This particular change was stimulated by technological change—the development of center-pivot irrigation. Changes are also brought about by market forces, such as current prices for land, grains, or beef, or other agricultural and forest products.

One land use continues to grow at the expense of other uses—urbanization. This includes the building of cities, towns, factories, and roads. In fact, this trend is speeding up. According to the USDA, in the period from 1958 to 1967, about one million rural acres were converted to urban use yearly. In the period from 1967 to

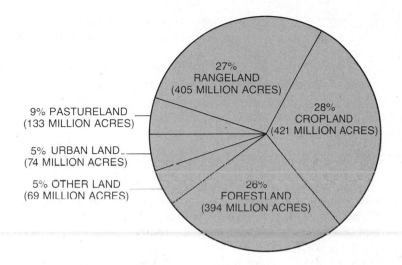

Figure 1-16. Use of nonfederal land in the United States in 1902. Urban land includes builtup and rural transportation (USDA, Preliminary Report, from 1982 National Resource Inventory, 1984)

1977, 3.3 million acres changed from rural to urban land each year. More than a third of the loss of rural land was prime farmland.

Figure 1-16 displays data gathered in 1982 during a USDA inventory of land in the United States called the 1982 National Resource Inventory (NRI). The inventory was repeated in 1987. Early data for the 1987 NRI shows that in the years 1982-1987, prime farmland continued to disappear, by 1.4 million acres. The loss was to urban development and loss of irrigation on some land. The data also shows that grazing land decreased the most (by 8.3 million acres) while urban land increased the most (by 4 million acres).

SUMMARY

The significance of soil is best explained by describing its function in three ways. First, soil is an interface between the earth's crust and the atmosphere. Soil is involved in interactions between the crust and atmosphere, including the recycling of resources such as energy, water, gases, and nutrient elements.

Second, soil supplies anchorage, water, and nutrients to the plant and oxygen to the plant roots. Soil is able to answer these plant needs because it is a three-phase matrix consisting of solid particles with water and air in the pore spaces between the particles.

Third, people inhabit the soil surface and have both agricultural and nonagricultural uses for soil. Agricultural uses include the production of food, fiber, timber, and ornamental plants. Engineering or nonagricultural uses include recreation, the building of foundations and roadbeds, and waste disposal. Soil also provides a source of building material. Properties that make soils suitable for engineering uses, such as a low shrink-swell potential, often differ from the properties needed for agriculture.

This chapter has stressed how important soil is and has explained why it is important. It also noted that the world's soil base is shrinking because of expanding urban areas. Other threats to the soil, noted later in the text, are erosion, shrinking supplies of water, and salt buildup. These problems emphasize the need to use soil correctly.

REVIEW

1. Atmospheric carbon is changed to biological carbon by respiration (True/False). *False - Photosynthesis*
2. Nitrogen comes from the atmosphere. *True*
3. Carbon in the soil is returned to the atmosphere by decay. *False*
4. All plant nutrients come from the soil. *False*
5. Most plant roots grow deeply into the soil. *False*
6. Annual crops require some yearly soil preparation. *True*
7. Tree roots grow to the edge of the canopy. *False*
8. The soil matrix consists of soil particles with pore spaces between them. *True*
9. Eighteen nutrients are required by plants. *False - 16*
10. Sewage sludge can be either harmful or helpful, depending on how it is used. *True*
11. The ideal soil consist of 25% *gas*, 25% *water*, and 50% *solid particles*
12. Roots grow most thickly where *oxygen* and *water* are in good supply.
13. *Aeration* is the exchange of soil and atmospheric air.
14. The mixture of soil water and nutrients dissolved in it is call the *soil solution*.
15. A layer in the soil that hinders roots from growing deeply is a *hardpan*.
16. Of the two forms of grazing land, the one found mainly in the drier climates is *rangeland*.
17. A building material made of sun-baked clay is *adobe*.
18. The process in a plant that makes sugars is called *photosynthesis*
19. *desertification* is the process by which land turns into desert.
20. A crop that stays in the ground for more than one year is *perennial*.
21. Name two engineering properties of soil. *shrink + swell, load bearing capacity*
22. Where are most roots found? *top 12 inches of soil*
23. What are the phases of soil? *solid, liquid + gas*
24. What is the source of energy for plants? *Sun*
25. Why stake a newly planted tree? *anchorage*
26. Explain why recycling is an important function of the soil.

27. Explain why soil usually has less oxygen and more carbon dioxide than the atmosphere.
28. If grain exports were to increase suddenly, how would the land use chart of figure 1-16 change?
29. Compare atmospheric and soil air. *atmospheric = more oxygen, soil air = more carbon dioxide*
30. Draw the water cycle.

SUGGESTED ACTIVITIES

List a number of additional examples of each of the following:
   (a) Situations in which plants are not adequately anchored
   (b) Situations in which humans create barriers to soil aeration
   (c) Recreational uses of the soil surface
   (d) Past or present examples of soil affecting human societies
   (e) Specific crops of each of the types mentioned under agricultural uses of the soil

chapter two
Soil Origin and Development

These children, playing in shallow waters covering the floodplain of the Mississippi River when it overflowed one recent spring, don't know that soil is being formed beneath their feet. They do know the mud left behind squishes nicely between their toes. Soil formation has been going on since the earth's beginning and continues today. This chapter will help you understand the processes that created the soils we use.

OBJECTIVES

After studying this chapter, you should be able to:

- define a soil body
- list examples of the five soil formation factors
- describe how soils form and age
- describe the master horizons of the soil profile

TERMS TO KNOW

alluvium	leaching	polypedon
alluvial fan	levee	residual soils
chemical weathering	loess	river terrace
colluvium	marine sediments	root wedging
eluviation	master horizons	sedimentary rock
eolian deposits	metamorphic rock	soil horizons
frost wedging	mineral soil	soil profile
glacial drift	organic soils	solum
glacial outwash	oxidation	solution
glacial till	parent material	subsoil
hydrolysis	pedon	topsoil
igneous rock	physical weathering	virgin soil
illuviation	plow layer	weathering
lacustrine		

Soil is a very slowly renewable resource. Although our soils originated long ago, the forces that created them continue to operate. Our soils still grow, change, and develop. The challenge is to conserve soils, so they are not depleted faster than they are renewed. This text will discuss later how to conserve soils; this chapter will describe how they form. Before describing how a body of soil forms, let's define what we mean by a soil body.

THE SOIL BODY

The soil is a collection of natural bodies of the earth's surface containing living matter, which is able to support the growth of plants. It ends at the top where the atmosphere begins. It ends at the bottom at the furthest reach of the deepest rooted plants. The soil varies across the landscape: in one area it may be mostly made of decayed plant parts, in another place it may be mostly sand.

It is not possible to learn everything about a soil just by standing on the surface. One must dig a hole to see what it looks like below the surface. Since a soil scientist cannot dig up acres of ground to study a whole body of soil, the soil is broken up into small parts that can be easily studied. This small body is called the pedon (figure 2-1). A *pedon* is a section of soil, extending from the surface to the depth of root penetration of the deepest rooted plants.

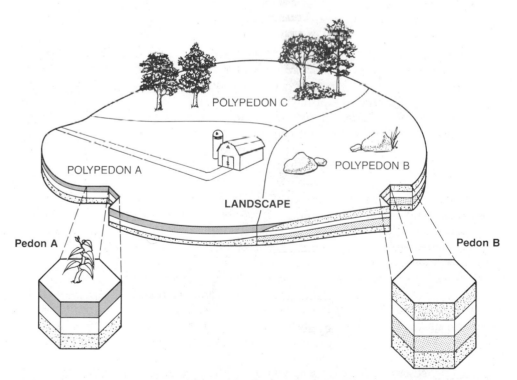

Figure 2-1. Two soil pedons show how each relates to the polypedons and the total landscape.

The area of a pedon varies from about 10 to 100 square feet, depending on how much the soil changes. Soil scientists use the pedon as a unit of soil that can be simply and easily studied by digging a pit in the ground with a shovel.

The traits of a pedon are set by the combination of factors that formed it. In the landscape near the pedon being studied are other pedons that are probably very similar. As one moves across the landscape, however, one will reach a pedon that is different, because the combination of factors that formed it were different. A collection of pedons that are much the same is called a *polypedon*. Later in this text we will learn how these polypedons are mapped into units called a soil series.

How does a soil pedon form? Picture a section of bare rock that will someday become a soil pedon. In the process of soil formation, this rock is changed into a layer of small, broken rock particles with some organic matter mixed in. Weather and plants are the major agents responsible for forming soil from rock, and the process is called *weathering*.

Physical weathering refers to the effects of such climatic factors as temperature, water, and wind on stone. One of the most important is *frost-wedging*. Water expands by nearly 10% as it freezes. Thus, if water runs into cracks in rock, when it freezes, the expansion will wedge the rock apart. Less importantly, simple heating of rock causes it to expand and cooling to shrink. The daily cycle of heating and cooling, by this means, can weaken and fracture rock. Rain, running water, and windblown dust can also wear away at rock surfaces.

Chemical weathering changes the chemical makeup of rock and breaks it down. The simplest process is *solution.* Rainwater is mildly acidic, and can slowly dissolve many soil minerals. In *hydrolysis,* water reacts with minerals to produce new, softer compounds. Some minerals react with oxygen in the atmosphere, and so *oxidation* further acts to decompose rock.

Plants also pay an important role in rock crumbling. Roots can exert up to 150 pounds per square inch pressure when growing into a crack in rock. *Root wedging* from the pressure pries apart stone.

Lichens, which are primitive plants, can grow on bare rock (figure 2-2). Lichens form mild acids that react with minerals to help break down the rock. When the lichen dies, its dry matter is added to the slowly growing mixture of mineral particles and organic matter. When a small bit of soil forms in rock crevices, plants begin to grow from seed that has blown into the crevice, continuing the cycle.

Soil formation does not stop when a layer of young soil covers the surface. The new soil continues to slowly age and develop over thousands of years. Soil scientists state that five factors operate during the process of soil formation and development: parent material, climate, life, topography, and time. One could say that over time, climate and living things act on parent materials with a certain topography to create soil.

This process begins with rock, which supplies the parent materials for most soils. Before studying the five factors, let's look at rocks of the earth's crust.

Figure 2-2. Plants help break down rocks to form soil. Here, patches of lichens grow on bare rock, and plants grow in soil forming in the cracks.

ROCKS AND MINERALS

The original source of most soils is rock—the solid, unweathered rock of the earth's crust. Solid rock breaks into smaller particles, which are the *parent materials* of soil. Rock is a mixture of minerals that, when broken down, supply plant nutrients. Geologists classify rock into three broad types: igneous, sedimentary, and metamorphic. Refer to figure 2-3 for help in understanding the following paragraphs.

Igneous Rock. The basic material of the earth's crust is *igneous rock*, created by the cooling and solidification of molten materials from deep in the earth (refer to figure 1-1. A most dramatic example resulted from the eruption of Mount St. Helens at 8:32 AM May 18th, 1980. Countless tons of material were thrown into the air. Much of the ash settled nearby, where it will be parent material for new soil.

Igneous rocks, such as granite, contain minerals that supply 12 of the 16 required plant nutrients (listed in chapter nine of this text). Granite, which is mined for monuments and building material, consists of about 50% feldspar minerals, 30% quartz, and 20% other minerals. Feldspar, a fairly soft mineral containing potassium and calcium, weathers easily to clay. Quartz, a very hard and resistant mineral, weathers slowly to sand. Figure 2-4 lists the nutrient content of two sample igneous rocks: a granite and a basalt.

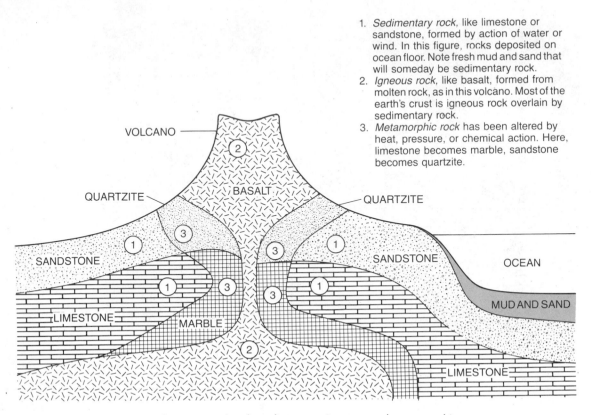

1. *Sedimentary rock*, like limestone or sandstone, formed by action of water or wind. In this figure, rocks deposited on ocean floor. Note fresh mud and sand that will someday be sedimentary rock.
2. *Igneous rock*, like basalt, formed from molten rock, as in this volcano. Most of the earth's crust is igneous rock overlain by sedimentary rock.
3. *Metamorphic rock* has been altered by heat, pressure, or chemical action. Here, limestone becomes marble, sandstone becomes quartzite.

Figure 2-3. Three types of rock: sedimentary, igneous, and metamorphic

Sedimentary Rock. Igneous rock comprises only about one-quarter of the earth's actual surface, even if most of the crust is igneous. This is because sedimentary rock overlays about three-quarters of the igneous crust. *Sedimentary rock* forms when loose materials like mud, sand, or dead ocean creatures are deposited by water, wind, or other agents. Then these loose materials are slowly cemented by chemicals and/or pressure into rock. Much of the sedimentary rock covering North America was deposited in prehistoric seas. Some of it contains not only mineral particles but also bodies and fossils of marine organisms.

The parent materials of many American soils derive from two important sedimentary rocks: sandstone and limestone. Sandstone, which consists mostly of cemented quartz grains, weathers to sandy soils. Generally, these soils are infertile and droughty. Limestone is high in calcium and weathers easily to soils high in pH, calcium, and magnesium. Limestone is also mined as a source of agricultural lime (figure 2-5). Figure 2-4 lists the contents of a typical sandstone and limestone.

MINERALS	Gray Granite	Basalt	Hinckley Sandstone	Platteville Limestone
% quartz – sand	64	49	94	7.5
% feldspars, others – clay	20	20	2	-
% calcite, dolomite	7	15	5	90
Elements				

Elements	Pounds per Ton of Rock			
Calcium	69	150	-	704
Potassium	66	17	-	-
Magnesium	36	66	-	18
Iron	23	35	15*	-
Phosphorus	5	5	-	-
Manganese	-	3	-	-

Figure 2-4. Composition of several igneous and sedimentary rocks. Quartz minerals weather to sand and contain few plant nutrients. Feldspars weather to clay and contain several plant nutrients. Calcite and dolomite are primary sources of calcium and magnesium. Most elements were not analyzed for the sandstone and limestone because only very small amounts appear. The starred number was estimated by the author.

Metamorphic Rock.   If igneous and sedimentary rocks are subjected to great heat and pressure, they change to form *metamorphic rock*. For instance, limestone is a fairly soft, gritty rock. When it is subjected to heat and pressure, it changes. The new metamorphic rock, marble, is harder, not gritty, and can be cut and polished. Soils arising from metamorphic parent materials resemble soils from the original sedimentary or igneous rock.

For a further listing of some soil-forming rocks and minerals, refer to the list in the activities section at the end of this chapter.

PARENT MATERIAL

The description of soil origin at the beginning of this chapter was of a soil formed directly from bedrock. These *residual soils*, as they are called, are actually less common than the soils from parent materials that were carried from elsewhere by wind, water, ice, or gravity. Residual soils (figure 2-6) form very slowly, as solid rock must be weathered first. *Transported soils*, however, grow from rock that has already been weathered and then carried somewhere else. These soils form more quickly. Let's look at the various parent materials. Figure 2-7 shows where soil parent materials are in the United States.

Glacial Ice.   Glacial ice carried parent materials over the northern part of North America (figure 2-8) during four separate periods of glaciation. Evidence suggests that these periods began about one million years ago and that the last glacier melted about 10,000 years ago. The glaciers expanded out of several centers in

Figure 2-5. A limestone quarry. Limestone is mined for gravel, agricultural lime, and other uses. Note the limestone bedrock visible in the upper left corner.

Figure 2-6. A residual soil. Four to eight inches of soil have formed on top of basalt. (USDA, Soil Conservation Service)

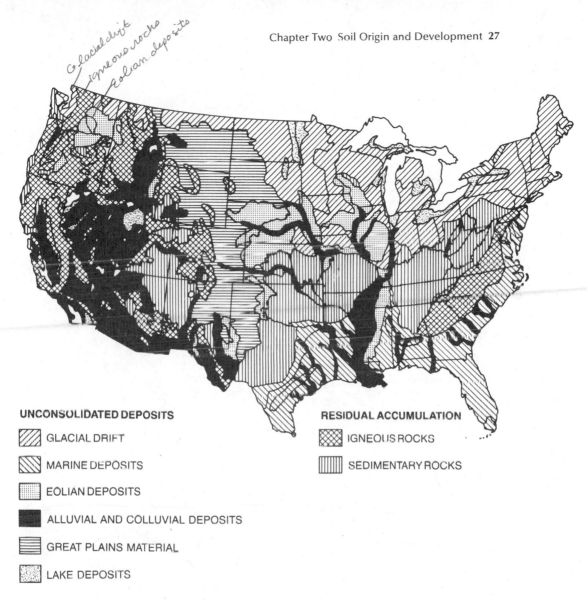

Glacial drift
Igneous rocks
Eolian deposits

UNCONSOLIDATED DEPOSITS

GLACIAL DRIFT

MARINE DEPOSITS

EOLIAN DEPOSITS

ALLUVIAL AND COLLUVIAL DEPOSITS

GREAT PLAINS MATERIAL

LAKE DEPOSITS

RESIDUAL ACCUMULATION

IGNEOUS ROCKS

SEDIMENTARY ROCKS

Figure 2-7. Parent materials of soils of the United States (Source: USDA)

Canada. As they grew, they carved and ground the earth, picking up soil, gravel, rocks, and other debris. As the glaciers melted and shrunk between glacial periods, transported material remained in deposits called *glacial drift*. In the process, they left behind a very distinctive landscape over much of the northern United States.

Glaciers deposited materials in many ways, so there are several kinds of glacial drift. During the melting process, some debris simply dropped in place to form deposits called *glacial till.* Since there was no sorting action in the deposition, glacial till is extremely variable, and so are the soils derived from it. Often till soils contain pebbles, stones, and even boulders (figure 2-9).

Figure 2-8. Glaciers have covered much of North America several times.

Other materials carried by the glacier washed away in meltwater to form sediments in streams and lakes. During the process, the materials were sorted by size. The coarser material, being larger and heavier, was deposited near the glacier and in streams and rivers to form *glacial outwash*. Outwash soils tend to be sandy. The smaller particles reached glacial lakes to form *lacustrine* deposits on

Figure 2-9. Glacial till. Till soils often contain rocks. (Courtesy of Howard Hobbs, Minnesota Geological Survey)

the lake bottoms (see later discussion of lacustrine soils). Glacial till, outwash, and lacustrine deposits dominate the agricultural soils of the northern tier of states.

Wind. Some parent materials were carried by the wind, leaving *eolian* deposits. For example, some soils in Nebraska formed from sand dunes, which are deposits of sand carried by rolling in the wind. Most of the eolian soils in the United States are actually a result of the last glacial period.

After the last glaciers melted and the meltwaters subsided, large expanses of land were exposed to a dry climate with strong westerly winds. The winds picked up silt-size (medium) particles and deposited them in the Mississippi and Missouri River valleys and elsewhere. These *loess soils*—wind-deposited silt—are important agricultural soils in much of Iowa, Illinois, and neighboring states.

Water. *Alluvial soils* are soils whose parent materials were carried and deposited in moving fresh water to form sediments (figure 2-10). Alluvial materials can be deposited in several ways. *Alluvial fans* form below hills and mountain ranges where streams flowing down the slope deposit material in a fan shape at the base. As the water speed slows abruptly at the foot of

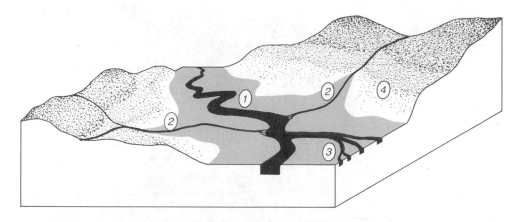

Figure 2-10. Water and marine deposited soils. (*1*) Flood plains form along rivers from materials deposited during flooding. (*2*) Alluvial fans form at the base of slopes when running water slows and large particles drop out. (*3*) Deltas form when smaller particles drop out as the river enters an ocean. (*4*) River terraces are old flood plains left above a new river level.

the slope, large particles drop out first. As a result, alluvial fans are generally sandy or gravelly. Finer materials are carried away in rivers.

Flooding rivers also leave deposits behind. Often coarser materials are deposited in low ridges, or *levees,* along the river bank. Away from the river, floodwaters spread over large flat areas called *floodplains.* Here the water will be shallow and slow moving; fine particles will settle out (figure 2-11). Flood plains tend to be fertile because new soil is added at each flood, but the soils tend to stay wet. Levees, being coarser and elevated, dry more quickly. Floodplain soils are especially important along the Mississippi and its tributaries and along rivers that flow into the ocean on the East and Gulf coasts. Many important soils of California are from river alluvium.

Sometimes a river will cut deeply into its flood plain and will flow at a lower elevation. This establishes a new riverbed and flood plain. The old flood plain is left higher as a *river terrace.* An example of river terrace soils is some soil of the San Joaquin Valley of California.

Lacustrine deposits form under still, fresh water. Most of our lacustrine soils remain from giant glacial lakes that have since dried up. Examples include Glacial Lake Agassiz of northern Minnesota and North Dakota, and Glacial Lake Bonneville of Utah. When glacial runoff water ran into the lake, the heaviest materials were left near the shore, while the smallest particles were carried to the center of the lake. Thus lacustrine soils are sandy near the old shoreline and grade to soils with smaller particles toward the old lake center.

Marine sediments form in the ocean. Many scattered soils of the Great Plains and the Imperial Valley of California are beaches of prehistoric seas that once covered the United States. Other beach soils are common along the Atlantic coastline and the Gulf of Mexico. These all tend to be sandy soils. *Deltas,* in

Figure 2-11. Mississippi mud. This alluvial deposit remained after floodwaters receded.

contrast, have very small particles and tend to be wet. Deltas form when rivers flowing into an ocean deposit sediments at the mouth of the river. The Mississippi River Delta of Louisiana is a prime example.

Gravity. Some parent materials move simply by sliding or rolling down a slope. This material, called *colluvium*, is scattered in hilly or mountainous areas. An example of a colluvial material is a *talus*—sand and rocks that collect at the foot of a slope. Avalanches, mudslides, and landslides are other examples.

Organic Deposits. Characteristics of the soils formed from parent materials described so far in this chapter are set by the mineral particles in the soil. These *mineral soils* contain less than 20% organic matter, except for a surface layer of plant debris. *Organic soils*, containing 20% or more organic matter, form under water as aquatic plants die and drop to the lake floor. Low oxygen conditions under water retard decay of these dead plants, so they tend to pile up on the lake bottom. Eventually the lake fills in and is replaced by an organic soil. Organic soils are extensive in Minnesota, Wisconsin, and Florida. They are often used to grow vegetables and sod.

CLIMATE

Climate first affects soil by causing physical and chemical weathering of rock, as described earlier in this chapter. However, climate continues to affect soil development long beyond this initial stage. The main effects are due to temperature and rainfall.

Temperature affects the speed of chemical reactions in the soil—the higher the temperature, the faster a reaction. Chemical weathering in soils occurs mostly when the soil is warmer than 60 degrees. Thus, in cold areas, like the tundra, soils develop slowly. In warm areas, like the tropics, soils develop more rapidly.

Another result of temperature is its effect on organic matter. Warmth promotes plant growth and greater vegetation, so more organic matter is added to the soil. However, warm temperatures also speed up the decay and loss of organic matter. Thus, soils of warm climates tend to be low in organic matter.

Rainfall affects the development of soil mainly by leaching. Leaching means that chemicals and other materials are carried deeper into the soil by water moving downward through the soil. Leached materials include lime, tiny mineral particles (clay), plant nutrients, and other chemicals. These materials are then deposited in lower parts of the soil.

High rainfall areas also tend to grow more vegetation, so the soils of humid areas tend to have more organic matter than soils of drier regions. To summarize, rainfall tends to cause leaching and the accumulation of organic matter.

The United States is a good example of the effects of climate on soil (figure 2-12). The climate of the United States cools from south to north. This is reflected in an increase in average organic matter content from south to north. Also, the most weathered soils in the United States are in the South. The average rainfall of the United States increases from west to east. As a result, the organic matter content of the United States soils also tends to increase from west to east.

ORGANISMS

Organisms that live in the soil—like plants, insects, and microbes—actively affect soil formation. Plants contribute to soil formation both by helping to break down rock and by adding organic matter to the soil. The actual properties of a developing soil are influenced by the type of plants growing on it. Figure 2-13 shows the parent vegetation of soils of the United States.

Mineral soils having the highest organic matter content form under grasslands. Grasses usually have a dense mat of fibrous roots, some of which die each year. This keeps the organic matter content high. Since organic matter is dark in color, grassland soils also tend to be dark. In a northern forest, much of the organic material is above ground in the trees. When the leaves fall or the tree dies, the material falls to the soil where it creates a surface layer of organic matter that does not mix with deeper layers. As a result, forest soils have less organic matter than prairie soils; forest soils are also lighter in color. The type of trees also influences the soil. For instance, softwoods (conifers) tend to be more acid in

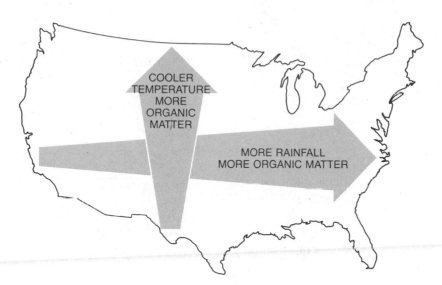

Figure 2-12. One effect of climate on United States soils. The average organic matter of soils increases to the east and north because of cooler temperatures and higher rainfall.

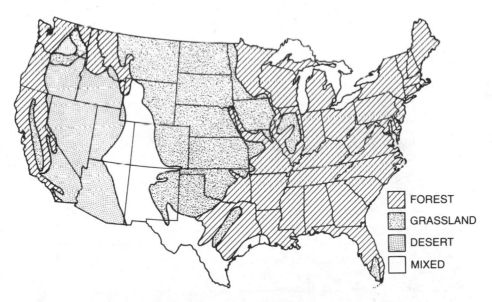

Figure 2-13. Native vegetation of the United States. (Source: USDA)

their effect than hardwoods (broadleaf trees). In many tropical forests, virtually all organic material is in the trees and very little in the soil.

Other organisms also influence soil formation. Soil microbes break down plant material into organic matter, while earthworms digest organic matter and leave behind tunnels. Ants and burrowing mammals mix the soil. Chapter seven covers these effects in more detail.

TOPOGRAPHY

Topography influences soil development mainly by affecting water movement. Water runs off slopes, so less filters into the soil. This action affects soil moisture, which in turn influences leaching, chemical reactions, and the types of vegetation. The direction a slope faces is also important—south-facing slopes are hotter and drier than north-facing slopes.

If enough water runs off a slope, it may carry away soil as fast as it is formed. Thus, soil may be thin on a slope and thick at the base of a slope. The effect of topography is most obvious in rolling fields where sloped areas are light brown from topsoil loss while lower areas are black from accumulating topsoil and organic matter (figure 2-14).

Low spots in the land are often very different from nearby areas. They tend to gather soil that runs off from other areas. They also tend to be wet much of the time. The result is often a deep soil that is wet and high in organic matter.

TIME

Soils change over time, undergoing an aging process. Initially, a thin layer of soil forms on the parent material. Such a young, immature soil takes as little as a

Figure 2-14. One effect of topography on soils. Topsoil has eroded off the knolls and has been deposited on the lower spots. (USDA, Soil Conservation Service)

hundred years to form from well-weathered parent materials under warm, humid conditions. Under other conditions, it may take hundreds of years.

Weathering of the young soil continues, and many generations of plants live and die, so the young soil becomes deeper and higher in organic matter. If there is enough rainfall, leaching begins to carry some material deeper into the soil, creating the soil profile described next in this chapter. Soils at this stage are considered mature and at the height of productivity. It takes from one to several thousand years to produce a mature soil.

The productive, mature soil continues to change, becoming severely weathered, highly leached, and low in organic matter. Old soils are weathered out and infertile. Such a soil may begin to appear after two to ten thousand years.

THE SOIL PROFILE

As soils develop and age, horizontal layers form in the soil (figure 2-15). These layers are known as *soil horizons*. The horizons are visible wherever the earth is dug deep enough to expose them. The *soil profile* is a vertical section through the soil extending well into the unweathered parent material and exposing all the horizons. Each horizon in the profile differs in some physical or chemical way from the other horizons.

Horizons develop as soils age. In a very young soil, weathering and plant growth produce a thin layer of mineral particles and organic matter on top of the parent material. The thin layer of soil is labeled the *A horizon* and is defined as a surface mineral horizon with an organic matter accumulation. The parent material below the A horizon of this young soil is termed the *C horizon*. It is defined as a subsurface mineral layer only slightly affected by soil-forming processes. Thus, this young soil has an AC soil profile.

As the young soil ages, the soil increases in depth. In addition, clay-sized particles and certain chemicals leach out of the A horizon (figure 2-16), moving downward in the profile to create a new layer, the *B horizon*. The A, B, C, and other horizons are described next.

Master Horizons. The A, B, and C horizons are known as *master horizons*. They are part of a system for naming soil horizons in which each layer is identified by a code. In 1981 the Soil Conservation Service (USDA) changed that system. Since the old system will probably remain in common use for a time, we will try to describe both here.

In the old system, there were five possible master horizons: the O, A, B, C, and R. In the new system, the master horizons are O, A, E, B, C, and R. These horizons are shown in figure 2-17. In the brief descriptions of each horizon, the first indicated code is from the new system.

O (O)   The O horizon is an organic layer made of wholly or partially decayed plant and animal debris. The O horizon generally occurs in a *virgin soil* (one unchanged by humankind), since plowing mixes the organic

1''

15''

30''

Figure 2-15. A soil profile (USDA, Soil Conservation Service)

Figure 2-16. This dark staining in the light horizon was leached (illuviated) from the "A" horizon. (Courtesy of Howard Hobbs, Minnesota Geological Survey)

material into the soil. In a forest, fallen leaves, branches, and other debris make up the O horizon.

A (A)   The *A horizon*, generally called the *topsoil* by most growers, is the surface mineral layer where organic matter accumulates. Over time, this layer loses clay, iron, and other materials to leaching. This loss is called *eluviation*. Materials resistant to weathering, such as sand, tend to concentrate in the A horizon as other materials leach out. The A horizon provides the best environment for the growth of plant roots, microorganisms, and other life.

E (A2)   The *E horizon* was a subdivision of the A horizon, called the A2 in the old system. This is the zone of greatest eluviation. The E horizon is very leached of clay, chemicals, and organic matter. Because the chemicals that color soil have been leached out, the E layer is very light in color. It usually occurs in sandy forest soils in high rainfall areas.

B (B)   The *B horizon*, usually called the *subsoil* by growers, is often called the "zone of accumulation" where the chemicals leached out of the A horizon accumulate. The word for this accumulation is *illuviation* (figure 2-16). The B horizon has a lower organic matter content than the topsoil and often has more clay. The A, E, and B horizons together are known as the *solum.* This part of the profile is where most plant roots grow.

C (C)   The *C horizon* lacks the properties of the A and B horizons. It is the soil layer not touched by soil-forming processes and is usually the parent material of the soil.

R (R)   The *R horizon* is the underlying bedrock, such as limestone, sandstone, or granite.

Subdivisions of the Master Horizons. As soils age, a soil may develop more horizons than the basic master horizons. Some of these layers are between the master horizons both in position and properties. In the new system, these layers are identified by the two master letters, with the dominant one written first. Thus an AB layer lies between the A and B horizons and resembles both, but it is more like the A than the B. Figure 2-17 shows these layers and indicates their labels under the old system.

A soil layer can be further identified by a lowercase letter suffix that tells some trait of the layer. Appendix Four lists some of these suffixes but two will serve as examples here—the Ap and Bt (figure 2-18). An Ap layer is a surface layer disturbed by humankind, so that the old layers were mixed up. For instance, plowing would mix up an O, A, and AB horizon if they were all in the top eight inches. The Ap horizon is the same as the *plow layer,* the top seven or eight inches of soil in a plowed field. A Bt horizon is a B horizon in which a lot of certain types of clay have accumulated, usually by illuviation.

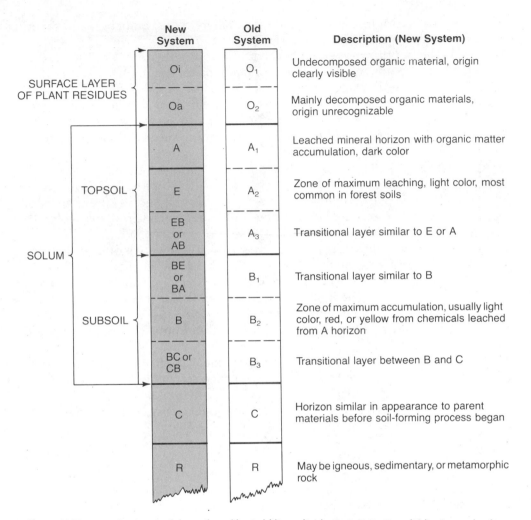

New System	Old System	Description (New System)
Oi	O₁	Undecomposed organic material, origin clearly visible
Oa	O₂	Mainly decomposed organic materials, origin unrecognizable
A	A₁	Leached mineral horizon with organic matter accumulation, dark color
E	A₂	Zone of maximum leaching, light color, most common in forest soils
EB or AB	A₃	Transitional layer similar to E or A
BE or BA	B₁	Transitional layer similar to B
B	B₂	Zone of maximum accumulation, usually light color, red, or yellow from chemicals leached from A horizon
BC or CB	B₃	Transitional layer between B and C
C	C	Horizon similar in appearance to parent materials before soil-forming process began
R	R	May be igneous, sedimentary, or metamorphic rock

SURFACE LAYER OF PLANT RESIDUES — SOLUM — TOPSOIL — SUBSOIL

Figure 2-17. Main horizons of the soil profile. Bold lines divide O, A, E, B, C, and R horizons; broken lines show the subhorizons.

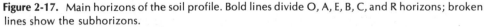

In the new system, further subdivisions are noted by a number following the letters. Thus, one could have a soil with both a Bt1 and a Bt2 horizon. This means that the Bt horizon of the soil has two distinct layers in it.

Now for an example. The soil profile in figure 2-18 is the same one shown in figure 2-15 but the various horizons are labeled. Figure 2-18 has the profile Ap-E-Bt-C. The top seven inches are an old plow layer or Ap. A strong, light-colored E (A2 in old system) horizon extends from the 7- to 14-inch depth, showing a leaching of clays and chemicals. Those clays then settled in the Bt horizon lying between 14 and 22 inches deep. Below that is the C horizon of sand and gravel. Notice the rodent hole in the E horizon.

HORIZONS

Ap 7″

E 7″– 14″

Bt 14″– 22″

C 22″+

Figure 2-18. Sample soil profile with horizons identified

SUMMARY

Soils form from minerals broken up by the action of weathering and plant roots and from the addition of decaying plant parts. Young soils continue to age—growing deeper, being leached by rainfall, developing layers, and changing over time. Five factors govern the soil development process: parent material, climate, life, topography, and time.

Residual soils develop directly from bedrock (igneous, sedimentary, or metamorphic). Most mineral soils come from parent materials moved from one area to another by ice, water, wind, or gravity. Organic soils are composed of decaying plants on lake bottoms. Each type of parent material is responsible for a different soil.

Parent materials are acted on by climate and living organisms. Soils develop quickly in warm areas with high rainfall, then age into heavily weathered soils low in organic matter. In cooler regions, organic matter accumulates and weathering is less extreme. In arid climates, sparse plant growth inhibits the formation of organic matter. Grassland soils tend to be high in organic matter, forest soils lower, and dryland soils lowest of all.

Topography affects soil formation by changing water movement and soil temperature. Low areas often have deep, rich soils that drain slowly. Erosion causes thin soils on slopes.

Time is a factor because soil development is a continuing process. Young soils tend to be thin with little horizon development. Mature soils are deeper and productive with several recognizable horizons. Old soils are severely weathered, highly leached, and less productive.

Soil profiles, which develop over time, are divided into master horizons. These, in turn, may also contain layers. Each layer is named by a code system that identifies its position in the profile and provides some information about it.

REVIEW

1. Solution is an example of physical weathering (True/False).
2. The surface layer of rock on the earth is usually sedimentary.
3. Glacial soils are found mostly in northern states.
4. Colluvium is moved by glaciers.
5. Forest soils tend to have more organic matter than prairie soils.
6. Soils on slopes tend to be thin.
7. The portion of the soil profile made of unaltered parent materials is labeled R.
8. A BC horizon most clearly resembles the B horizon.
9. Soils age until the master horizon forms, then continues unchanged.
10. Soil forming forces no longer operate.
11. Climate pulverizes rock by a process called _____.
12. _____ soils form directly in place from solid rock.
13. Glacial deposits are called _____ _____, sediments deposited in moving fresh water are _____, and wind deposits are _____.
14. Soil tends to be deep on low spots, thin on _____.
15. The prying apart of rocks by freezing water is called _____ _____.
16. Sandy soils tend to be derived from the weathered mineral _____.

17. Wind deposited silt is called _____ .
18. Other factors being equal, soil organic matter tends to _____ from south to north.
19. The _____ horizon usually is the topsoil.
20. A horizon that would always be absent in a plowed soil is _____ .
21. Name the five soil forming processes.
22. Name the parent material that is not derived from rock.
23. What causes eluviation?
24. Draw a soil profile and label the horizons. Indicate the topsoil, subsoil, parent material, bedrock, and solum.
25. Using appendix 4, what would be the code for a topsoil that has an accumulation of sodium?
26. Describe a horizon labeled Bmy.
27. Explain why the concept of a soil pedon is useful.
28. What do alluvial fans, flood plains, deltas, and terraces have in common? Explain how and why each is different.
29. How does soil change over time?
30. Describe the main parent materials and native vegetation that contributed to the soils of your state.

SUGGESTED ACTIVITIES

1. Study the history of the soils in your state or vicinity.
2. Dig a soil pit and study the soil profile. See if you can name the layers.
3. Obtain samples of common soil-forming rocks and minerals, such as the examples listed on the next page. Find more information about each from a simple field guide to rocks and minerals. What plant nutrients does each contain (see chapter nine for a list)? Using one of the several available laboratory exercises, experiment with the various weathering processes. For instance, try to scratch feldspar with quartz, and vice versa. Which is harder? Relate this to what the two minerals weather to.
4. To observe the effects of freezing on physical weathering, pat a handful of clay soil into a ball. Inject water into the ball with a syringe, then freeze overnight. Observe the results.

Minerals	
Quartz SiO_2	Feldspars* $KAlSi_3O_8$
Mica* $KAl_2(Si_3\,Al)O_{10}(OH)_2$	Apatite $Ca_5(PO_4)_3(OH, F, Cl)$
Gypsum $CaSO_4 \cdot 2H_2O$	Calcite $CaCO_3$
Dolomite $CaMg(CO_3)$	Olivines* $Mg_2\,SiO_4$
Limonite $Fe(OH)_3$	Hematite $Fe_2\,O_3$

*Formula is for one of several types

Rocks			
Sedimentary	*Igneous*	*Metamorphic*	*Main Components*
Sandstone		Quartzite	Quartz sand
Limestone		Marble	Calcite
Shale		Slate	Feldspar clays
	Granite	Gneiss	Quartz, mica, feldspar
	Basalt	Schist	Feldspar, mica, olivine

chapter three
Physical Properties of Soil

A germinating seedling needs moisture, temperature, and oxygen to grow through the soil and emerge into air and light. How warm and moist the soil is, how much oxygen there is, the presence or absence of a hard crust that will hinder emergence; these all depend partially on soil traits we call physical properties. Knowing about physical properties will form a basis for much of what you learn about and do with soils. Physical properties can be seen or felt, so be prepared to get your fingers dirty!

(Courtesy of *Crops and Soils Magazine*, American Society of Agronomy)

OBJECTIVES

After studying this chapter, you should be able to:

- describe the concept of soil texture and its importance
- identify the texture of a sample of soil
- describe soil permeability and related properties
- describe structure and its formation and importance
- explain other physical properties
- discuss soil compaction

TERMS TO KNOW

aeration pores	friable	particle density
blocky structure	gleying	peds
bulk density	granular structure	percolation
caliche	hydraulic conductivity	permeability
clay	infiltration	physical properties
claypan	macropores	platy structure
clods	massive soil	plinthite
compaction	mechanical analysis	prism-like structure
cone penetrometer	micropores	puddling
crumb structure	mottling	sand
duripan	Munsell system	silt
fragipan	oven-dry soil	single-grain soil

soil aggregates	soil texture	total pore space
soil consistence	soil triangle	
soil separates	subsoiling	

Physical properties are soil characteristics a grower can see or feel. Physical properties greatly affect how soils are used to grow plants or for other activities. Is the soil loose so roots can grow easily through it or water seep in easily? Or is the soil tight, preventing root growth and water absorption? How well does the soil supply air, water, and nutrients? A knowledge of physical properties helps to answer these questions.

SOIL TEXTURE

The most fundamental soil property, one that most influences other soil traits, is texture. *Soil texture* is a term that refers to the size of the mineral particles in the soil. It does so by giving the proportion of three sizes of soil particles—sand (large), silt (medium), and clay (small). The size of soil particles, in turn, affects such soil traits as water-holding capacity and aeration. Let us first describe why particle size affects these properties.

Effect of Particle Size. Soil particle size affects two important soil features: internal surface area and the numbers and sizes of pore spaces. The internal surface area of a soil is the total surface area of all the particles in the soil. Figure 3-1 uses children's alphabet blocks to demonstrate that *the smaller the particles in a soil, the larger the internal surface area.* Since soil contains many small particles, a handful of soil may hold many thousand square feet of internal surface area.

Internal Surface Area. Soil surface area is important because reactions occur on the surface of soil particles. Picture pouring water over a pile of marbles. Most of the water runs quickly away. Droplets clinging to the surface of the marbles are the only water retained in the pile, since water cannot soak into the marbles. Following the rule about particle size, a pile of small beads holds more water than a pile of marbles because it has more surface area for water to cling to. Because soils with the smallest particles, like silt and clay, have the largest surface area, they hold the most moisture.

Reactions that hold plant nutrients in the soil also occur on particle surfaces. Therefore we can make the rule that *the smaller the particles in a soil the more water and nutrients the soil can retain.*

Soil Pores. Pore size and number depend on particle size. Figure 3-2 suggests that more pores are found between small particles than between large ones. However, the figure also shows that pores are larger between larger particles. Thus soils high in clay have many small pores, while soils high in sand have fewer

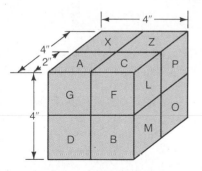

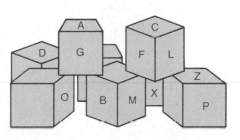

(A) Eight blocks have been put together to make a single large block. Each side measures 4 × 4 inches. The total surface area of this large block is 96 square inches:

Total area = area of each side × number of sides
96 sq. in. = 4 × 4 × 6

(B) The large block has been cut into eight equal blocks. The total surface area of all the blocks is now 192 square inches:

total area = area of each block × number of blocks
192 sq. in. = 2 × 2 × 6 × 8

By halving the size of the blocks, the total surface area is doubled. In the soil, small particles create a large surface area for water and nutrients to hold on to.

Figure 3-1. The smaller the soil particles, the greater the internal surface area.

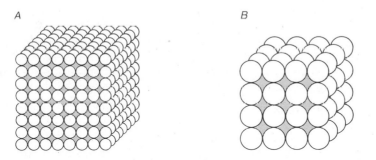

Figure 3-2. Size of soil particles affects pore size. (A) Small soil particles create many small pores. (B) Pores are larger but fewer in number between large soil particles. Micropores usually hold water, macropores hold air.

but larger pores. Water drains rapidly through large pores, which are called *macropores* or *aeration pores.* As the water drains, it pulls air in behind it, filling up the spaces. Small pores, which are known as *micropores,* tend to retain water. Both pore sizes are important, since the soil needs micropores to hold water and macropores for air.

Soil Separates.   Soil scientists divide mineral particles into size groups called *soil separates* and define three broad classes: sand, silt, and clay. The largest size

Separate	Diameter (mm)	Comparison	Feel
Very coarse sand	2.00–1.00	36″	Grains easily seen, sharp, gritty
Coarse sand	1.00–0.50	18″	
Medium sand	0.50–0.25	9″	
Fine sand	0.25–0.10	4½″	Gritty, each grain barely visible
Very fine sand	0.10–0.05	1¾″	
Silt	0.05–0.002	7/16″	Grains invisible to eye, silky to touch
Clay	<0.002	1/32″	Sticky when wet, dry pellets hard, harsh

Figure 3-3. The United States Department of Agriculture System of Soil Separates. The diameter of particles is in millimeters. The comparison shows the differences by setting a very coarse sand grain equal to three feet in size.

separate, sand, is further divided into four subcategories. Figure 3-3 names the separates and gives their sizes according to the system adopted by the United States Department of Agriculture. Figure 3-4 gives some idea of the relative sizes of the separates.

Sand, the largest soil separate, is composed mainly of weathered grains of quartz. Individual sand grains, except for very fine ones, are visible to the eye. All are gritty to the touch. Sand grains do not stick to one another, so they act as individual grains in the soil. Enough sand in a soil creates large pores, so sand improves water infiltration (rate at which water enters the soil) and aeration. On the other hand, large amounts of sand lower the ability of the soil to retain water and nutrients.

Silt is the medium-sized soil separates (figure 3-5). Silt particles are silky or powdery to the touch, like talc. Like sand, silt grains do not stick to one another. Silt, of all the soil separates, has the best ability to hold large amounts of water in a form plants can use.

Clay is the smallest soil separate. While sand and silt result simply from rock crumbling into small particles, clay results from chemical reactions between weathered minerals to form tiny particles of new minerals. These new minerals are able to bond nutrients chemically to their surfaces. These reactions, described in chapter nine, help hold plant nutrients in the soil.

Clay particles stick to one another and so do not behave as individual grains in the soil. Wet clay is sticky and can be molded. Some types of clay swell when wet and shrink as they dry.

As mentioned earlier, internal surface areas influence a number of soil properties. A handful of sand may have a surface area the size of a ping-pong table, while a handful of clay could reach the area of a football field.

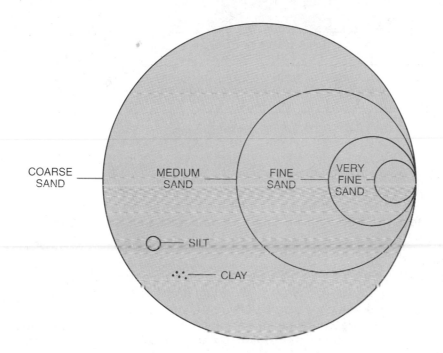

Figure 3-4. Comparing the relative sizes of soil separates. Very coarse sand is not shown.

Figure 3-5. A silty soil. The soil layer above the weathered bedrock is a silty loess. (Courtesy of Howard Hobbs, Minnesota Geological Survey)

Class	Diameter Range (mm)	Diameter Range (in.)
Gravel	2–75	1/12–3
Cobbles	75–254	3–10
Stones	More than 254	More than 10

Figure 3-5. USDA size classification for stones in the soil

It is not surprising that soils high in clay better retain water and nutrients. Conversely, clays are less well aerated and water seeps into them more slowly.

Gravel and other pieces of stone larger than 2 mm are not considered to be part of soil texture. They often are, however, part of the soil and affect its use, as anyone who has picked rocks out of a field can testify. Figure 3-6 lists the USDA size classifications of rock fragments in the soil.

Textural Classification. Soils usually consist of more than one soil separate; all three separates are found in most soils. The exact proportion, or percentage, of the three separates is called *soil texture*. Obviously, any number of combinations of the three are possible, so soil scientists simplify texture by dividing soils into textural classes. Soils in the same textural class are similar.

The 12 textural classes are shown in the *soil triangle* in figure 3-7. Each side of the triangle represents the percentage of one soil separate. A person can measure the amount of sand, silt, and clay in a soil sample and simply read the class off the triangle. An example included in figure 3-7 shows how to read the triangle.

Examine the soil classes carefully. Each corner of the triangle is a class dominated by one soil separate: sand, silt, or clay. The largest class is clay soil, because clay has the most powerful effect on soil properties. With as little as 40% clay a soil is classified as a clay soil. Another important textural name is *loam,* a soil in which sand, silt, and clay contribute equally to the soil's properties. The remaining classes have properties between those of the four major classes and their names suggest the difference. For example, a loamy sand is a sandy soil containing enough clay or silt to make it more loamy.

Growers can usually manage soils without knowing the exact soil texture. A broader classification is often adequate. One can simply classify soils as sandy, loamy, or clayey, as described above. Figure 3-8 shows another approach. The 12 classes are divided into three broad categories: coarse, medium, or fine, based on the size of the soil separates.

Determining Soil Texture. The amount of sand, silt, and clay in a soil can be measured by *mechanical analysis*. Mechanical analysis is based on the fact that the larger a soil particle, the faster it sinks in water. For instance, it takes only 45 seconds for very fine sand to settle through four inches of water, while it takes about eight hours for large clay particles. In mechanical analysis, one stirs soil into water, and notes how fast the soil particles settle out. Appendix two provides instructions for a simple type of mechanical analysis called a sedimentation test.

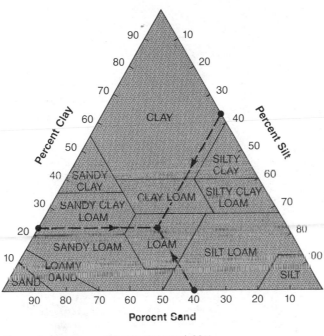

Example: Identify a soil that is 40% sand, 22%
            clay, and 38% silt
            1. Find 40 on the side for sand.
            2. Draw a line in the direction of the arrow.
            3. Do the same for clay (22%) and silt (38%).
            4. The spot where the three lines come together is the soil
               texture. In this case, the soil is a loam.

            A textural name may include a prefix naming the
            dominant sand size, as in "coarse sandy loam."

Figure 3-6. The soil triangle. Each side of the triangle is a soil separate. The numbers are the percentage of soil particles of that type. For example, the bottom line is the percentage of sand, ranging from 0% on the right to 100% on the left.

An even simpler test, which can be done on site, is the ribbon or feel test. The test is based on the feel of damp soil and how easily it can be molded. All those who work with soil should be able to do ribbon testing. The procedure for the test is as follows:

Step 1   Obtain a large enough sample of soil to form 1/2" ball. The sample should contain no gravel or bits of leaves, straw, or other debris. If needed, one can run the sample through a sieve to remove such material.

Step 2   Moisten the sample to a medium moisture level, like workable putty. Work the soil between the fingers until it is uniformly moist and dry lumps are wetted. Note any grittiness that indicates sand or the stickiness of clay. Clay also stains the fingers.

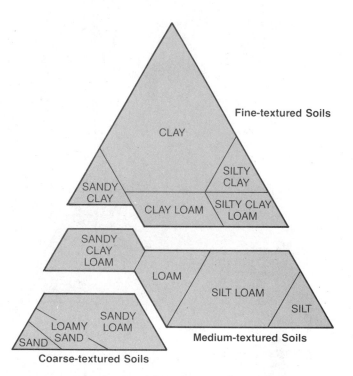

Figure 3-8. The soil triangle is redrawn to show fine-, medium-, and coarse-textured soils. An exception is very fine sandy loam, which is considered medium textured.

Step 3   Mold the sample into a 1/2" ball and try to lightly squeeze the ball. If it breaks at the slightest pressure, the soil is a sand or coarse sandy loam. If the ball stays together but changes shape easily, it is a sandy loam, loam, or silt loam. Finer-textured soils resist molding.

Step 4   Squeeze out a ribbon between the thumb and forefinger, noting how long a ribbon can be formed before it breaks (figure 3-9). Use this guide to narrow down the choice of textures:
no ribboning: loamy sand
ribbon shorter than one inch: loam, silt, silt loam, sandy loam
ribbons one to two inches long: sandy clay loam, silty clay loam, clay loam
ribbons two to three inches long: sandy clay, silty clay, clay

Step 5   Put all the observations together to decide the textural class. Sand feels gritty, silt feels smooth, clay feels sticky. So, for instance, sandy clay forms a long ribbon yet feels slightly gritty. A short ribbon that feels smooth is a silt loam.

The ribbon test is only useful if one has practiced it enough to "get a feel" for it. Try it out a few times.

Characteristics of Textural Classes.   Soil scientists place soils in textural classes because each class has properties important to its management. One can only generalize about textural effects because other properties also affect the soil. However, here are a few useful guidelines.

Texture governs the way water behaves in the soil. For instance, water enters *(infiltration)* and drains through *(percolation)* coarse soil most rapidly because of the large pore spaces. Thus, a coarse soil dries out most quickly after a heavy rain or in the spring, allowing a grower to get into the field more quickly. Similarly, coarse soils are more likely to need frequent irrigation. Growers with fine soils are likely to worry about the opposite problem—dealing with excess water.

Fine soils retain plant nutrients better than coarse soils. This is true partly because the rapid percolation of water through coarse soil leaches out nutrients. Also, clay particles have the best holding ability for nutrient chemicals.

Soil texture influences how easily a soil can be worked. Because clay particles stick together, it takes more horsepower and fuel to pull tools like plows through a fine soil. Landscapers also find that it is harder and slower to dig holes in fine soils for tree planting.

Figure 3-9. Ribbon test to determine soil texture

The stickiness of clay also affects the physical condition of the soil. For instance, fine soils often form clods when they are tilled. A crust may also form on the surface and interfere with seedling emergence. A fine soil tends to be "tight," meaning it has mostly small pores that are difficult for air and roots to penetrate. In contrast, coarse soils are "loose" and well aerated.

For most purposes, growers consider medium soils to be ideal. They hold water, but they don't stay wet too long. They are not sticky and are not hard to work. In general, medium soils have the good traits of both coarse and fine soils without their bad traits.

Modifying Soil Texture. Growers and engineers use soils for many purposes. For each purpose, a different soil texture may be best. For example, corn tends to be most productive on a loam, potatoes on a sandy loam, and black walnuts on heavy soils.

Can a grower change soil texture to improve it for the crop being grown? Except in very small areas, like golf greens or potting soils, changing texture is impractical. The amounts of clay or sand to be added are too large. Figure 3-10 shows the effect of adding sand to a clay soil to loosen it—clay particles surround the sand grains and fill in any pores that may be created. As a result, the soil continues to behave much like clay.

There are three ways for growers to take texture into account. First, select a crop to fit the soil, or purchase land that suits the crop. For example, an apple grower may purchase land with the fine soil on which apples grow best Second, manage the soil in a manner that fits the texture. For example, with proper fertilization and irrigation, coarse soils can be very productive. Third, organic matter can improve texture extremes by making sandy soils less droughty and by loosening clay soils.

SOIL DENSITY AND PERMEABILITY

As stated earlier, important physical properties relate to the spaces between soil particles. For a better understanding of soil pore space, this discussion will work through a series of related properties, beginning with soil density. We begin here because the density of soil—its weight per volume—is related to the amount of empty space in the soil.

Particle Density. One could ask how much soil would weigh if there were no pore space. This is called *particle density*, which is the density of the solid particles only. As an example, the particle density of a soil made wholly of quartz sand would be the same as the density of a solid block of quartz or 2.65 grams per cubic centimeter (166 pounds per cubic foot).

Particle density varies according to the type of minerals in the parent material and the amount of organic matter in the soil. Figure 3-11 lists the density of several soil-forming minerals. Note that the densities are very similar. In fact, there is surprisingly little variation in the particle densities of most mineral soils. Most

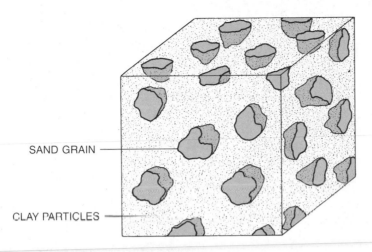

SAND GRAIN

CLAY PARTICLES

Figure 3-10. When sand is mixed into clay soil to loosen it, clay particles surround the sand grains and large pores are not formed. Very large quantities of sand are needed to improve clay soils.

Mineral	Density (grams/cm³)	Density (lbs/ft³)
Water	1.0	62.5
Quartz	2.65	166
Feldspars	2.5-2.7	156-169
Micas	2.7-3.0	169-188
Clay minerals	2.0-3.0	125-188

Figure 3-11. Densities of several soil-forming minerals

soils average about 2.65 grams per cubic centimeter, a value used as a standard density in soil calculations. High amounts of organic matter reduce the value because organic matter is much lighter than mineral matter.

Bulk Density.   Because soil does contain pore spaces, the actual density of a soil is less than the particle density. This measurement is *bulk density* or the weight of a volume of oven-dry soil.

To measure bulk density, a core of soil of known volume is carefully removed from the field. The soil core is then dried in an oven at 105 degrees Celsius until it reaches a constant weight. This is called *oven-dry soil*. The core is then weighed, and the bulk density is calculated. The example that follows is for a core of 500 cubic centimeters (cm³) that weighs 650 grams (g):

$$BD = \frac{\text{weight dry soil}}{\text{volume dry soil}} = \frac{g}{cm^3} \text{ or } \frac{\text{pounds}}{ft^3}$$

$$BD = \frac{650 \text{ g}}{500 \text{ cm}^3} = 1.3 \text{ g/cm}^3$$

The bulk densities of mineral soils depend mostly on the amount of pore space in the soil, since particle weight is fairly constant. Bulk densities of mineral soils usually range from 1.0 grams per cubic centimeter (62.5 lbs/ft^3) for "fluffed-up" clay soils to 1.8 grams per cubic centimeter (113 lbs/ft^3) for some sandy soils. Organic soils are much lighter, with values of 0.1 to 0.6 grams per cubic centimeters (6-38 lbs/ft^3) being common.

Soil Porosity.   *Total pore space* is a measure of the soil volume that holds air and water. This value is usually expressed as a percentage and is known as the *porosity.* Thus, a soil with a 50% porosity is half solid particles and half pore space.

Porosity can be measured by placing an oven-dry soil core in a pan of water until all of the empty pore space is filled with water. The difference in weight between the dry and the wet cores is the total pore space. This number is converted to a percentage to get porosity. The soil core used as an example before had a volume of 500 cubic centimeters and weighed 650 grams when dry. When wet, the same core weighs 900 grams. Porosity is calculated as follows:

$$\text{Porosity} = \frac{\text{wet weight (g)} - \text{dry weight (g)}}{\text{soil volume (cm}^3)} \times 100 = \frac{900 - 650}{500} \times 100 = 50\%$$

Porosity can also be calculated from bulk density and particle density. If there were no pore space, then bulk density *(BD)* would be the same as particle density *(PD).* The ratio *BD/PD* would be equal to one. The more pore space, the smaller the bulk density and the smaller the ratio *BD/PD.* In fact, the ratio *BD/PD* is simply the percentage of the soil that is solid matter. If one subtracts that percentage from 100%, the difference is the percentage of pore space. To make the calculation, one can usually assume that *PD* is 2.6 grams per cubic centimeter. The following equation can be used to calculate porosity:

$$\text{Porosity} = 100\% - (\frac{BD}{PD} \times 100)$$

If we substitute the values for the bulk density just calculated:

$$\text{Porosity} = 100\% - (\frac{1.3}{2.65} \times 100) = 50\%$$

The porosity of sand (about 30%) is lower than the porosity of clay (about 50%). Figure 3-12 shows that porosity increases at finer textures. Yet common

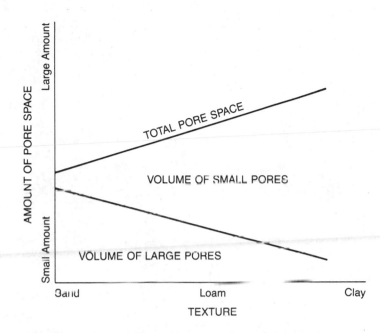

Figure 3-12. Texture affects soil pores. The top line shows the total amount of pore space in the soil. Clay has the greatest total pore space. The lower line shows how much space is in large pores. Sand has the most large pore space. The amount of small pore space is between the two lines. Clay has the most small pore space. Loam has a balance of large and small pores.

sense tells us that water seeps into sand very rapidly, but it seeps only slowly into clay. The next section explains why.

Permeability. *Permeability* is the ease with which air and water and roots move through the soil. In a highly permeable soil, water infiltrates soil rapidly, and good aeration keeps roots well supplied with oxygen. Roots grow through permeable soil with ease. We can think of permeable soil as being "loose" and impermeable soil as being "tight."

Permeability depends partially on the *number* of soil pores, but it depends more on the *size* and *continuity* of the pores. The movement of air, water, and roots can be likened to walking a maze. If the paths are too narrow, progress is difficult. Progress is even more difficult when paths come to a dead end. Like a maze with dead ends, soils lacking large, continuous pores limit the flow of air and water.

Large, continuous pore spaces in the soil, or macropores, occur between large particles. Therefore, the number of macropores depends on texture, as shown in figure 3-12, and permeability must also depend partially on texture. Permeability is not a soil property that can be measured directly. However, the movement of water, which reflects permeability, can be measured.

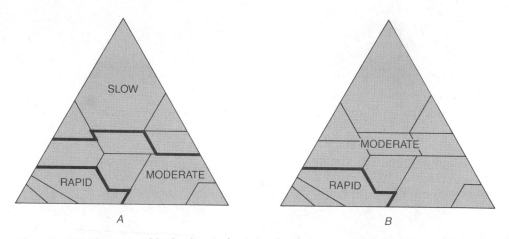

Figure 3-12. Infiltration and hydraulic conductivity of soil textures. Slow is less than 0.014 in./hr; moderate is 0.014-1.44 in./hr; rapid more than 1.44 in./hr. *(A)* Rates for soils with weak structures or soils in which structure has less influence. *(B)* Rates for soils with strong structure or for soils in which structure has little influence.

Hydraulic conductivity is a measure of the rate of water movement through a soil. Figure 3-13 shows a suggested classification system for hydraulic conductivity. Figure 3-13 (A) shows how texture influences water movement.

The *A* in figure 3-13 suggests that heavy soils are rather impermeable. Fortunately for those who crop fine soil, another physical property, structure, influences permeability.

SOIL STRUCTURE

Heavy soils would make very poor root environments, except that structure can alter the effects of texture. Structure refers to the way soil particles clump together into large units (figure 3-14). These large units are called *soil aggregates.* Aggregates that occur naturally in the soil are *peds,* while clumps of soil caused by tillage are called *clods.*

Peds are relatively large, ranging from the size of a large sand grain to several inches. Spaces between clay particles may be tiny, but the spaces between peds may be large. Good structure can therefore improve air and water movement and make it easier for roots to grow while maintaining good water-holding capacity within the peds. Figure 3-13 (B) shows, for instance, how structure improves water movement in fine soils.

There are many different kinds of structure, and some are better at improving permeability than others. Soil scientists classify structure according to three groups of traits.

- *Type* refers to the shape of the soil aggregates (figure 3-15). These shapes are described in detail below.

Figure 3-14. Soil aggregates. The peds create large spaces in the soil to improve infiltration, aeration, and root growth.

- *Class* is the size of the peds, which can be very fine, fine, medium, coarse, or very coarse.
- *Grade* refers to how distinct and strong the peds are. One grade, structureless, applies to soils that have no peds. Weak grades are barely visible in a moist soil, whereas strong peds are quite visible and can be easily handled without breaking.

Structureless Soil.   Sand behaves as individual grains, so sandy soils seldom have much structure. These soils are called *single-grain*. Sandy soils are naturally permeable, so single-grain soils have good infiltration rates and aeration.

Finer soils lacking structure are a solid mass stuck together like molding clay. These *massive* soils, as they are called, lack permeability. Massive soil is typical of some C horizons. Tillage of wet soil may result in massive soil in the A horizon.

Types of Soil Structure.   *Granular structure* or *crumb structure* is commonly found in A horizons. The peds are small, usually between 1 to 10 millimeters (1/25

Structure   Horizon of Common Occurrence

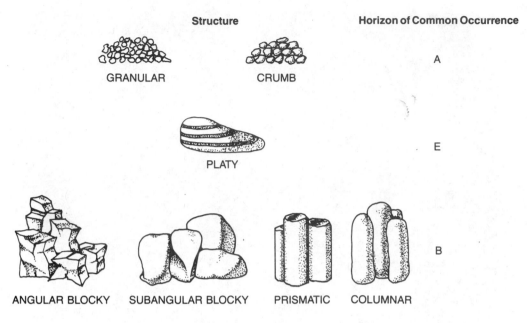

Figure 3-15. Types of soil structures

and 2/5 inch), rounded in form, and considered the most desirable of structures. A granular ped is not very porous within the ped. A crumb is porous, almost like a bread crumb. Such structure increases total pore space and lowers bulk density compared with a soil lacking structure.

Platy structure is usually found in E horizons. Peds are large but thin, platelike, and arranged in overlapping horizontal layers. The arrangement makes discontinuous pores that reduce penetration of air, water, and roots. Soil compaction can create platy structure in the A horizon when granules of topsoil are crushed into thin layers.

Blocky structure is typical of many B horizons. The peds are large, 5 to more than 50 millimeters (1/5 to 2 inches) and blocklike in shape. If the ped is very angular, it is termed *angular blocky*. More rounded peds are described as *subangular blocky*. Blocky structure has medium permeability.

Prism-like structures also occupy the B horizon of some soils. Peds are large, usually from 10 to more than 100 millimeters (1/5 to 4 inches), forming angular columns that stand upright in the soil. If the top of the ped is pointed or flat the structure is called *prismatic*. If the top of the ped is rounded, it is termed *columnar*. Prismatic structure is moderately permeable, columnar structures slowly permeable.

Formation of Soil Structure.   Soil structure is formed in two steps. First, a clump of soil particles sticks loosely together to form a loose aggregate. These clumps are very weak and easily crushed. Freezing and thawing or wetting and drying can break soil up into aggregates. Roots, soil fungi, and tillage also create loose aggregates.

Second, the weak aggregates are cemented together to make them distinct and strong. Clay, certain mineral chemicals, and organic matter may each act as cements. In most soils, microorganisms provide the best cement. When soil microbes break down plant residues, they produce gums that glue the peds together. Therefore, the best way to enhance soil structure is to frequently add organic matter to the soil.

Large amounts of sodium in the soil reverse the process, causing soil aggregates to break up. Sodium occurs naturally in some soils or may build up because of use of high-sodium irrigation water. This will be discussed in more detail in chapter ten.

SOIL CONSISTENCE

Consistence refers to the behavior of soil when pressure is applied. It relates to the degree that soil particles stick to one another and mostly results from certain types of clay.

The effect of consistence can be best explained by some examples. Loose sand, for instance, shifts easily under pressure, so that vehicles may get stuck in sand along a beach. Preparing a seedbed for planting is another example. A grower wishes to break apart large chunks of soil to get a fine surface to plant seeds in. It is the consistence of the soil that determines how easily those chunks can be broken down.

Consistence depends on how moist a soil is, so it can be measured at three different moisture levels. Each level has its own descriptive terms (figure 3-16), which are described as follows:

Wet Soil.   Wet soil is checked for stickiness and plasticity. Plasticity is how easily soil can be molded between the fingers. To determine stickiness, some soil is pressed between thumb and forefinger, and the amount that sticks to the fingers is noted.

Moist Soil.   The terms friable and firm apply to soils in the moist state. *Friable* means that soil materials can be crushed easily under pressure.

Dry Soil.   Determine consistency of dry soil by trying to crush an air-dry mass of soil in the hand.

By rating a soil for consistence, one can infer such information as suitability for plowing, likelihood of erosion, or texture. In fact, the feel test for texture works because of the consistence of different soil textures. A loose soil is coarse textured, a friable soil medium textured or well aggregated, and a firm soil is a tight, fine textured soil. A firm soil may lack good structure or be compacted.

	Wet		Moist	Dry
Stickiness		**Plasticity**		
Nonsticky	Nonplastic		Loose	Loose
Slightly sticky	Slightly plastic		Very friable	Soft
Sticky	Plastic		Friable	Slightly hard
Very sticky	Very plastic		Firm	Hard
			Very firm	Very hard
			Extremely firm	

Figure 3-16. Consistence terms for soil at three different moisture levels

SOIL TILTH

Tilth is a general term for the physical condition of a tilled soil. It suggests how easy the soil is to till, how good a seedbed can be made, how easily seedlings can come up and the ease of root growth. Tilth is actually a combination of other physical properties, including texture, structure, permeability, and consistency.

Tillage improves soil tilth for a time, improving soil-air-water relations for new seedlings. It does so by loosening the soil and stirring air into it. Fine-textured soils are most improved by tillage because coarse soils are already well-aerated.

Tillage tends to cause a year-by-year decline in soil structure, however. Compare the topsoil of a cultivated field with that of a nearby fencerow that has not been cultivated. Peds from the fencerow will be more numerous, stronger, and of a better type. The weakening structure of the tilled soil, in turn, lowers water infiltration, aeration, and ease of root growth.

Compaction. Compaction results when pressure is applied to the soil surface. The pressure squeezes together soil particles, shrinking soil pores. The number of air-providing macropores decreases, while micropores increase in number. Compaction reduces porosity and increases bulk density. Permeability declines, so that aeration and water infiltration becomes more difficult. Roots have more difficulty growing through the denser soil.

A number of agricultural activities induce compaction. For instance, cultivation and other equipment operations during the growing season compact the soil between crop rows (figure 3-17). If severe enough, it can restrict root growth between the rows.

Annual plowing breaks up this compaction. However, just below the plow layer, a compacted zone develops. This *plow pan*, or *tillage pan* (figure 3-17), restricts the growth of roots and the drainage of water deeper into the soil profile.

The worst compaction occurs when heavy equipment compresses the subsoil, leading to subsoil compaction. Harvest and transport equipment can

Figure 3-17. Tillage compaction. This cross-section of a corn field pictures compaction from both between-row cultivation and a deeper plow pan (Courtesy of *Crops and Soils Magazine*, American Society of Agronomy)

exert loads up to 40 tons per axle (figure 3-18). Such a heavy load, if applied to wet soil, can cause deep compaction that seriously reduces crop yields. Tillage equipment, which generally exerts only about 5 tons per axle, does not create subsoil compaction.

The degree of compaction relates to the natural compressibility of a soil, its moisture content, axle weight, and how often equipment is driven on the soil. Fine textured soils and wet soils are most compressible.

Compaction has varying effects on crop yield. Research indicates these effects:

- any degree of compaction decreases yields when conditions are wet.
- slight compaction improves yields under normal conditions, especially on sandy soils which could use a few more micropores.
- moderate compaction can improve yields in dry soils, by increasing micropore space to hold water.
- severe compaction always inhibits production
- the finer the soil texture, the more damaging is compaction.

Figure 3-18. Harvest equipment can cause deep subsoil compaction. Grain carts such as this can carry loads of up to 40 tons per axle. (Courtesy of *Crops and Soils Magazine*, American Society of Agronomy)

Compaction is also a problem for other soil users. Virtually all landscape sites are badly compacted by construction equipment. Parks and recreational areas suffer from the pressure of countless footsteps (figure 3-19), mowers, and off-road vehicles. Logging operations compact forest soils. Often, compaction is most severe in these nonfarming areas, because the land is not plowed each year.

Compaction can be measured two ways. The most accurate is to compare the bulk density (BD) of the compressed soil to that of nearby unaffected soil. An average uncompacted tilled loam might have a BD of about 1.3 gms/cc. Compaction by farm equipment can raise the BD to 1.8. This amounts to a reduction of pore space from 50% to almost 30%.

More simply, but less precise, is the use of a *cone penetrometer*. A rod with a cone-shaped tip is pushed into the soil, and a dial reads the pressure it took. The result is an index that can be compared to a nearby unaffected soil. Unfortunately, soil type and moisture level also affect the result. Use cone penetrometer indices only to compare nearby plots of the same soil type and moisture level.

Aggregate Destruction. Plowing tends to create large aggregates as it flips the soil over. However, other tillage operations, such as cultivating, tend to crush soil

Figure 3-19. Footpaths in parks are a sign of compaction.

aggregates. High-speed rototillers are especially destructive because they batter the aggregates apart (figure 3-20). Many gardeners favor the fluffy, loose bed created by rototillage, but, over the long term, structure is hurt.

There are two reasons why tillage destroys aggregates. First, by stirring oxygen into the soil, tillage speeds up organic matter oxidation. This loss, in turn, reduces the amount of "organic glue" that holds the peds together. Second, the tillage tools smash the now weakened peds.

Puddling and Clods. Working wet soil greatly harms tilth, especially in soils high in clay. When pressure is applied to very wet soil aggregates, they fall apart. This results in a condition known as *puddling*—the conversion of aggregated soil into massive soil. The puddled soil is very dense and tight. In fact, it is done on purpose in some rice paddies, canals, and reservoirs to keep water from leaking away.

Working soil that is either too wet or too dry can also break up soil into large, seemingly indestructible clods. Soils with a hard consistence are most likely to form such clods.

Figure 3-20. A high-speed rototiller being used in a school nursery. Tillage tends to cause a deterioration of structure.

Surface Crusts. Most forms of tillage bare the soil until the crop grows large enough to cover the soil. When raindrops hit this bare surface, their impact breaks apart peds on the surface. The free soil particles then splash around, washing into spaces between large particles and sealing the surface. When the soil dries, a crust forms that sheds water and inhibits the emergence of seedlings. The more clay in the soil, the harder the crust.

Improving Tilth. As noted earler, tilth relates to the properties of texture, structure, permeability, and consistence. However, texture and consistence cannot, in most cases, be changed. Therefore, improving tilth is largely a matter of improving structure and avoiding compaction. The following practices can help improve tilth:

- Never work wet soils, and avoid quite dry ones as well.
- Avoid unneccessary traffic over the soil. If possible, set aside paths through fields to limit driving on the soil. If possible, drive only on dry soils.
- Employ controlled traffic in the field by setting the wheel base on all equipment to the same width. Then always drive in the same rows. While the wheel tracks will compress more severely, the remaining rows will be compaction free.
- Use equipment with the lowest practical axle weight. Large or dual tires do not seem to greatly reduce compaction, but spread it over a wider area.
- Reduce the number of tillage operations. One can reduce trips across the field by combining operations or by simply not repeating them so many times. Some modern tillage practices, called *minimum* or *reduced tillage*, use less tillage. This is covered in chapter fifteen.

Figure 3-21. A subsoiling chisel plow. A tool like this can break up hardpans in the soil to temporarily improve penetration of water and roots. (Courtesy Year-A-Round Cab Corp.)

- Deep plowing, or *subsoiling* (figure 3-21), can break up tillage or natural hardpans, resulting in deeper penetration of water and plant roots and improved yields. The benefits may be short-lived, however, because further tillage reforms the compacted layer.
- Wherever possible, keep the soil covered by vegetation or mulch. Crops that fill in between rows quickly and crops that form complete cover, like alfalfa, protect soil from raindrop impact. Tillage that leaves a lot of crop residue on the soil surface also helps by creating a mulch.
- Frequently add organic matter to the soil. Growers can leave crop residues in the field and spread manure. Gardeners and some organic farmers use compost. Lawn clippings, leaves, and any other sources of organic matter can be useful.
- Plow under "green manure" crops of grasses or legumes. The plant's roots help create the loose aggregates, while decaying organic matter glues them together. Taprooted plants like alfalfa also help break up hardpans.
- Add lime where needed. Many soil organisms that decay organic matter require adequate levels of lime.
- Treat high-sodium soils with gypsum to remove excess sodium (see chapter ten) and manage high-sodium irrigation water correctly (chapter six).

SOIL PANS

The preceeding section mentioned plow pans. Any layer of hardened soil is called a pan. Pans restrict the deep rooting of crops and the deep percolation of water, and can be a serious hindrance to cropping.

Growers create plowpans, but other types are natural. Examples include:

- *Claypans* occur where extreme illuviation has caused a very high clay content in a subsoil layer. The layer is quite dense.
- *Fragipans*, like claypans, result from clay accumulation. Here the clay binds soil particles into a hard, brittle layer.
- *Plinthite*, formerly known as laterite, layers are cemented by a special type of clay common to the tropics. When plinthite dries, it hardens to a bricklike substance; the process cannot be reversed by later wetting. Plinthite commonly renders tropical soils poor for agriculture.
- *Caliche* and *duripans* are layers of soil in which chemicals cement soil particles together. Lime cements caliche, typically a white, hardened layer found in arid regions. Many soils of the American Southwest contain caliche.

SOIL TEMPERATURE

As indicated in chapter one, soils in the growing regions of the world keep a temperature balance over the year that is satisfactory for plant growth. Short-term temperature changes, between seasons or night and day, can be dramatic. For instance, in the summer, soil temperatures can easily rise thirty degrees

Fahrenheit in the top inch of soil during the day. Below twelve inches, temperature varies little day to day.

Soil temperature is critical to the grower. Seed germination, for instance, is affected by soil temperature. Each crop has a temperature range that is best for seed growth. Thus, the crop is planted when the soil has warmed enough in the spring for rapid germination.

Root growth also depends on soil temperature. Warm season crops, such as corn or tomatoes, grow best at relatively high soil temperatures, while peas and other cool season vegetables grow best at cooler temperatures. Root injury due to very high soil temperatures often occurs in plants grown in containers when sunlight strikes the pot on hot summer days.

Soil temperature depends not only on weather conditions but on several soil factors. Sunlight striking the earth is partially absorbed by the soil and partially reflected into the atmosphere. Dark soils absorb more sunlight, so they tend to be slightly warmer than pale-colored soils.

Sunlight absorbed by the soil raises the temperature of both mineral particles and soil water. It takes five times as much energy to warm water than to warm an equal volume of mineral particles. As a result, it takes far less energy to warm dry soils. Sandy soils, which hold the least amount of water, tend to warm up most quickly in the spring and remain warmer during the season.

Managing Soil Temperature. The most obvious effect of soil temperature is in determining planting dates, which, in turn, affect harvest timing. The value of perishable crops, like vegetables, is at its highest when supplies are low, usually very early in the season. Vegetable growers, therefore, often favor coarse soils that allow early planting.

Gardeners and growers change soil temperature by using mulches. Light-colored organic mulches like straw insulate the soil and reflect sunlight. This lowers average soil temperature, improving the growth of many crops. Crops that grow best in warm soil, like melons, are often mulched with black plastic (figure 3-22). The plastic absorbs sunlight and raises the average soil temperature. The warm soil improves production and speeds growth so the melons come to market early enough to demand premium prices.

Recent trends in tillage have made soil temperatures a greater concern. Conservation tillage, a newer style of tillage, leaves crop residues on the soil surface. The residues both reflect sunlight and reduce soil drying. Thus, soil temperature can be several degrees cooler under conservation tillage. The difference can affect the production and harvest dates of several crops. Conservation tillage will be discussed in chapter fifteen.

Frost damage to crops can sometimes be avoided by taking advantage of heat stored in the soil. At night, when damaging frosts occur, heat absorbed during the day warms the air immediately above the soil. Growers can heighten the effect by keeping the soil bare of debris that cools and insulates it. Irrigation is also helpful, since moist soil stores far more heat than a dry soil.

Figure 3-22. Demonstration of a black plastic mulch. The mulch warms the soil to improve production of warm-season crops. (USDA, Soil Conservation Service)

SOIL COLOR

While soil color is easily noted, it does not itself greatly affect the soil. However, color is an indicator of soil conditions, so growers can learn about a soil by its color.

Brown to Black. Dark soil colors result from organic matter or dark parent materials, usually the former. Black soils can arise from three situations, which can often be distinguished by smell:

- organic matter can reach high levels in soils that are usually waterlogged. Such soils often have a sour, oily smell.
- organic matter can also reach high levels in an adequately aerated soil. These soils have the earthy smell of good soil.
- dark parent materials will affect the color of young soils. A faint chalky odor often describes these soils.

White to Light Gray. This color may indicate that the chemicals that color soil, especially organic ones, have leached out. It may be seen in heavily leached sandy soils and E horizons. White color may also be due to accumulations of lime, gypsum, or other salts.

Yellow to Red.   These are the colors of iron oxides, most commonly seen in warm climates. Red color is from iron oxide, essentially rust. Red color indicates good drainage because there is enough oxygen in the soil to form the oxide. Yellow is from an iron oxide that includes some water (the mineral limonite), indicating the soil is slightly less well drained than a red soil.

Bluish-Gray.   This is the color of unoxidized iron and indicates a lack of oxygen in the soil. The lack of oxygen results from water-logging, so bluish-gray color indicates poor soil drainage. The occurrence of this color is called *gleying.*

Mottled Colors.   The soil shows patches of different colors, often spots of rust, yellow, and grey. *Mottling* suggests that the soil is waterlogged for part but not all of the year.

Color as a Guide to Soil Use.   Soil color can be a useful guide to the suitability of the soil for various uses. As a guide, it is most dependable within a region. Comparisons between regions with different climate or mineralogy may not be valid.

White or light-colored soils usually have low fertility, either because they are leached or high in salts. Proper irrigation, fertilization, or treatment to remove salts may render these soils usable. Very dark topsoils that are high in organic matter may be quite fertile. However, one should check the subsoil for gleying—the high organic matter content may result from lack of oxygen needed to decay organic matter.

Mottling and gleying of the subsoil is a good indication of poor soil drainage. Thus a grey or mottled subsoil indicates the need for artificial drainage for agriculture. Figure 3-23 shows a simplified classification system for soil drainage based on soil color.

Drainage Class	Description
Very poorly drained	In level or low spots, black topsoil with gray color under the A or AB horizon
Poorly drained	High water table or impermeable sub-surface layer, gray or black surface, gray B horizon with brownish mottles
Somewhat poorly drained	Gray or brown A horizon with brownish upper B horizon, gray and rust mottles between 10- to 18-inch depth
Moderately well drained	Fairly bright colors in upper B horizon, few gray mottles between 18- to 30-inch depth
Well drained	Free of mottles, may be a few mottles below 30 inches
Excessively well drained	Sandy soils with rapid permeability, shallow soils on steep slopes

Figure 3-23. Guide for determining natural soil drainage class using soil color (Source: USDA)

Describing Soil Color. A simple description of a soil as "dark" would not be adequate for a soil survey. Soil surveys rely on a system that provides a precise description of soil color, the *Munsell system* of color notation. Using the Munsell method, the surveyor matches the soil to standard color chips. The Munsell system identifies each chip with three variables:

- *Hue* is the color, such as red or yellow.
- *Value* is the lightness or darkness of the hue. Value is denoted by the numbers 1-10, where 0 is black, grades through lighter colors to 10 for white.
- *Chroma* is the purity of the dominant color, and is also denoted by a number. Low chroma suggests muddy colors.

Using the Munsell system, a soil might be labeled 10YR 3/6. This soil has the hue 10YR, a yellow-red; the value 3 (dark), and the chroma 6. This soil could be described as a fairly pure, dark yellowish brown.

SUMMARY

The most basic physical property of soil is texture—the proportion of sand, silt, and clay. All the possible variations of the three are divided into 12 textural classes. These classes range from the coarsest, which is sand, to the finest, which is clay. Medium-textured soils are called loams.

Texture strongly affects how growers use soil. Coarse soils are easy to work and dry out quickly. They warm up early in the spring, but they do not hold nutrients well. Therefore coarse soils perform best with irrigation and proper fertilization. Fine soils hold water and nutrients well, but they are more poorly aerated unless of good structure. Fine soils tend to stay wet later in the spring and are more difficult to work.

The ease with which air, water, and roots move through soil is called permeability. Coarse soils, because of their large sand content, are naturally permeable. Finer soils depend on structure, the aggregation of soil particles in peds, to create large pore spaces.

Consistence measures traits such as stickiness, plasticity, and friability. Soil tilth, the physical condition of the soil for growing plants, results from the interaction of consistence, texture, structure, and permeability. Tillage, done correctly, improves the tilth of a seedbed. In the long term, tillage can cause compaction, crusting, puddling, or deterioration of structure. Minimizing tillage, adding organic matter, and working soil at the proper moisture level preserve tilth. Some soils contain hardened soil pans such as plow pans or caliche.

Seed germination and plant growth are strongly affected by soil temperature. Some crops, like corn, germinate best in warm soil while other crops, like peas, accept cooler soils. Growers can change soil temperatures by

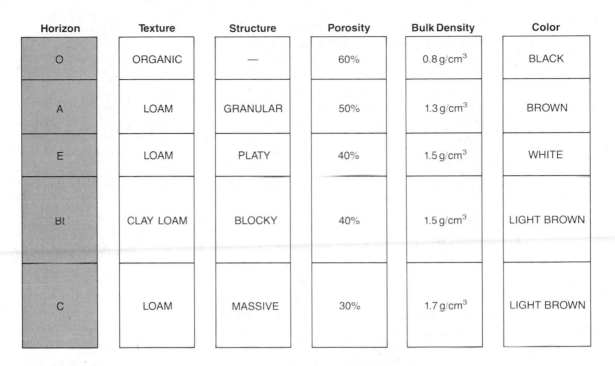

Horizon	Texture	Structure	Porosity	Bulk Density	Color
O	ORGANIC	—	60%	0.8 g/cm³	BLACK
A	LOAM	GRANULAR	50%	1.3 g/cm³	BROWN
E	LOAM	PLATY	40%	1.5 g/cm³	WHITE
Bt	CLAY LOAM	BLOCKY	40%	1.5 g/cm³	LIGHT BROWN
C	LOAM	MASSIVE	30%	1.7 g/cm³	LIGHT BROWN

Figure 3-24. Variation of physical properties through a soil profile. This figure represents the trends one might see in some soils. Other soils would show a different pattern.

using mulches. Organic mulches lower the average soil temperature while plastic mulches raise it.

Soil color is an indicator of soil conditions. For instance, dark color in the topsoil suggests a high amount of organic matter. Gray or mottled colors suggest slow drainage. Soil scientists employ the Munsell system to identify soil color.

Color and other physical properties vary through the soil profile. Figure 3-24 summarizes variations in one sample soil profile.

REVIEW

1. Macropores are important to retain air in the soil (True/False).
2. The easiest way to improve a tight, clay soil on a farm is to add sand.
3. Stickiness in a soil comes from silt.
4. Particle density and bulk density are the same.

5. When bulk density rises, porosity declines.
6. Platy structure is considered undesirable.
7. Adding organic matter to soil can improve structure.
8. A friable soil is one that has suffered compaction.
9. A damp soil is likely to be cooler than a dry one.
10. Gleying indicates good drainage.
11. Compaction is most severe when one drives on _____, fine-textured soil.
12. _____ is a hard soil layer cemented by lime, common to arid regions.
13. A coarse sandy soil would likely be described by the consistence term _____.
14. Patchy subsoil colors that indicate poor drainage are called _____.
15. If a soil has a prosity of 30%, the soil is _____% solid matter and _____% pore space.
16. Many large, continuous pores aid the movement of air and water, a property called _____.
17. Soils in which structure has been destroyed by working them wet are called _____.
18. The primary heat source for soil is _____.
19. Sand feels _____, silt feels _____, and clay feels _____.
20. A soil with 20% clay and 20% sand, has _____% silt and is a _____.
21. A soil sample forms a 2″ ribbon with no feeling of grittiness, but is quite sticky. What soil class is it?
22. A soil core has a volume of 400 cubic centimeters. After drying in an oven, it weighs 600 grams. Then the core is allowed to soak up water until all the pores are full. The core now weighs 772 grams. Calculate the bulk density and porosity of the sample.
23. Does puddling change the soil texture?
24. Name three ways a grower can preserve soil tilth.
25. A loam with good structure, that makes a good seedbed, would be described by what soil consistence term?
26. Describe how soil aggregates are held together.
27. Why would some vegetable growers prefer a coarse soil?
28. Describe how a crust forms on the soil surface.
29. How could you tell something about soil drainage from its physical properties?
30. Analyze the effects of slight compaction on a sandy soil. What happens to particle density, bulk density, porosity, soil pore size, water holding capacity, and crop production?

SUGGESTED ACTIVITIES

1. Obtain soil samples of different known textures and practice the ribbon test. Then try to identify some unknown samples.
2. Try measuring soil texture using the sedimentation test in appendix two.
3. Obtain soil samples from an old fencerow and the neighboring field. Observe the difference in structure.
4. Fill three 12″ deep boxes with 8″ of moist soil. Place soil probe thermometers two and eight inches deep. Mulch one with an organic mulch, one with black plastic, and leave one bare. Place the boxes in the full sun and record soil temperatures over time. Compare the effects of different mulches.
5. Determine the bulk density of a soil sample. Remove both ends of a soup can and calculate its volume. (The volume is 3.14 times the radius of the base squared times the depth of the can.). Drive the can into a soil, then remove by cutting around the can. Make sure the core fills the can exactly. Now remove the soil core, put it in a tray, and dry in an oven. Weigh the dried sample and calculate the bulk density.

chapter four

Soil Water

Furrows in this Texas field hold a most critical substance for plant growth: water. Places like this appreciate the importance of water because crops depend on a sparse water supply. This chapter will consider how water behaves in the soil, plant/water interactions, and how to measure soil water.

(Courtesy of USDA-ARS)

OBJECTIVES

After completing this chapter, you should be able to:

- identify the role of water in plant growth
- define the forces that act on soil water
- classify types of soil water
- discuss how water moves in the soil
- explain how plant roots remove water from the soil
- describe how to measure soil water content

TERMS TO KNOW

adhesion
adhesion water
available water
bar
capillary
capillary rise
capillary water
cohesion
cohesion water

field capacity
gravitational flow
gravitational water
hydraulic conductivity
hygroscopic water
matric potential
permanent wilting point
potentiometer

resistance block
saturated flow
saturation
soil moisture tension
soil-water potential
temporary wilting point
unsaturated flow
vapor flow

HOW PLANTS USE WATER

On the average, crop plants use 500 to 700 pounds of water to produce a single pound of dry plant matter. Water deficiency commonly limits plant growth; in several agricultural areas water is the most important need in farming. Water is vital to growers because of the several functions it serves in plant growth:

- Plant cells are largely made up of water. Plant tissue is 50% to 90% water, depending upon the type of tissue.

- When plant cells are full of water, the plant is stiff (turgid) or semi-rigid because of water pressure in plant tissue. This keeps stems upright and leaves expanded to receive sunlight.
- Photosynthesis uses water as a building block in the manufacture of carbohydrates (figure 1-4).
- Transpiration, or evaporation of water from the leaf, helps cool the plant.
- Plant nutrients are dissolved in soil water and move toward roots through the water. Water is thus important in making nutrients available to plants.
- Water carries materials such as nutrients and carbohydrates throughout the plant.

Effect of Water Stress.   Water stress is caused by a shortage of water in plant tissue. As will be explained later, stress can occur even at moisture levels that do not cause wilting. Part of the reason for such stress is that as the soil dries, it becomes increasingly difficult for a plant to absorb moisture. This sets off the following sequence of events. As the plant becomes deficient in water, guard cells begin to close the stomata, slowing down the exchange of oxygen and carbon dioxide. As a result of the reduced exchange of the two gases, photosynthesis must also slow down. With less photosynthesis, plant growth is inhibited.

As the soil dries further, or if the weather is hot and dry, the plant becomes even more deficient in water. The plant begins to lose water faster than it can be absorbed and the plant temporarily wilts. At this *temporary wilting point,* the plant will recover when conditions improve. Wetter soil, cooler temperatures, a more humid atmosphere, shade, or less wind (figure 4-1) can help the plant recover. Although the plant recovers, episodes of water stress can reduce plant growth and crop yields. With further drying, the *permanent wilting point* is reached. Now the plant will not recover even if conditions improve.

Plants suffering from chronic water stress are small and sparse with small, poorly colored leaves. Old leaves often turn yellow and drop off. Some plants show specific symptoms of water stress. For example, the leaves of corn plants curl when they need water.

Seed germination is very sensitive to water shortage. While seeds efficiently absorb moisture through the seed coat, the emerging seedling is easily injured by dry soil.

FORCES ON SOIL WATER

A number of forces influence the way water behaves in the soil. The most obvious is gravitational force, which pulls water down through the soil. Other forces, called adhesion and cohesion, work against gravity to hold water in the soil. *Adhesion* is the attraction of soil water to soil particles, while *cohesion* is the attraction of water molecules to other water molecules.

(A)

(B)

Figure 4-1. The temporary wilting point. In (A), this azalea in dry soil has wilted on a hot day. After watering with sprinklers, the plant revives quickly (B). A moist soil and cooling by wetting leaves have reversed the water balance.

Adhesion and cohesion happen because of the shape of the water molecule and the way an electron is shared in the oxygen-hydrogen bond. Examine the two water molecules pictured in figure 4-2. Hydrogen consists of one proton and one electron. When the two hydrogen atoms combine with oxygen to form water, each shares its one electron with the oxygen atom. Each electron sits *between* the oxygen and hydrogen atoms, leaving the positively charged protons positioned on one side of the molecule. As a result, that side has a slightly positive charge. To balance that positive charge, the oxygen side assumes a slightly negative charge. The water molecule is then like a bar magnet—positive on one end, negative on the other. Like bar magnets, the opposite ends of water molecules attract. The bond between the hydrogen of one water molecule and the oxygen of another, called a hydrogen bond, accounts for cohesion.

Hydrogen bonding also accounts for adhesion. The main chemical in soil minerals is silica (quartz is pure silica). Silica, with the chemical formula SiO_2, has oxygen atoms on the surface that can form hydrogen bonds with soil water.

Together, adhesion and cohesion create a film of water around soil particles. The film has two parts. A thin inner film is held tightly to the particle by adhesion. The *adhesion water* is held so tightly it cannot move. A thicker outer film of water is held in place by cohesion to the inner film. *Cohesion water,* sometimes called *capillary water,* is held loosely and can be absorbed by plants. Thus, plants use cohesion water that is clinging loosely to soil particles.

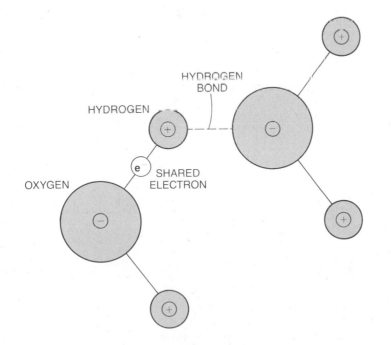

Figure 4-2. Two water molecules attract each other because the electrical charge is unequally distributed. The "plus" side attracts the "minus" side, forming a hydrogen bond.

Capillarity.   Soil water exists in the small spaces in soil as a water film around soil particles. The small pores can act as capillaries. A *capillary* is a very thin tube in which a liquid can move against the force of gravity. Capillary action is shown in figure 4-3. Water is attracted to the glass tube by adhesion and a thin film flows up the side of the tube. The force of cohesion between water molecules causes more water to be added to the flow, figure 4-4. The liquid rises to the point where gravity balances the adhesive and cohesive forces. The narrower the tube, the higher the water column can rise. A common blood test is another example of capillary action. A thin tube is touched to a drop of blood on the fingertip and blood rises into the tube.

   Capillary action, the additive effect of adhesion and cohesion, holds soil water in small pores against the force of gravity. The fact that soil water can move in directions other than straight down is also due to capillary action. The smaller the pores, the greater that movement can be.

Soil-Water Potential.   Plants obtain moisture by drawing off water from the films surrounding soil particles. The difficulty of the process depends on the strength of the force attracting the water molecules to the soil particles. Until recently, this force was measured by the *soil moisture tension* (SMT), which stated how much "suction" is required to pull the water away from the soil particles.

   Currently, *soil-water potential* is the concept used to measure soil forces. This is defined as the work water can do when it moves from its present state to a pool of water in a defined reference state. More simply, one can think of it as the

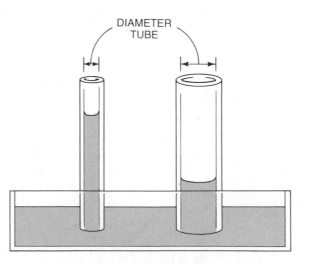

Figure 4-3. Capillary action is shown by the movement of water upward against gravity in a capillary tube. The thinner the tube, the higher the column of water rises. Small pores in the soil can act as capillaries.

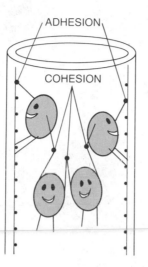

Figure 4-4. In capillary action, water rises against gravity because adhesion pulls a few water molecules up the side of the glass. Cohesion causes other water molecules to be pulled along.

potential energy contained in soil water. As is the rule in all energy transactions, water always tries to achieve a lower energy state or decrease its potential energy. This rule dictates the behavior of water in the soil.

As an example, consider raindrops falling on dry soil. This water is capable of movement, and since movement is itself a form of energy, the water is in a high-energy state. This means that the water can do work, such as eroding the soil. Thus, it has a high water potential. The water can move to a lower energy state (lower potential) by getting stuck (adsorbed) to a soil particle. Its movement becomes limited and it can no longer do work, such as erosion. The water loses energy in the form of heat (soil being wetted does warm slightly) and moves to a lower water potential.

Now consider a soil water molecule located far from a soil particle (figure 4-5). At this distance from the particle, capillary forces are weak, so the water molecule can move about. It still contains some potential energy and is at a higher water potential than it could be. Water molecules touching the soil particle directly, on the other hand, are very tightly fixed and cannot move. These molecules have a very low water potential. One can make the rule that *the lower the soil-water potential, the more tightly water is adsorbed to soil particles.*

Soil-water potential consists of the sum of several separate forces. Three of these forces are sufficiently important to discuss here. In most soils, the main force is the one just described, which is called the *matric potential.* The matric potential is always zero or less (a negative number). This is because of the definition of potential. Adsorbed water has less ability to do work than free water

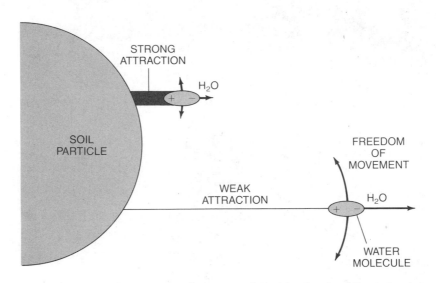

Figure 4-5. Energy level of water molecules near a soil particle. The closer the molecule is to a soil particle, the more restricted its movement and the lower its potential.

in a pond, which is defined as zero potential. Put another way, rather than being able to do work, work must be applied to adsorbed water to move it.

A second force is the *gravitational potential.* This arises from the fact that soil water is elevated above the water table and so carries potential energy from gravity. To achieve a lower energy state, the water simply percolates through the soil to a lower elevation.

The third force is important only in salty (saline) soils. Salt ions in the soil solution can also attract water molecules, since ions also carry a charge. The oxygen end of a water molecule, for example, will tend to stick to a postively charged ion. Thus work has to be done to pull water off the ion.

Water potential can be expressed in several different units. Currently the official unit of water potential, acceptable for scientific publications, is the *megaPascal* (MPa). Still in common usage is the older term *bar*, equivalent to 10 MPa, and slightly less than one atmosphere pressure. For simplicity, we will use the term bars.

The potential in soil water is a negative number because the largest part of the potential, the matric potential, is a negative value. The negative number means that work must be done to move the soil water. The larger the absolute value of the negative number, the more work must be done. Figure 4-6 shows a water film with bars of soil potential. Note that the force varies through the film—the closer to the particle, the larger the negative potential, and the more work must be done to pull the water away. The potential for water in the soil, as a whole, is expressed as the most weakly held water in the soil.

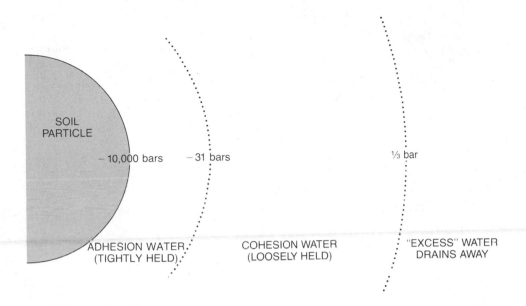

Figure 4-6. A water film, showing bars of potential. A plant root must work to overcome the potential.

TYPES OF SOIL WATER

Consider what happens after a heavy rain. At first, all soil pores are completely filled with water. This is called *saturation*; see B in figure 4-7. All of this water does not normally remain in the soil pores. In larger pores, some water is too far from the nearest surface to be attracted by the particles to overcome gravity; the gravitational potential exceeds the matric potential. The extra water, called *gravitational water,* drains through the soil profile, usually within 24 to 48 hours in a well-drained soil. As the soil drains, it pulls air in to fill the large soil pores. This action provides new oxygen-rich air to plant root systems.

Eventually drainage ceases. The soil moisture level at that point in time is called *field capacity.* At field capacity, the remaining water is close enough to the surface of a particle to be held against the force of gravity. The soil-water potential is about -1/3 bar (see C, figure 4-7). Air fills the large pores, and thick water films (cohesion water) surround each soil particle. Plant growth is most rapid at this ideal moisture level, because there is enough soil air yet sufficient water is held loosely at high potential.

Once drainage stops, plant and evaporation continue to remove cohesion water, shrinking the soil water films. As the water films become thinner, the remaining water clings more tightly, being held at lower potential (larger negative value). It becomes increasingly difficult for plant roots to absorb water. Eventually, at the *permanent wilting point,* most of the cohesion water is gone and the plant can no longer overcome the soil-water potential. The plant wilts and dies. The potential at this point varies according to plants and conditions but is generally about -15 bars.

Beyond the wilting point, some capillary water remains but is unavailable to plants. The capillary water may also evaporate, leaving only the thin film of adhesion water (see *D*, figure 4-7). This point is called the *hygroscopic coefficient,* the point at which the soil is air dry. This *hygroscopic water,* as it is called, is held to particles so tightly, between -31 and -10,000 bars, that it can only be removed by drying the soil in an oven. In fact, the strength of the soil water potential

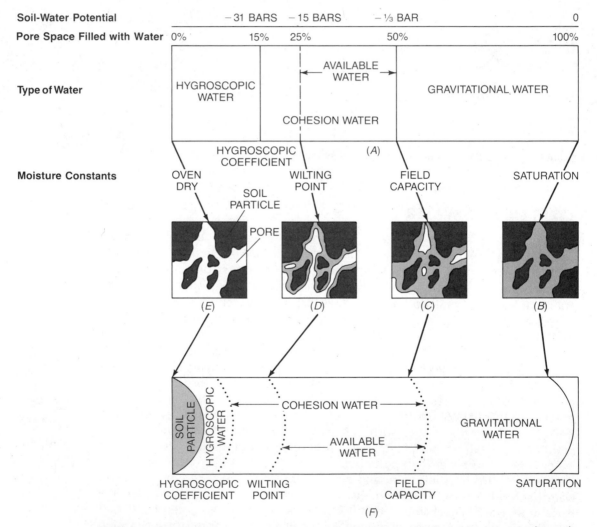

Figure 4-7. In *A*, types of soil water are shown, plus the moisture constants, the corresponding soil-water potential in bars, and the amount of average pore space. Parts *B*, *C*, *D*, and *E* illustrate the soil pores at each constant, and *(F)* illustrates the water films.

is so great at this stage that if oven-dry soil is exposed to air, it will bind water vapor from the air until the soil moistens to the hygroscopic coefficient.

Available Water.   *Available water* is that part of soil water that can be absorbed by plant roots. Soil scientists consider gravitational water to be largely unavailable, because it moves out of the reach of plant roots. If the excess water is unable to drain away, roots become short of oxygen and fail to function. Hygroscopic water cannot be removed by roots, so it is also unavailable to the plant. Only some cohesion water can be used by plants. Available water is defined as lying between the field capacity and the wilting point or between -1/3 and -15 bars. In a medium soil, available water amounts to about 25% of the water held at saturation.

WATER RETENTION AND MOVEMENT

Both the retention of water and the movement of water in the soil are governed by the energy relations just described. We can begin by looking at water retention.

Water Retention.   How much water can a particular soil retain and make available to plants? Actually, these are two separate questions. Not all the water film surrounding a soil particle can be drawn on by plants, so only a portion of the total water-holding capacity of a soil can be said to be plant-available. Both the total water-holding and the available water-holding capacity are based mainly on soil texture. Let's look at the effect of each soil separate.

Sand grains are large, so the internal surface area of a soil high in sand is quite low. Thus, there is little surface to hold water films. In addition, the pores are large enough that much of the volume of each pore is too far from a surface to retain water against gravity. The opposite is true of clay soils— they have small pores and a large internal surface area. Thus, soils high in sand have a low tottal water-holding capacity, while soils high in clay have a large water-holding capacity.

Not all of this water is available to plants, however. In a soil high in clay, clay particles are crowded together tightly, leaving tiny pores. Any one water molecule occupying one of the pore spaces will be close to a clay surface, therefore tightly bound. Most of the water in a high-clay soil is held at low water potential. Sand is the opposite. With large pores, much of the water can be fairly distant from a grain, and therefore be held at high potential.

This leads to two rules. First, water in fine soils is held at low potential and water in coarse soils is held at high potential. Thus, it is easier for plants to remove water in coarse soils than in fine soils. Second, since most of the water in high-clay soils is held at low potential, much of the water is not available to plants. In contrast, most of the water in a sandy soil is available.

Silt particles, and to some degree very fine sand, are a special case. They are small enough that there is a high surface area to hold water. The pores are small enough to hold large amounts of water by capillary force but large enough that much of it is held loosely at high water potentials. Thus soils high in silt hold large amounts of plant-available water.

To hold the largest amounts of plant-available water, then, a soil should have a mixture of large and small pores with many of the medium-sized pores caused by silt and very fine sand. Figures 4-8 and 4-9 show the effects of texture on soil-water retention. There are several important points to note in the figures:

- The fine sandy loam holds much more water than regular sand loam, reflecting the influence of very fine sand.
- The finest soil, clay, has the highest total water-holding capacity. But note that it holds no more *available* water than a sandy loam.
- Medium-textured soil has the highest available water-holding capacity. Note that the best soil is a silt loam.

Water Movement. Horticulturists suggest that trees be watered by letting a hose trickle on the ground under the tree for a few hours. How will the water move into the soil?

Figure 4-10 shows the penetration of water over time for two soil textures. First, water infiltrates the soil, then it percolates downward through the soil profile. The distance, direction, and speed of travel are set by gravity, matric forces, and hydraulic conductivity.

Directly below the nozzle of the trickling hose is a column of percolating water. This water is gravitational water, moving under the influence of gravity (or moving in response to the gravitational potential). It is called *gravitational flow*. Gravitational flow only occurs under saturated conditions, when the matric potential is so low that it cannot hold water against gravity. Because it occurs under such conditions, it is also called *saturated flow*. Saturated flow resembles the flow of water through water pipes.

Texture	Field Capacity (in./ft)	Permanent Wilting Point (in./ft)	Available Water (in./ft)
Fine sand	1.4	0.4	1.1
Sandy loam	1.9	0.6	1.4
Fine sandy loam	2.5	0.8	1.8
Loam	3.1	1.2	1.95
Silt loam	3.4	1.4	2.03
Clay Loam	3.7	1.8	1.95
Clay	3.9	2.5	1.4

Figure 4-8. Water retention of several soil textures. (Adapted from Water: *The Yearbook of Agriculture*, USDA, 1955)

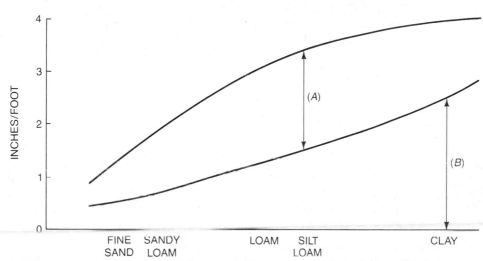

Figure 4-9. Water-holding capacity of soils at field capacity and wilting point. Available water lies between the two curves. *(A)* Silt loam holds the most available water. *(B)* While clay holds the greatest amount of water, most of it is unavailable. (Adapted from Water: *The Yearbook of Agriculture*, 1955, USDA)

In a water pipe, water creates friction as it flows along the pipe wall. The friction causes a drop in water speed and pressure. The narrower the pipe, the greater the friction losses. Similarly, water flowing through soil suffers friction losses, and the smaller the soil pores, the greater the losses. Thus coarse soils permit more rapid water movement, or have greater hydraulic conductivity. Figure 4-10 reflects this fact.

Since the force of gravity is directed straight down, how does water spread sideways, as it obviously does in figure 4-10? Lateral movement is by *capillary flow*, the flow of cohesion water through capillary pores in the soil. Recall that water tries to achieve a low energy state. It can do this by moving from areas of moist soil (high water potential) to areas of less moist soil (low water potential). One can picture it as water flowing over a dam from an elevated position (high potential) to a lower position (low potential). Capillary flow can occur in any direction—up, down, or laterally (figure 4-11). In fact, the downward movement of water is aided by capillary flow. Dry soil below pulls water downward from the wetted soil above.

Capillary flow occurs in unsaturated soil, and so is called *unsaturated flow*. It depends on an unbroken film of water spreading through a series of connected capillary pores. This is like siphoning water—if a bubble gets in the siphon tube, water stops flowing. In sandy soils, many large pores often contain "air bubbles." In unsaturated conditions, these bubbles break the continuous water film. Thus, the clay loam soil in figure 4-10 is capable of carrying water farther laterally than the sandy loam.

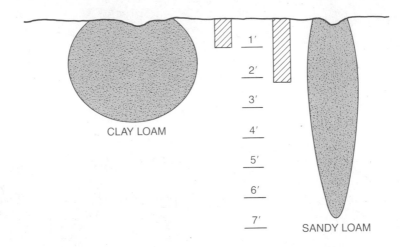

Figure 4-10. Wetting pattern in soil. This represents the way water penetrates the soil from a furrow or trickling hose. The shaded areas show penetration after 24 hours. The bars show how deep two inches of water penetrates for each soil texture.

Figure 4-11. Subsurface drip irrigation in a California tomato field. Water trickles out of the device being shown. From here, water moves by capillary action in all directions to moisten crop roots. (Courtesy of USDA-ARS)

As long as the hose in the figure 4-10 keeps trickling, the core of saturated soil under the hose can supply free water to keep water films thick where capillary flow is occurring. What happens if the hose is shut off? The column of saturated soil drains to field capacity, and the soil begins to dry. Then the water films thin out and begin to become discontinuous. Under these circumstances, unsaturated flow slows dramatically. As a result, unless there is a nearby source of saturated soil, capillary flow occurs very slowly over very short distances, often a fraction of an inch per day.

To summarize, saturated flow downward is the movement of gravitational water, usually percolation. It occurs most rapidly in coarse soils with large pores. Unsaturated flow is the movement of water from moist to dry soil by capillary action, in any direction. Large pores in sand tend to inhibit unsaturated flow.

The Wetting Front. Figure 4-10 shows that water advances into the soil with a definite wetting front—wet behind a distinct line, dry ahead of it. In the dry soil, pores contain much air, and water films are thin and discontinuous. These pores have little ability to "pull" water deeper into the soil. The water cannot move forward until water films behind the front are so thick that soil particles let go of the water. Behind the front, soil is nearly saturated; immediately ahead of the front, soil is still unwetted. One cannot "half wet" a soil volume. Either all of a soil is wet, or else part is wet and part is dry. To water crops properly, growers must understand this fact.

Capillary Rise. Water moves upward in the soil as the surface layers dry, moving from areas of high potential to areas of low potential. This upward movement is called *capillary rise.*

Capillary rise has important consequences. Because of it, soil water can evaporate from the soil surface. However, capillary rise does not continue until the entire soil column dries. When the surface dries, cohesive films become too thin for capillary flow, so upward migration almost halts. This creates a sharp boundary between the dry surface and a moister soil below. The boundary protects against further rapid moisture loss.

Generally, capillary rise does not bring much water from deeper in a soil to plant root systems, because unsaturated flow works slowly over a short distance. The plant still extends its roots into lower horizons to draw water. However, where there is a water table near the surface that is a source of saturated soil, capillary rise does carry moisture to plant roots. One method of irrigation makes use of capillary rise. Some potted plants are even watered from below by dipping the pot bottoms into a saturated material like wet sand.

In dry climates, as water rises to the surface it carries dissolved salts with it. These salts are left behind when water evaporates, so they build up in the root zone. Salts may reach levels that injure the plant. In arid zones, the problem is counteracted by overwatering slightly during irrigation, washing excess salts deeper into the soil.

Effect of Soil Horizons.   Figure 4-10 suggests that water flows differently in soils of different textures. Normal soil profiles contain horizons that may differ in texture. What happens when percolating water meets the boundary between two horizons of very different characters?

In figure 4-12, water percolating through a fine-textured horizon encounters a coarser layer of soils. One might expect that percolation would speed up because of the larger pores of the coarse layer (greater hydraulic conductivity). The soil would become more droughty. In fact, percolation instead slows down. As the figure shows, the water does not enter the coarser layer at first, but instead it spreads out along the boundary. Why?

Recall that water is being pulled down by capillarity as well as being pushed down by gravity. The large, noncapillary size pores of the coarse layer exert much less "pull" on the advancing water than do the fine pores above the boundary, so the clay will not "let go" of the water. The front spreads rather than being pulled across the boundary. When the topsoil is almost saturated, its potential is high enough that sand can pull the water downward.

Commonly, the A horizon is coarser than the B horizon, so a sharp boundary exists between a coarse upper layer and a finer lower layer. As expected, the finer pores pull the water down. However, since water naturally drains more slowly through fine soils, water will still build up above the fine-textured layer.

As a rule, then, any sharp boundary between two layers of dissimilar texture retards drainage. This is often helpful because it can improve the water-holding capacity of a topsoil. In some situations, though, the slowed drainage may keep the soil saturated long enough to injure roots. Landscapers encounter, or even create, these interfaces between different textures. An example is topdressing a yard with "black dirt" before laying sod. This will be examined in more detail in chapter sixteen.

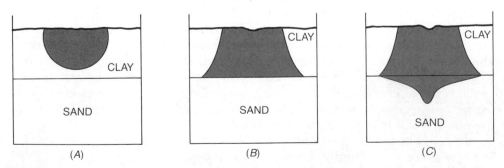

Figure 4-12. Effect of textural boundaries. *(A)* Water infiltrates heavy surface soil normally. *(B)* Wetting front strikes layers of sand, and water spreads out along boundary. *(C)* When the clay is saturated, water begins to move into the sand.

Vapor Flow.   So far, this chapter has covered the saturated flow of gravitational water and the unsaturated flow of cohesion water. What happens when the soil becomes too dry for either mode of water movement? Under dry conditions, vapor flow becomes important. *Vapor flow* is simply the movement of water vapor from moist soil to drier soil, where it may condense if the particles are colder. As a rule, vapor flow is very slow and moves very little water.

HOW ROOTS GATHER WATER

The absorption of water by plant roots is also governed by the soil-water potential. At field capacity, water films are as thick as they can be. Therefore, the soil-moisture potential is high, and water can move easily into a root hair. A root hair pulls a bit of water from this thick film (figure 4-13), so the film becomes thinner at that point and the water potential decreases. This causes capillary flow toward that point from nearby thicker areas of the water film. Therefore the root hair can continue to draw on the water around it until the water film becomes too thin. The distance in which this capillary flow can occur is less than an inch.

Most water absorbed by plants is "sucked" into the plant to replace water lost by transpiration. A useful comparison is to a person drinking out of a straw. As the person removes water from the top of the straw, water flows upward in an unbroken column from the pool of water in the glass. A plant can be compared to an unbroken column of water, beginning at the soil volume that supplies water to a root hair and ending in the air outside the leaves. As the leaves lose moisture to the air by transpiration, leaf tissues enter a state of low water potential.

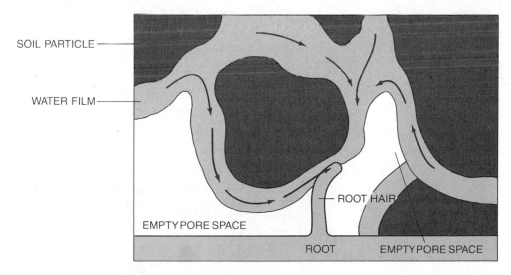

Figure 4-13. Water withdrawal by plants. As a root hair removes water from a pore, nearby water flows by capillarity through the film toward the root to equalize the potential at the point the root hair is withdrawing water.

With all the leaves transpiring, the entire water column to the root hair acquires a low potential. As a result, water flows into the root.

Normally it is only the matric potential that resists the flow of water into a root. As the soil dries, two things happen. First, as water films thin, soil potential decreases. Thus, the soil is "pulling back" harder on the water. Second, thinning water films slow capillary movement of water to the root. As a result of both actions, as the soil dries it becomes increasingly difficult for plants to extract water from the soil.

In saline soils, the force of the matric potential is joined by the salt potential. In this case, both salt ions and particles are holding back soil water. As a result, even moist soil could be at quite a low potential, keeping water from entering roots. Thus saline soils present special water-management problems.

Pattern of Water Removal. The limited capillary flow cannot supply all the water needs of a plant. Thus, it is very important that roots be able to spread easily, seeking moist soil. Further, roots tend to grow best where oxygen and moisture are highest. Because the oxygen level is highest near the soil surface, plants first use the water near the surface. As that zone is depleted, absorption shifts downward to zones where the soil is still moist (figure 4-14).

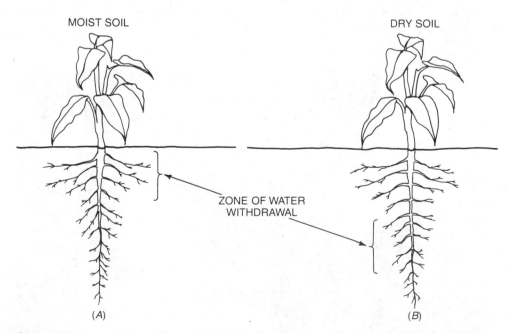

MOIST SOIL

DRY SOIL

ZONE OF WATER
WITHDRAWAL

(A)

(B)

Figure 4-14. Water withdrawal from soil. *(A)* When the whole rooting zone of a plant is moist, it draws first from soil near the surface. *(B)* As that dries, the plant begins to draw more heavily from deeper soil.

The pattern of water absorption suggests how irrigation water should be applied. Plants develop only shallow roots and lack deeper roots when irrigation is too shallow. The plant is unable to exploit lower soil levels quickly when the surface dries, and is, therefore, vulnerable to drought. Thus, proper irrigation should wet most of the rooting depth of the crop.

MEASURING SOIL WATER

People who design or use irrigation systems need to be able to measure the amount of water in a soil. They also need terms to name the amount of water present. Four methods are common: gravimetric measurements, potentiometers, resistance blocks, and neutron probes. At the base of all these are gravimetric measurements.

Gravimetric Measurements. Gravimetric methods measure the percentage of soil weight that is water. The percent moisture by weight can then be converted to other useful quantities.

Weight Basis. To measure the percent moisture of a soil sample by weight, the sample is weighed and the weight recorded. The sample is then oven-dried, and the dry weight is noted. The difference between the two weights is the weight of water in the soil. The percent moisture is the amount of moisture divided by the oven-dry weight:

$$\% \text{ moisture} = \frac{\text{moist weight} - \text{dry weight}}{\text{dry weight}} \times 100$$

As an example, suppose one needs to measure the moisture percentage of a soil at field capacity. A sample is taken two days after a heavy rain. If the sample weight was one pound when wet and 0.80 pound when dry, the moisture percentage would be

$$\% \text{ moisture} = \frac{1.0 \text{ pound} - 0.80 \text{ pound}}{0.80 \text{ pound}} \times 100 = \frac{0.20}{0.80} \times 100 = 25\%$$

Volume Basis. It is often more useful to calculate the percent moisture on a volume basis. However, it is impractical to measure a volume of water in the soil. This problem can be solved by making a percent weight determination and converting it to percent volume using soil and water densities (density being weight per volume). The equation for the conversion is

$$\% \text{ water by volume} = \% \text{ water by weight} \times \frac{\text{soil density}}{\text{water density}}$$

If we are using U.S. Customary Units, then density can be expressed as pounds per cubic foot. Water has the density of 62.5 pounds per cubic foot, so the formula then would be

$$\% \text{ water by volume} = \% \text{ water by weight} \times \frac{\text{soil density}}{62.5}$$

In the example above, if the bulk density of the soil sample were 90 pounds per cubic foot, the percent water by volume is:

$$\% \text{ water by volume} = 25\% \times \frac{90}{62.5} = 36\%$$

Soil Depth Basis. A meteorologist states how much rain falls in inches of water, and irrigation is measured in inches of water as well. In saying that one inch of water fell, the meteorologist means that if rainfall were caught in something like a cake pan, it would fill the pan to a one-inch depth. Inches of water is a convenient, easily visualized unit that can also be used to measure the amount of water in a soil.

Let's say one could take one cubic foot of soil and squeeze all the water out of it into a one square foot cake pan. How many inches of water would be in the pan? If there were two inches of water, the soil had two inches of water per foot of soil. This can be calculated simply by the equation:

$$\text{inches water per foot soil} = 12 \text{ inches} \times \frac{\% \text{ water by volume}}{100}$$

In above sample, then:

$$\text{inches water per foot} = 12 \text{ inches} \times 36\%/100 = 4.32$$

In the sample, each foot of soil depth contains 3.36 inches of water. If a soil profile were three feet deep, and each foot was the same, then the total of the entire profile would be 10.08 inches of total water.

Inches per foot is a common measurement used in irrigation. Irrigation also uses the *acre-inch*, the volume of water that would cover one acre of soil one inch deep. In the metric system, the measurement equivalent to an inch per foot is centimeter of water per centimeter of soil.

From these calculations, one can determine how much water a soil holds at each moisture constant or how much of each type of water a soil can hold. For instance, if a soil contained three inches of water per foot of soil at field capacity and one inch per foot at the permanent wilting point, then the soil holds two inches per foot of soil of available water.

Potentiometers. In practice, it would be a bother to make gravimetric measurements each time one wanted to decide whether or not the soil needs watering. Besides, from the plant's point of view, the important thing is not how much water is in the soil, but the water potential it is being held at. A device called a *potentiometer,* or *tensiometer,* acts like an "artificial root." It measures soil-moisture potential and so gives a "root's eye view" of how much water is available.

A potentiometer is a plastic tube with a vacuum gauge at one end and a porous clay cup at the other (figure 4-15). The tightly closed tube is filled with water, then buried with the gauge sticking out. The dry soil outside the tube pulls water out through the clay cup, creating a partial vacuum inside the tube. The gauge records the vacuum. Potentiometers function best at potentials between 0 and -.8 bar—a narrow range but an important one to plants.

If a grower wanted to use a potentiometer to measure the percentage of water it would be necessary to calibrate the device. Calibration involves making gravimetric measurements at different gauge readings to prepare a calibration chart. Calibration must be done for each soil because the same amount of water will be held at a different matric potential in different soils.

Resistance Blocks. Another device for measuring soil moisture is the *resistance block* (figure 4-16) or *Bouyoucos block,* after the person who introduced the device. Two electrodes are imbedded in a block of gypsum, fiberglass, or other material. When in the soil, the device measures resistance to electrical flow

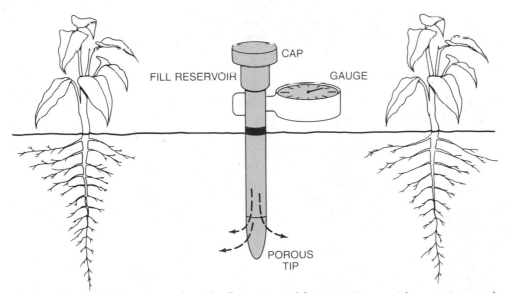

Figure 4-14. A potentiometer. As dry soil pulls water out of the porous tip, a partial vacuum is created inside the tube that is measured by the gauge.

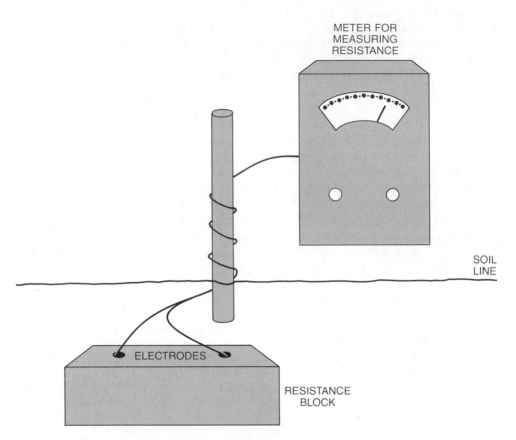

Figure 4-16. When using the resistance block, the meter reads the resistance to electrical flow between two electrodes buried in the block. The drier the soil, the greater the resistance.

between the two electrodes. It is more difficult for electricity to flow in dry soil than moist soil, so the reading indicates moisture level. Resistance blocks work well between the field capacity and wilting point.

Actually, pure water conducts electricity very poorly. It is ions in solution that carry the electrons of electrical flow. Therefore, most resistance blocks sense both water content and salt content of the soil. These blocks must be calibrated for each soil to obtain moisture readings, because of the different salt content and matric potential of different soils.

Neutron Probe. A *neutron probe* is a long tube containing a radioactive material that emits a stream of neutrons. The tube is inserted into a steel tube in the soil so the neutrons stream into the soil. The behavior of the neutron stream depends on the amount of water in the soil and can be measured. Like the other devices, the neutron probe must be calibrated for different soil types.

Figure 4-17. A neutron probe. The sensing probe is lowered into a tube imbedded in the soil. Note the symbol for radioactivity on the face of the unit. (Courtesy of USDA-ARS)

While the neutron probe is an effective, simple device to use, it is rather expensive. Its radioactive content also subjects it to legal restrictions. Therefore, the neutron probe is less practical for general use than the potentiometer and the resistance block.

SUMMARY

Soil-water potential dictates the behavior of water in the soil. Based on their differing potential, three types of soil water can be defined. Gravitational water drains away under the force of gravity. Capillary water is loosely held by cohesion and is mostly available to plants. Hygroscopic water is too tightly held by adhesion to be used by plants.

When all soil pores are occupied by water, the soil is saturated. Most plants do not live well in saturated soils because of the lack of air for the roots. At the field capacity, the last bit of excess water has drained away, and pores are occupied by air and water. Plants grow best at field capacity. As the soil dries further, plants must work harder to absorb moisture, so water stress increases until the permanent wilting point is reached. When the last of the water that can be removed by plants or evaporation is gone, the hygroscopic percentage is reached.

Water moves through soil by both gravity and capillarity. Both forces contribute to downward movement, but only capillarity carries water upward or laterally. Downward movement is most rapid in coarse soils, while lateral movement is most extensive in fine soils.

Water penetrates soil along a wetting front. Below the advancing front, soil remains dry. Whenever a wetting front meets a sharp boundary between soil layers of different texture, percolation slows. This may either improve the water-holding capacity of a soil or cause drainage problems.

Fine-textured soils have the highest water-holding capacity but medium-textured soils retain the greatest available water.

Water is absorbed into plants because transpiration causes low potential inside the plant. As the root draws from a water film, more water moves towards the root from nearby by unsaturated flow. Since such flow is slow, the extension of roots into moist soil is critical. The water absorbed by plants is used in photosynthesis, to transport materials within the plant, and to keep plant tissues stiff.

REVIEW

1. Once a plant wilts, it dies (True/False).
2. Hydrogen bonding is responsible for adhesion and cohesion.
3. The movement of water is governed by the rule that it will flow from areas of high potential to low potential.
4. In most soils, matric potential dominates the other forces on soil water.
5. As soil dries, its matric potential decreases.
6. The soil faction that holds the most available water is clay.
7. Capillary movement can occur in any direction but up.
8. A layer of sand below a layer of clay will slow percolation.
9. As long as the whole soil profile is moist, plants will pull water mostly from the top soil layers.
10. The absorption of water by roots is powered by transpiration loss of water from the leaves.
11. The soil texture that holds the most available water is _____
12. The inner film of water around a soil particle is _____ water, the outer film is _____ water.
13. The _____ a capillary tube is, the more tightly it holds water and the farther water can rise.

14. The type of water considered to be available to plants is called _____ water.
15. The portion of the soil water potential caused by gravity is the _____ potential. The portion caused by adhesion/cohesion is _____ .
16. Water that moves by percolation is _____-water.
17. The most water a soil can hold against gravity is at the _____ .
18. Comparing equal amounts of water being dripped onto a clay and a sandy soil, it will percolate most deeply in the _____ .
19. A _____ can be used to measure soil water potential.
20. Water moves sideways in the soil by means of _____ action.
21. At what moisture level will roots suffer from lack of oxygen?
22. Which texture soil would have to be watered most often, coarse of fine? Which would need more water each time?
23. A resistance block provides information about soil moisture by actually measuring what?
24. A 112 gram sample of soil, after drying in an oven, weighs 100 grams. Assume a bulk density of 1.4 gms/cc. Calculate percent moisture by weight, by volume, and inches of water per foot soil.
25. Samples of soil are weighed several times after a rain. The weights are as follows:

immediately after the rain:	132 grams
two days after the rain:	116 grams
when plants wilt:	108 grams
after drying in oven:	100 grams

    Using these values, calculate the inches of available water per foot soil for this sample at field capacity. Assume a bulk density of 1.4 gms/cc.
26. Potted houseplants often have a white crust on top of the soil. Explain why.
27. In order to avoid staining the table, a homeowner is careful to never put so much water into a potted plant that it drips out the bottom. Explain why the plant is not being effectively watered.
28. Explain why shallow irrigation should usually be avoided.
29. Explain why dissolved salts in the soil increase the soil water potential. How does this affect plants?
30. In the metric system, density can be measured as grams per cubic centimeter. The metric density of water is 1.0 gms/cc. Show how one can simplify the formula for percent soil water by volume in the metric system.

SUGGESTED ACTIVITIES

1. Test the effects of cohesion and adhesion using two glass microscope slides. Hold the two slides flat against each other, then pull (not slide) them apart. Now put a drop of water between them, and repeat. Feel the difference.
2. Grow several tomato plants in pots until each has several leaves. Then divide the plants into three groups, watering each differently. Water one group so often the soil stays wet. Water the second group when the soil surface dries. Water the third group only when the plants wilt. Observe the differences in growth.

chapter five

Water Conservation

(Courtesy of USDA-ARS)

Here's water going to waste off a field in Tennessee. The rainfall there is adequate, but for many parts of the country, keeping this water on the field can greatly improve yields and profitability. Regional droughts in the past decade have driven home the importance of our dwindling water supplies. Water conservation can only become more critical. What are our water supplies, and how can growers best use them? This chapter has some suggestions.

OBJECTIVES

After completing this chapter, you should be able to.

- explain the source of our fresh water supplies
- explain the need for water conservation
- describe ways to make better use of water
- discuss water quality

TERMS TO KNOW

aquifer	groundwater	strip-cropping
arid climate	hydrologic cycle	stubble-mulching
buffer strip	mulch	subsoiling
consumptive use	precipitation	surface water
contour tillage	runoff	terracing
crop rotation	semi-arid climate	water table
evapotranspiration	soil pitting	water-use efficiency
furrow diking		

THE HYDROLOGIC CYCLE

Of the nine planets occupying our solar system, only earth is known to contain large amounts of liquid water. Geologic evidence suggests water came from inside the earth while it was still evolving and was spewed out of volcanoes as water vapor. The condensed water naturally flowed to the lowest point on the earth's crust; even today 97% of our water remains in oceans. Since then little water has been created. Existing water is simply recycled by the hydrologic cycle (figure 5-1).

The *hydrologic cycle* is like an engine that transports water from the ocean to land and back again. The engine is fueled by the sun's energy. Air moistened by evaporation from the surface of the ocean passes over the continents, where it is shifted upward by warm air rising from the land mass. When the moist air has risen high enough, water vapor condenses into *precipitation* (rain, snow, and hail). Some of the rainwater runs into streams and lakes. Most of this water finds its way into rivers that finally flow into the ocean. Other water is absorbed into soil to later evaporate or be used by plants. Finally, some rainwater percolates into the *water table*, which is the upper surface of saturated underground material.

Let's look more carefully at the point in the cycle when precipitation lands on the soil. The water proceeds in the cycle by one of four paths:

● Rainfall or snowmelt that cannot be absorbed into soil fast enough runs into streams or lakes. This water is called *runoff*.

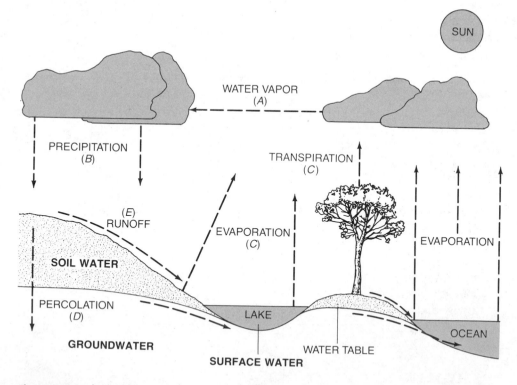

Figure 5-1. In the hydrologic cycle, water cycles between the ocean and land, turning saline ocean water into fresh water available for our uses. Water evaporates from the ocean surface and falls on the land as precipitation. The water then finds its way back to the ocean by a number of routes. The letters labeling the cycle refer to the quantities listed in figure 5-2.

- Gravitational water percolates below the root zone of plants. Some may enter the water table.
- Some of the remaining water stored in the soil evaporates from the surface back into the atmosphere.
- Most water taken up from soil by plants is transpired into the atmosphere. The total water loss due to transpiration and evaporation is called *evapotranspiration*.

WATER RESOURCES IN THE UNITED STATES

Figure 5-2 shows where the United States's water resources are in reference to the hydrologic cycle. The reservoirs of water we draw on include water vapor in the atmosphere, surface water, and groundwater. Note from the figure that enormous amounts of water vapor float over the United States daily. Some work has been done trying to use this water source by "cloud seeding," most notably in North Dakota. Until we find practical ways to tap this supply without disturbing weather patterns elsewhere, we rely on natural precipitation to satisfy our water needs.

An average of 30 inches of precipitation falls each year over the continental United States, but the supply of water is unequally distributed. Figure 5-3 shows the average annual rainfall of each state. However, the amount of rainfall does not by itself control soil moisture. Soil type, average temperature, and the season when the greatest rainfall occurs are equally important factors.

More important than simple annual precipitation is the balance of precipitation and evapotranspiration. In the eastern half of the country and much of the West Coast, more rain falls than is lost to evapotranspiration. Here water does not usually limit crop production, except on excessively well-drained soil or during occasional periods of dryness. The shaded area of figure 5-3 shows where the opposite is true. In a *semi-arid climate*, evapotranspiration slightly exceeds precipitation and moisture limits production. Special dryland farming methods (see chapter 15) or irrigation overcomes the problem (figure 5-4). In *arid climates*, where evapotranspiration greatly exceeds precipitation, crop growth

Yearly Amount	Inches/Year	Billion Gallons per Day
(A) Water vapor	—	40,000
(B) Precipitation	30	4,200
(C) Evapotranspiration	21	2,900
(D) Percolation	3	411
(E) Runoff	6	822
Consumptive use	-	106

Figure 5-2. Each year in the United States tremendous amounts of water are involved in the hydrologic water cycle. (*USDA Appraisal, Part I: Soil, Water, and Related Resources in the United States,* 1980)

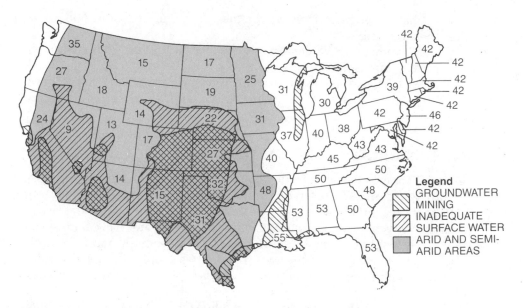

Figure 5-3. Average annual precipitation levels in the continental United States. The value given for each state is the number of inches of average annual rainfall. The arid or semi-arid areas, where evapotranspiration exceeds precipitation, are shaded. Also indicated are areas where groundwater is being mined and areas where there is not enough surface water to meet all demands. (*USDA Appraisal, Part I: Soil, Water, and Related Resources in the United States,* 1980)

depends on irrigation. Despite the importance of irrigation nationally, the USDA says that natural rainfall on fields remains the most important source of water. Seventy-five percent of our food and fiber is grown on "rainfed" fields.

The remaining three sources of water, as shown in figure 5-5, comprise stored water that society draws on: saline ocean water, fresh surface water, and groundwater. As figure 5-5 shows, saline ocean water supplies only a small percentage of the water used annually. Fresh surface water and groundwater are critical sources for irrigation and other nonagricultural uses.

Surface water occupies lakes, rivers, and ponds and covers approximately 60 million acres of American land. The water is distributed unequally; most surface water is in the Great Lakes and a few other states, especially Alaska, Texas, Minnesota, Florida, and North Carolina. Figure 5-3 shows areas where there is not enough surface water to meet both urban and agricultural demands.

Groundwater is stored in underground rock formations called *aquifers.* There is far more groundwater than surface water. Experts estimate that some 8,000 to 10,000 trillion gallons of water are contained in mainland United States aquifers. However, this water renews very slowly, averaging about three inches per year (figure 5-2). In some parts of the country, groundwater is "fossil water," deposited long ago and not supplemented since then. According to the USDA, about 25% of the nation's groundwater supplies are being "mined." This means

Figure 5-4. Rangeland in Idaho. Raising animals, like these sheep, is one way to use soil in drier climates. (Courtesy of USDA-ARS)

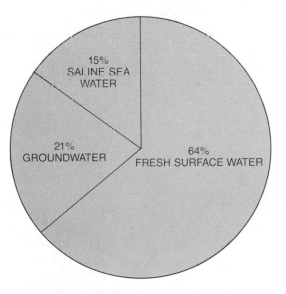

Figure 5-5. Annual water needs are met by the use of stored water. Saline seawater contributes only 15% of the annual consumption of water. We depend on fresh surface water and groundwater to meet our water needs. (*USDA Appraisal, Part I: Soil, Water, and Related Resources in the United States,* 1980)

that the water is being withdrawn more rapidly than it is being renewed. This is occurring mostly in the Southern Plains and the Southwest (figure 5-3), where the water loss per day (in an average year) is 15 billion gallons of groundwater. As vast as our groundwater resources are, problems are already occurring. In some parts of central Arizona, for example, water tables fall 7 to 10 feet a year.

However, improvements are being made. One of the nation's largest and most important aquifers, the Ogallala, which underlies the High Plains of northern Texas and adjoining areas, has seen drastic drawdowns due to irrigation. In recent years, improved irrigation methods, and the return of some farms to dryland agriculture, has slowed the overdraft.

Reasons for Conservation. Agriculture consumes more than 80% of the water used annually in this country (figure 5-6). It takes forty gallons of water to produce one egg, 150 gallons for one loaf of bread, and 2,500 gallons for one pound of beef. If forecasts are correct, growers will compete on an increasing scale with other users for the nation's supply of water. As the largest user, agriculture has a special responsibility to conserve, both for national well-being and in its own self-interest. The nation and its growers benefit from agricultural conservation efforts in three ways: preservation of water resources, increased crop yields, and fewer problems from runoff.

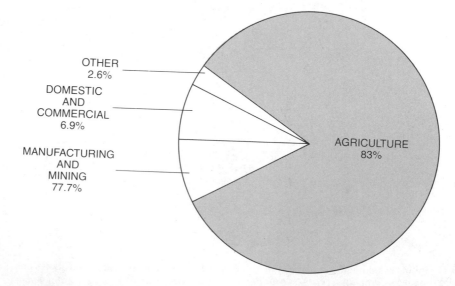

Figure 5-6. Agriculture is the largest consumer of water in the United States. (*USDA Appraisal, Part I: Soil, Water, and Related Resources in the United States*, 1980)

Preservation of Water Resources. Water shortages in the United States loom as a major problem, both for agriculture and the nation as a whole. For example, the USDA forecasts that within 30 years 6.6 million acres of land now irrigated by groundwater mining will no longer be able to be used for agriculture because of dwindling water supplies.

Agriculture itself can help reduce water shortages. The USDA projects from current trends that in the next 30 years irrigation efficiency will improve from 41% to 47%. This would mean a savings of some 24 million acres/foot of water.

Increased Yields. Water conservation methods result in improved soil moisture and better yields. Semi-arid areas respond especially well to conservation efforts. For example, about four inches of available water is needed to mature a wheat crop. Reports from USDA research indicate that each additional inch of available moisture stored in the soil can raise production from two and a half to six bushels per acre. Figure 5-7 shows some of the improved yields for wheat and the economic value of the yield.

Reduced Runoff. Some of the strategies for conserving water involve reducing runoff from farm fields. Methods for reducing runoff do more than improve soil moisture. They have beneficial side effects, such as:

- Reduced erosion and topsoil loss.
- Reduced downstream flooding because less water runs into rivers (figure 5-8).
- Reduced water pollution from fertilizers and pesticides carried off farm fields by runoff.

Water-Use Efficiency. Water-use efficiency is a good place to begin a discussion of conservation of agricultural water. Of all the water that lands on a field, little actually becomes part of the dry matter of a crop plant. Most of the water is lost to runoff, percolation, or evapotranspiration. The total amount of water needed to produce a unit of dry plant matter, such as a bushel of oats, on a given field is a measure of the *water-use efficiency.* Of course, not all of the water that fails to become a part of a plant is wasted. Much of it returns to streams or aquifers where it can be reused. However, as it passes through farm fields, the quality of the water may be reduced because it picks up sediments or chemicals.

Location	Yield Improvement (bu/acre)	Value ($/acre)
Eastern Washington	5.8	20.30
Northern Plains	2.9	10.15
Great Plains	2.72	9.52

Figure 5-7. Yield improvements in wheat for each inch of additional water provided, based on three USDA studies. The value of the additional yield is based on the average price of about $3.50 per bushel quoted on the Chicago Board of Trade on July 6, 1984. (*USDA Appraisal, Part I: Soil, Water, and Related Resources in the United States,* 1980)

Figure 5-8. Growers can help avoid the problem of downstream flooding by reducing runoff from their fields. (USDA, Soil Conservation Service)

There are three primary means of improving water-use efficiency. One is to capture more of the water from precipitation in the root zone of crop plants. This means improving the infiltration rate and reducing percolation. Having captured more water, the second step is to reduce consumptive use. *Consumptive use* is the sum of the water lost by evapotranspiration and the amount contained in plant tissues. About 1% to 10% of total consumptive use actually becomes part of the plant; 90% to 99% of the total is evapotranspiration. The third way to improve water-use efficiency is by improving irrigation systems. This subject is discussed in chapter six.

CAPTURING WATER IN SOIL

To capture into a crop's root zone more of the water landing on a field, a grower can improve infiltration and/or lower percolation. Of the two options, improving infiltration is easier. Let's review the factors that influence the amount of infiltration:

- Soil texture dictates the size of soil pores. Since water moves most easily through large pores, infiltration is most rapid in coarse soil and least rapid in fine soil.

- Soil structure can create large pores. Thus, infiltration is most rapid in granulated soil and least rapid in massive soil.
- Organic matter improves soil structure, so infiltration improves with the addition of organic matter.
- Compaction shrinks the size of soil pores. Thus, a compacted soil has a slower infiltration rate than the same soil without compaction.
- Hardpans, bedrock, or other subsoil layers can slow infiltration. For instance, a soil with a plow pan has a slower infiltration rate than the same soil without.
- Slope affects how much water seeps into the soil. There is less infiltration on sloped land than on flat land.

Soil structure is worth a second look. The structure of the entire soil profile is important, but the structure at the very surface of the soil is most critical to infiltration. As noted earlier, the soil surface is sealed by the shattering of surface aggregates by raindrops or puddling. As a result, a crust may be formed that reduces infiltration significantly.

The list of factors influencing infiltration suggests two problems that lower the amount of infiltration. One problem is low water-intake rates and the other is runoff due to slope. Let's look at methods for dealing with slope first.

Capturing Runoff. *Terraces* have long been used to capture runoff water. In some parts of the world, whole mountainsides are terraced, allowing the mountain slope to be farmed. Terraces consist of a series of low ridges and shallow channels running across the slope, or on the contour, as shown in figure 5-9. When water begins running down the slope, it runs into the terraces, where it gathers while it seeps in.

All terraces are built to control runoff. However, in humid areas, the main concern is to control erosion. In drier areas, the primary purpose of terracing is to increase moisture in the soil. Terraces for erosion control are designed partly to channel water safely off the field (Chapter 18). To save moisture, terraces are designed to cause ponding of water on the terrace, giving water time to infiltrate. Figure 5-9 shows the cross section of a flat-channel terrace built for this purpose.

Terraces are very effective at controlling runoff. They are also expensive to build and maintain, although some modern tools have improved the efficiency of terrace construction. The extra crop yields that result from the water conservation may not be enough to justify the cost of terracing in all cases. A newer style of terrace, the mini-bench terrace, is less costly to install and works well on slopes of 1%–2%. These small terraces measure about one to two equipment widths wide. In any event, the added benefit of erosion control makes terraces practical in many areas.

Furrow-diking is another means to capture water in drier areas of the country. To furrow-dike, special equipment creates furrows with small ridges, or dikes, across them (figure 5-10). This creates basins that capture and hold water. Somewhat similar is *soil pitting,* a practice that creates tiny pits on rangeland to capture water.

(A)

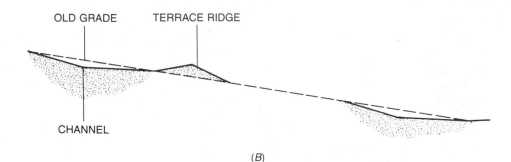

OLD GRADE   TERRACE RIDGE

CHANNEL

(B)

Figure 5-9. *(A)* Measuring water captured on a terrace in the Great Plains. In drier areas, terraces not only help control erosion, but retain water for crop use. (Courtesy of USDA-ARS) *(B)* The cross section of a flat-channel terrace designed for water conservation.

Contour tillage is practiced by operating all equipment across the slope of the field, on the contour. This practice makes many tiny ridges across the slope. Water ponds beyond these ridges, giving it time to infiltrate the soil. In contrast, tilling up and down hills creates actual channels for water flow, contributing to runoff and erosion.

Figure 5-10. Furrow dikes in a Texas field. Note that in the undiked furrows on the right, water has already run off. The diked furrows capture water for crop use. (Courtesy of *Crops and Soils Magazine*, American Society of Agronomy)

Strip-cropping slows runoff water by alternating bands of different crops across the slope (figure 5-11). One band may be a row crop that leaves much of the soil bare, such as corn or soybeans. The next band would be a close-growing crop (small grains), or a crop that completely covers the soil, such as hay. The close-growing strips slow the water, keeping it from achieving the speed it would on a continuous slope of corn.

Improving Water Intake Rate. A second way to decrease runoff is to improve the rate at which soil absorbs moisture. The soil's physical properties of texture, structure, and permeability set the infiltration rate. Chapter three lists a number of ways to improve soil permeability and structure in the section on tilth. Because compacted soil tends to shed water, growers should be particularly careful to avoid severe compaction. The following practices also help soil absorb moisture.

Subsoiling, or deep plowing, can shatter plowpans resulting from years of tillage. This may improve infiltration by letting water seep deeper into the soil. At the same time, it allows roots access to deeper soil levels. A high-power requirement for subsoiling equipment makes this method expensive, especially if it must be repeated later.

Figure 5-11. This farmer practices strip-cropping by planting alternating strips of corn and hay. Close-growing crops slow the downhill flow of water. (USDA, Soil Conservation Serivice)

Mulches, such as a layer of straw or woodchips on the soil, have a very powerful effect on absorption. When raindrops strike bare soil, the impact shatters the surface aggregates. As a result, a thin "massive" layer seals the surface and blocks water absorption. Mulches protect the soil surface from the impact of raindrops. Some studies show that mulched soil absorbs water two to four times better than bare soil. Gardeners or growers of small, high-value crops may mulch with straw, leaves, or grass clippings. Landscapers use gravel or woodchips.

Conservation tillage is a fairly new method of working soil that preserves both soil and water. Conservation tillage does not bury crop residue; rather, it leaves much of it on the soil surface to act as a mulch (figure 5-12). Figure 5-13 shows how leaving crop residues on the soil surface improved the infiltration rate in one soil. Conservation tillage often results in less compaction, which also improves moisture infiltration.

Crop rotation means that different crops are planted on a piece of ground each year or in succeeding years. Crop rotation usually has good effects on the organic matter level of the soil and on its soil structure. For that reason, fields that are rotated usually absorb water more quickly than fields continually planted to row crops such as corn.

Figure 5-12. In conservation tillage, the layer of plant debris left on the soil surface acts as a thin mulch. (USDA, Soil Conservation Service)

Crop Residue (tons/acre)	Runoff (percent)	Infiltration (percent)
0	45	54
1/4	40	60
1/2	25	74
1	0.5	99
2	0.1	99
4	0	100

Figure 5-13. Effect of crop residue on water runoff and infiltration on a 5% slope. (Source: Purdue University)

Capturing Snowfall. In some areas, capturing winter snowfall is an important way to retain natural precipitation. The key is to keep snow from blowing away from fields. Some growers leave strips of a tall crop standing at right angles to the prevailing winter wind. These *buffer strips*, which act as snow fences, should be about 50 feet apart (figure 5-14). In a North Dakota study, five foot tall wheatgrass planted in double rows 48 feet apart raised wheat yield from about 16 to 23 bushels per acre. Some strawberry growers in cold climates plant such strips between stawberry rows to capture snow for cold protection.

Figure 5-14. Wheatgrass buffer strips on a Montana farm. These barriers capture snow that would otherwise blow off the field. (Courtesy of USDA-ARS)

Crop stubble left standing over winter also captures some snow (figure 1-3). Because the stubble also lowers runoff, more of the snowmelt water is then able to soak into the ground. This practice is called *stubble-mulching*.

Water Harvest. Figure 5-15 is one example of a most elaborate way to capture water, called water harvesting. In areas where rainfall is scant, it can be harvested from small watersheds and directed to the roots of crop plants.

In figure 5-15, taken in Arizona, the land is terraced into a series of shallow peaks and valleys. Christmas trees are planted in the valleys, while the rest of the soil is treated with wax or salt to repel water. By this means, all the rainfall is directed to the crop.

Water harvesting has the potential to improve the dryland farming to certain crops like trees. Researchers also hope that the method can be used in the battle against desertification in the United States and Africa.

Reducing Percolation. Water drains very quickly through coarse soils, causing high percolation losses. It is difficult to reduce these losses. The most practical way is to improve the water-holding capacity by maintaining the organic matter level. Since topsoil is usually the most moisture-retentive soil layer, reducing erosion is also important. In fact, the USDA estimates that the loss of moisture-holding capacity due to excess erosion cost American growers two billion dollars annually.

Figure 5-15. Water-harvesting on an Arizona Christmas tree plantation. Special soil treatment funnels all the sparse rainfall to the root zone of these pines. (Courtesy of *Crops and Soils Magazine*, American Society of Agronomy)

The moisture retention of coarse soils is improved by a clay layer in the B horizon. Some researchers have imitated the clay layer by installing a thin asphalt layer two feet under the soil surface. The layer doubles the water available for crop growth by greatly slowing percolation. Though improved production of high value crops would repay the cost of installing the asphalt, few growers practice the method at this time.

REDUCING CONSUMPTIVE USE

Consumptive use is the total water "used" to produce a crop—including evaporation, transpiration, and water that becomes part of the plant. Consumptive use varies dramatically from place to place and crop to crop. It depends on several factors:

- Warm-air temperature increases transpiration from leaves and evaporation from the soil.
- Relative humidity also affects evapotranspiration. Dry air increases water losses.

- Wind increases water losses. Normally, a film of humid air surrounds leaves and covers the soil surface. Wind strips away that film of humid air.
- The plant efficiency for using water depends on the health and spread of the root system, the species and variety, and nutritional status.

Reducing Evaporation.   The factors that affect evapotranspiration are largely beyond the grower's control. However, there are some methods that can help reduce evaporation.

Growers can lower evaporation from soil by covering the soil surface either by vegetation or mulch. This has the effect of shading the soil and reducing wind velocity at the soil surface. To shade the ground with crops means to grow them so the crop canopy quickly covers the soil.

Mulches have the additional benefit of acting as a barrier to moisture movement. Loose, organic mulches like straw form only a partial barrier, and reduce but do not eliminate evaporation. After a long period of drying, the soil under such a mulch will be as dry as an unmulched one. Plastic mulches form a complete barrier and reduce evaporation for longer periods of time. Many growers of high value crops, such as berries, nursery stock, or vegetables, mulch not only to preserve moisture but for weed control and other purposes (see chapter eight). For other growers, conservation tillage leaves some crop residues that act as a mulch on the soil surface.

Reducing the number of tillage operations helps control evaporative losses. Each time the soil is worked, moist soil is dragged to the surface where it can dry out. Conservation tillage, use of herbicides for weed control, and combining tillage operations are ways to reduce tillage.

Reducing Transpiration.   It is difficult to control transpiration in plants. Three weather conditions cause high transpiration: high temperature, low humidity, and wind. Frequent light sprinkling of plants with water improves conditions by wetting the leaves and raising humidity. This practice is called "crop cooling." However, this method can hardly be called a moisture conservation measure because it uses moisture.

Another method of reducing transpiration is to plant windbreaks to cut the wind (figure 5-16). Orchardists often plant tree windbreaks. A few growers protect low-growing crops by planting rows of taller plants across the field. For example, a truck farmer may plant a few rows of sweet corn every 50 feet at right angles to the prevailing summer wind. Between these rows are planted cucumbers or another low growing crop. The tall sweet corn rows create areas of less wind and so protect the cucumbers from the effects of the wind.

An important side effect of lowering transpiration is the reduction of water stress. Even when soil moisture is adequate, plants can experience water stress in extreme weather. This happens when roots gather water at a rate that is very close to the rate of water loss through leaves. A recent study in Nebraska showed that winter wheat yields rose an average of 15% when windbreaks protected crops from drying winds and winter extremes. The study further concluded that, on the average, the cost of windbreaks would be repayed within 15 years.*

Weeds in a field also transpire moisture, so weed control is a basic moisture control measure. Weed control is a central part of *summer fallow*, a technique commonly used in the farming of grains in semi-arid areas like the Great Plains. This will be covered in detail in chapter fifteen, but it basically involves leaving the soil bare every other year to store moisture for the following season. Weeds are controlled by cultivation or chemicals so they do not pull moisture out of the soil.

Improving Plant-Use Efficiency. So far this chapter has concentrated on how to increase the amount of water available for plants in the soil. Another way to conserve moisture is to help plants make better use of soil water. This allows greater production with the same amount of water.

The efficiency of plant water use can be measured by the *transpiration ratio*. This ratio is the amount of water transpired divided by the amount of dry matter

Figure 5-16. Windbreaks lower wind speed in the field. This reduces the loss of water from evaporation and transpiration and also helps relieve water stress. (USDA, Soil Conservation Service)

*Brandle, J. R. and Johnson, Bruce B. "Windbreak Economics: The Case of Winter Wheat Production in Eastern Nebraska." *Journal of Soil Water Conservation* 39(5):339-342, 1984.

produced. The ratios are affected by climate, the specific plant considered , and the soil conditions. Figure 5-14 lists sample transpiration ratios. The higher the ratio, the greater is the amount of water needed to produce a pound of dry matter. Note that there is great variation between plants. For example, alfalfa needs about three times as much water to produce a pound of material as sorghum.

How can growers improve transpiration efficiency? Windbreaks and crop cooling can help by lowering transpiration. One important method suitable for all growers is to improve the rooting zone of plants. Compaction and plowpans limit the soil volume that a plant can exploit, leading to an increase in the transpiration ratio. Therefore, the practices listed in chapter three for improving structure and controlling compaction are important techniques.

Good soil fertility also improves water use by plants. Adequate levels of nitrogen and phosphorus increase the size and depth of roots systems and lower the transpiration ratio. For example, work in Texas showed that unfertilized (nitrogen) sorghum produced a 190-pound yield per inch of water used, while sorghum fertilized with 240 pounds/acre of nitrogen produced a 348-pound yield per inch of water used.

Some plants are well-adopted for growing under dry conditions, and the use of such plants can improve water use. New crops can be one answer, but breeders also search for more drought-tolerant varieties of established crops (figure 5-18).

Plants avoid water stress several ways. Some have extensive or deep root systems to improve water gathering ability. Some grow leathery, glossy, or thick leaves that reduce water loss. Others have special ways to store water in plant tissues, like cacti.

Crop	Transpiration Ratio	Weeds	Transpiration Ratio
Alfalfa	858	Lambsquarter	658
Field pea	747	Russian thistle	314
Sweet clover	731	Pigweed	300
Oats	635	Purslane	281
Potato	575		
Cotton	562		
Wheat	505		
Corn	372		
Sugar beet	377		
Millet	287		
Sorghum	271		

Figure 5-17. Transpiration ratios of several crops and weeds at Akron, Colorado. The ratios are the average over a five-year period.

Urban dwellers can exercise the same concept by replacing humid climate grasses, shrubs, trees, and flowers with dryland adapted species. *Xeriscaping,* as this has been called, is growing rapidly in cities of the American Southwest where landscape irrigation strains water resources. Chapter sixteen will describe xeriscaping in greater detail.

WATER QUALITY

As important as water quantity is water quality—we depend both on enough and good enough water. Like conserving water, agriculture has a special role in the preservation of our nation's water supplies. Fish-kills in trout streams of the author's home state, caused by run-off from insecticide treated corn fields, are an example. So are the widely-publicized problems with waterfowl in the Kesterton Reservoir outside Fresno, California. There selenium-laden runoff-water from irrigated fields found its way into the reservoir, causing death and deformity in waterfowl.

Figure 5-18. Drought-resistant lima beans. These lima beans, bred for drought resistance, performed well when the corn in the foreground became shriveled and brown. (Courtesy of USDA-ARS)

Agricultural sources of water pollution are difficult to pin down, compared to a pipe coming out of a factory or sewage plant. *Point sources* of pollution, like factories, are easier to identify than *non-point sources* like farmland. Much progress has been made in the control of point sources of pollution, so American agriculture can expect more attention being paid to its non-point sources.

Current agricultural issues include the seepage of fertilizers and pesticides into ground water (figure 5-19), and the pollution of surface waters. Contaminants of the latter include soil particles, pesticides and fertilizers, organic debris, and disease-causing organisms. Some of these problems will be discussed in chapter fourteen, after fertilizers have been presented.

GROUNDWATER

It has been assumed for some time that the soil matrix filters chemicals from percolating water. In too many cases, the assumption has turned out to be false. One case is very coarse soils with a high water table, especially if irrigated. Examples are wells of such areas as Central Wisconsin and Long Island, New York, that have been contaminated by the toxic pesticide aldicarb used on potatoes.

Figure 5-19. Testing groundwater in Maryland. These folks are testing groundwater for nitrogen from farm fields. (Courtesy of USDA-ARS)

Parts of the United States with limestone bedrock exhibit *Karst topography,* or land in which solution of the bedrock by water creates caves, sinkholes, and other channels from the surface to bedrock. Agricultural chemicals traveling in run-off can flow directly into aquifers through these channels.

Even in areas lacking Karst topography and excessively drained soils, chemicals can leach to surprising depths. Recent research shows that soils contain many natural channels, such as old worm holes, decayed root tunnels, and cracks. Chemicals in water can flow quickly through these channels, bypassing the filtering action of soil.

AVOIDING WATER POLLUTION

The ways to avoid water pollution are largely covered in various sections of this text. Here is a brief summary:

- Reduce runoff that can carry contaminants (earlier this chapter).
- Reduce erosion that can carry contaminated soil (chapter eighteen).
- Reduce fertilizer losses in percolating water (chapter fourteen).
- Reduce the use of pesticides, especially soil-applied, by crop rotation and other strategies. Much progress is being made by Integrated Pest Management (IPM), the carefully planned use of many pest control methods such as biological or cultural ones. IPM promises a way to reduce the need for pesticides.
- Apply and store manures properly (chapter fourteen).

SUMMARY

Water conservation is important to American society because usable water is not an unlimited resource. Our major water supplies are rainfall and the "stored" water in fresh surface and groundwater. In many areas of the nation, rainfall is inadequate and stored water is being depleted. As the major user of water, agriculture has a special obligation to make more efficient use of water and to conserve water supplies.

Each grower should have an interest in conserving water as well. Making the best use of water increases yields. Further, many water conservation measures control erosion. These measures thus help growers save the soil, their most important resource.

Many techniques for improving water-use efficiency have been outlined in this chapter. One technique is to capture more of the water that lands on a field. This can be done by terracing, contour tillage, or furrow-diking. Improving the soil infiltration rate by means of preserving soil structure, mulches, conservation tillage, and subsoiling also helps. In some areas it is helpful to trap snow in the fields.

Consumptive use can be reduced by lowering the amount of evapotranspiration. Methods include reduced tillage, windbreaks, and weed control. A large, healthy root system, possible in soils of good physical condition, helps the plant to use more of the soil moisture and reduces the transpiration ratio. Good soil fertility also helps plants make the best use of soil moisture. The use of dryland adopted plants in agriculture and landscape can greatly reduce water use.

Growers need to pay attention to their contributions to polluted water. Fertilizers, pesticides, eroded soil, and organic debris can harm environment and health.

REVIEW

1. While growers in many areas of the country rely on irrigation, most fields still depend on rainfall (True/False).
2. Most of the water absorbed by plant roots become a permanent part of the plant.
3. Soil compaction reduces infiltration.
4. Mulches reduce transpiration.
5. Adequate fertilization improves water use by plants.
6. Agriculture consumes the largest share of water in the U.S.
7. In arid climates, water losses from the soil exceed precipitation.
8. Furrow dikes can help hold water on a field.
9. The only way for water to move deeper into the soil is by percolation through the soil matrix.
10. Aquifers are water-bearing underground formations.
11. The combined loss of water from evaporation and transpiration is called _____ .
12. The most important source of water for growers in most of the U.S. is _____ .
13. The two major sources of stored water are _____ and _____ .
14. A gardener can _____ to reduce evaporation.
15. On slopes, _____ can be used to capture water until it seeps into the soil.
16. The amount of water actually used by a plant is called _____ .
17. A method that farmers can use to reduce evaporation from the soil surface is _____ _____ .
18. The upper surface of saturated underground layers is called the _____ _____ .
19. Buffer strips can help capture _____ .
20. Windbreaks help reduce _____ .
21. List three reasons for conserving water.
22. List two ways that mulching conserves moisture for crops.

23. If you saw an unfamiliar plant with thick, glossy leaves, what might you conclude about the plant?
24. Where is most of the world's water?
25. How can you treat soil pans?
26. Summarize the water cycle.
27. Distinguish between water-use efficiency, consumptive use, and transpiration ratio.
28. Summarize the water supplies of the United States.
29. As has happened to many growers in semi-arid areas of the United States, a certain grower has had to stop growing irrigated corn because it has ceased to be profitable. Develop a strategy for the grower to continue farming without irrigation.
30. What role, however small, could animal manure play in improving water use? What would you have to watch out for?

SUGGESTED ACTIVITIES

1. Observe the effect of mulching by filling two deep jars with soil to within three inches of the top. Carefully moisten the soil in each jar with the same amount of water. Mulch the top of one jar, leave the other bare, and compare the drying. Weighing each jar before and after a period of drying may also provide interesting results.
2. Analyze the water resources of your area. How much is the annual precipitation, and when does it occur? Is your climate arid, humid, or semi-arid? Where does your drinking and irrigation water come from?
3. Visit local farms that use water-conserving practices.

chapter six

Irrigation and Drainage

(Courtesy of USDA-ARS)

Were you flying over these fields in Oregon on a clear day, you would see big circles of green. Airline passengers can see these anywhere in the world where center-pivot irrigation creates islands of green in land too dry to grow good crops. Irrigation is one of man's oldest ways of improving crops. Read this chapter to learn about irrigation—and soil drainage—the way it is practiced today.

OBJECTIVES

After completing this chapter, you should be able to:

- define drainage and explain its importance
- explain the difference between wetlands and wet soils
- identify methods of artificial drainage
- identify methods of irrigation
- decide when and how much to irrigate
- name water quality problems for irrigation

TERMS TO KNOW

border-strip irrigation	saline soil	surface irrigation
capillary fringe	seep	traveling-gun irrigation
center-pivot irrigation	solid-set irrigation	trickle irrigation
drainage	soluble salts	wheel-move irrigation
furrow irrigation	subirrigation	
hand-move irrigation	subsurface drainage	
leaching requirement	surface drainage	

Many acres of American farmland suffer from one or two moisture problems—either the soil is too wet or the soil is too dry. With proper treatment, however, some of these acres have become our most productive cropland. For example, some of the richest farmland of the Corn Belt states is drained to remove excess water.

Irrigation or drainage practices have been used by humans for thousands of years. This chapter describes how and where irrigation and drainage are used today.

THE IMPORTANCE OF DRAINAGE

The word *drainage* can have two meanings. The technical meaning is how rapidly excess water leaves the soil by runoff or draining through the soil. The term is also used to describe a condition of the soil—how much of the time soil is free of saturation.

In a well-drained soil, excess water leaves the root zone of plants quickly enough that plant roots do not suffocate. Poorly drained soils, or "wet soils," remain waterlogged for long periods of time. Soils may be wet because they are naturally impermeable, like a compacted heavy soil. On such soils rainwater collects on the surface while it slowly drains into the soil. Another form of saturation comes from a high water table. In this case, soil is kept wet from below. A third cause of poor drainage is frequent flooding from rivers or runoff from higher elevations. While most drainage problems occur in depressions or other low areas, parts of a hillside may be moist where a groundwater layer meets the land surface. Water leaks out into the soil at these locations, called *seeps*.

Effects of Poor Drainage. In poorly drained soils, gravitational water fails to drain away rapidly. As a result, large pore spaces remain water filled, blocking aeration. Because soil organisms use up oxygen and release carbon dioxide, soil air becomes very high in carbon dioxide and low in oxygen. Soon roots suffocate, especially the rapidly growing parts like root tips and root hairs. The roots die and rot away. Toxic levels of carbon dioxide and other materials may also build up in a poorly aerated soil.

If the entire root zone is saturated, common upland crops like corn cannot grow. Even where the water table only approaches the soil surface, aeration can be a problem. Water moves upward toward the surface by capillary action from the saturated zone. This means there will be a wet zone above the water table termed the *capillary fringe*. The fringe further restricts root growth.

Poor drainage has other effects on plant growth as well. Plants find it difficult to absorb many nutrients from wet soils, so poor drainage limits nutrient use. Also, in arid areas, poor drainage can cause the accumulation of salts in the root zones of plants.

Slow drainage interferes with tillage and other farming operations. The soil tends to stay wet late into the spring, preventing planting operations. Because wet soil warms slowly, poorly drained soil also stays cold late into spring. Draining these soils has the effect of lengthening the growing season. For example, the USDA has reported that one drainage project in Maine lengthened the growing season three weeks. Slow drying on poorly drained soils also keeps growers out of fields after a rain (figure 6-1).

Figure 6-1. Heavy rains on a poorly drained soil have halted this haying operation. (USDA, Soil Conservation Service)

WETLANDS AND WET SOILS

While there are systems for classifying wetlands, this text will make a simple distinction between wet soils and wetlands. Wet soils are simply soils that are poorly drained. Wetlands, like marshes, maintain enough surface water to provide a wildlife habitat. Both can be drained and farmed. Wetlands are sufficiently valuable to the nation that their drainage may be a questionable practice.

Wildlife Habitat. The most obvious value of wetlands is as wildlife habitat. Twenty-six species of game or fur animals are associated with wetlands, especially waterfowl. Hundreds of other nongame animals depend on wetlands for food, protective cover, or sites to raise young. About one-third of the endangered species in the United States is associated with wetlands. In addition, wetlands provide spawning areas for many fish species and shellfish. Coastal marshes, for instance, are breeding grounds for shrimp, crab, oysters, and many important fish. The Prairie Pothole region of Minnesota and the Dakotas is the nation's main breeding ground for waterfowl.

Water Control. Wetlands may play an important role in the water cycles of some areas of the United States. It has been commonly accepted by many that they are sites for the recharge of groundwater. Wetlands are also thought to store water during snowmelt or heavy rains, reducing the amount of flooding by buffering rivers from an immediate influx of water. While some researchers have questioned if this is really true, the National Fish and Wildlife Service quotes a number of examples of flood control benefits.

Water Quality.   Wetlands play a role in improving water quality by trapping and filtering nutrients, sediments, and pollutants. When running water laden with eroded soil enters the marsh, sediments drop out. Marsh plants absorb some of the nutrients and even some toxic chemicals. Peat on the floor of a marsh also absorbs toxic pollutants.

Recreation and Education.   Wetlands also provide both recreational and educational opportunities. Much hunting and fishing depends on wetland habitat, either directly or as nursery habitat for game or gamefish young. Science classes often visit wetlands as part of their curriculum. Serious research also takes place in wetlands. For instance, study of the pollen imbedded in different layers of peat on the marsh floor can re-create past climatic and vegetative conditions.

Wetland Crops.   Some wetlands are literally "farmed" for wetland crops, which include wild rice and cranberries. In the future, wetlands may also be used to grow cattails and other plants that can be converted to fuel alcohol or burned as fuel.

Unfortunately, while society benefits from wetland preservation, landowners bear the cost of the preservation without directly realizing all of its benefits. This makes wetland drainage common. Currently, about 200,000 acres of wetland are drained yearly in the United States. In some areas of the country, little of the original wetlands remain. Iowa has lost 95%, California and Nebraska 90%, and the lower Mississippi Valley 80%.

Because of increased costs, the rate of wetland drainage has declined in recent years. In addition, state and federal programs, like the federal Water Bank program and more recently, the Swampbuster provisions of the 1985 Farm Bill, have been enacted to save wetlands. Some states also grant tax credits to growers who preserve wetlands.

Wet Soils.   Wet soils have fewer ecological functions as compared with wetlands. Nearly 25% of all farmland would be too wet to support production if it were not drained. Figure 6-2 shows how wet soils are distributed in the United States.

The most apparent sign of poor drainage is the presence of standing water several days after a rain. Often water-loving plants are found growing on wet soil. A *percolation test* serves to precisely define drainage. In the test, holes are dug and filled with water to wet the soil around the hole. After emptying, the holes are refilled, and the time it takes for drainage noted.

Soil color can be another indication of poor drainage. Typically, poorly drained soils have a very black A horizon and grey or mottled B horizons. Figure 3-20 presented a chart for determining soil drainage class by color. Soils classed as *very poorly drained* must be artificially drained for all upland crops. *Poorly drained soil* must be treated for crops like corn or cotton. Even *somewhat poorly drained soils* should be drained for best results with alfalfa.

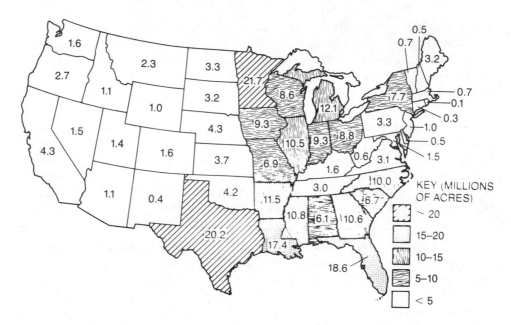

Figure 6-2. Wet soils of the continental United States. This map shows the acreage of wet soils for each state in millions of acres. (USDA, *Resource and Conservation Appraisal, Part I: Soil, Water, and Related Resources in the United States*, 1980)

Before draining these soils, however, it is wise to have an expert consider the effect on the local water system. Drainage may lower the water table enough to leave local wetlands dry.

ARTIFICIAL DRAINAGE

Soils that stay saturated can be artificially drained. Artificially drained soils can be remarkably productive, as they are in much of the Corn Belt. Two types of drainage are available: surface and subsurface drainage.

Surface Drainage. *Surface drainage* systems collect and remove excess water from the soil surface. Surface drainage is best suited to three situations:

- Collecting excess surface water on impermeable soil that cannot absorb it readily.
- Channeling away water flooding fields from higher elevations.
- Collecting irrigation water applied in excess of the soil's ability to absorb it.

Surface systems are inexpensive to install, since they are simply ditches dug through the field (figure 6-3). Water flows off the land into the ditch and is discharged off the field. Random ditches may be dug to drain a few depressions.

Figure 6-3. A drainage ditch on this farm carries off excess water.

In large, uniformly wet fields, a series of parallel ditches collect water from the whole area. Some grading may be done to improve the slope of the field to ensure that the water runs into the ditches.

The primary function of surface systems is to remove surface water. However, they can also lower the water table if the soil is permeable. In this case, the ditches are dug deep enough to enter the water table. Water leaks into the drainage ditch from the soil. In most situations, however, a subsurface system is needed to remove subsurface water.

Subsurface Drainage. *Subsurface drainage* collects water that has seeped into the soil and discharges it into surface ditches. A subsurface system consists of buried "pipes" into which the water seeps and then flows to an outlet. Eventually, the water is discharged away from the field.

Subsurface drains are made of a number of materials. The most common material in the past was short drain tiles made of fired clay or concrete. The tiles are hollow cylinders one or two feet long. The contractor digs a trench and lays the tiles end-to-end (figure 6-4). The tiles are not sealed together, so water seeps into the line through the joints.

Today, perforated plastic pipe is largely replacing tiles in subsurface drainage systems. A long continuous flexible plastic tube with holes (perforations) spaced along its length can be installed much more quickly than tiles (figure 6-5). Plastic

Figure 6-4. Short concrete drain tiles are laid end to end. Water seeps into the line through the joints and is carried off the field. (USDA, Soil Conservation Service)

Figure 6-5. Perforated plastic drainpipe can be installed more quickly and cheaply than clay or concrete drain tiles.

is less expensive to install than tile and is especially useful in soils that may break up tile lines by shifting.

Like surface drainage, subsurface drainage design depends on the situation. It cannot be used at all if the soil is so impermeable that water will not flow into the lines. Lines may be installed randomly to drain isolated wet spots. On larger fields, a definite pattern is followed (figure 6-6). The spacing of the lines depends on how rapidly water moves through the soil, which depends on texture and structure. The depth at which the drainage lines are installed depends on the crop (deep-soil rooted crops need the deepest lines) and texture. Figure 6-7 gives a suggested depth and spacing for each of several soil textures.

Maintenance. Drainage systems can fail to perform if not properly maintained. Make sure that all systems are inspected at least twice a year. Here are some suggestions:

- All ditches should be clear of brush, weeds, sediment, and debris
- Maintain an uncultivated zone along ditches to keep soil from washing into them
- A verticle hole in the field may mean a broken tile
- Soil may be wet along a tile line plugged with soil, debris, or tree and shrub roots
- Regularly clean tile outlets

Figure 6-6. Aerial view of a recently installed subsurface drainage system; 57,600 feet of tile were installed in 160 acres. (USDA, Soil Conservation Service)

Soil Texture	Depth (ft)	Spacing (ft)
Coarse	4.5	300
Medium	4.0	100
Fine	3.0	70
Organic	4.5	300

Figure 6-7. Guide to maximum depth and spacing for tile lines

IRRIGATION SYSTEMS

Irrigation has a long history in world agriculture. We know that irrigation has been practiced for at least 4,000 years. The success of civilizations along rivers such as the Nile River in Egypt and the Ganges River of India has been partially attributed to irrigation. In the United States, thousand-year-old irrigation canals can be found along the Gila River in Arizona.

Today, 11% of the world's cropland is irrigated, especially in China, the United States, Mexico, and the Mideastern nations. About 8% of American farmland is irrigated. Figure 6-8 shows where farmland in the United States is irrigated.

In arid parts of the country (figure 5-3), water management is an important part of farming. Farmers either irrigate or use dryland farming techniques. Even in the less arid areas, irrigation can be profitable to use on droughty soils or to relieve temporary rainfall shortages.

Irrigation water can be applied in a number of ways. Figure 6-9 lists nine systems and some of their traits. These nine systems can be divided into four categories: subsurface, surface, sprinkler, and trickle.

Subsurface Irrigation. *Subirrigation* is watering from below, using capillary rise from a zone of saturated soil lower in the soil profile. The zone must be high enough that water can rise into the root zone. However, the water must not be so high that it saturates the root zone. In some places, subirrigation occurs naturally. In other areas, pipes can be used in a manner opposite to drainage to produce an artificial water table. Subirrigation is rather inefficient and is useful only in special situations.

Surface Irrigation. *Surface irrigation* of fields involves flooding the soil surface with water released from canals or piping systems. Surface irrigation is most suitable to level or slightly sloping land of moderate permeability. In preparing land for surface irrigation, fields are carefully leveled to the slight slope needed for water to flood the land. A system of canals uses gravity to carry water to the farm and among the fields. Surface irrigation is especially suited to areas where region-wide canal systems can be built, as in parts of the American West. Once the water has reached the fields, it is distributed mainly by one of two ways— border strips or furrows.

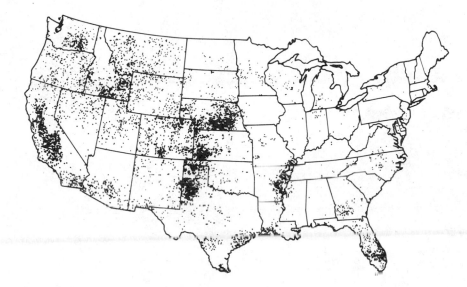

Figure 6-8. Irrigation in the United States. Each dot on the map represents 8,000 acres of irrigated land. (USDA, *Appraisal, Part I: Soil, Water, and Related Resources in the United States,* 1980)

Type	Cost to Install	Cost to Operate	Water-Use Efficiency	Flexibility
Subsurface	High	Low	Low	Very low
Surface				
border strip	High	Low	Low	Low
furrow	Medium	Low	Medium	Low
Sprinkler				
hand-move	Low	High	Medium	High
solid-set	High	Low	Medium	High
traveling-gun	Medium	Low	Low	High
center-pivot	Medium	Very low	Medium	Medium
wheel-move	Medium	Low	Medium	High
Trickle	Medium	Low	Very high	High

Figure 6-9. Several irrigation systems are compared. The flexibility rating is based on how easily the system can be adapted to various soil types and terrain.

Border-strip Irrigation. This is practiced by covering the entire soil surface of a field with a sheet of water. Each field is divided into smaller parts by the use of low dikes. Each of these sections is flooded in turn from a ditch or pipe running along the head of the field (figure 6-10). Because of the large surface area of the water flooding the ground, evaporation causes some waste of water. Runoff and percolation can also be extensive.

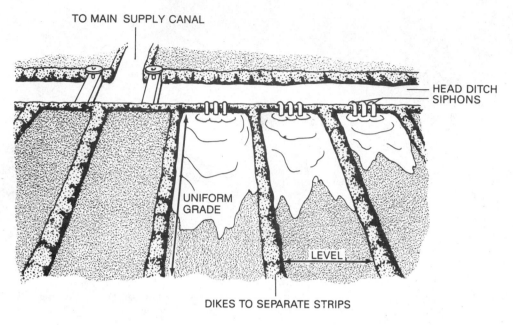

TO MAIN SUPPLY CANAL

HEAD DITCH
SIPHONS

UNIFORM
GRADE

LEVEL

DIKES TO SEPARATE STRIPS

Figure 6-10. Border strip irrigation (USDA)

Furrow Irrigation. This sytem distributes water through furrows, with crops planted in the ridge between two furrows. Furrows are best suited to row crops like vegetables, corn, or cotton. Evaporation is less of a problem than in border strips because less surface area is exposed to the air. Nearly half of all irrigated land in the United States is furrow irrigated.

Sprinkler Irrigation. Sprinkler systems pump water under pressure through pipes to sprinklers that spray water out in a circular pattern (figure 6-11). Sprinkler irrigation can be used where it is not possible to surface irrigate. For example, soil that is too permeable or too impermeable can be sprinkled. Ground that is not level can be sprinkled without leveling. However, wind can disrupt the sprinkling pattern so irrigation in windy weather may not be uniform.

Sprinkler irrigation equipment can be used for other purposes in addition to watering crops. For example, some growers have recently begun to apply agricultural chemicals like fertilizer or herbicides by sprinklers. This technique has been termed "chemigation." Sprinklers can also be used as a substitute for rainfall for the activation of herbicides. One sprinkler system can be used for frost control, which is described later.

Hand-Move Irrigation. This is the least expensive sprinkler system to install. The system consists of lightweight aluminum pipe that can be easily moved from place to place by a single person (figure 6-12). This type of system is very labor-

Figure 6-11. Sprinkler irrigation in a California field. Notice the main line running along the edge of the field. Compare to the layout drawn in figure 6-12.

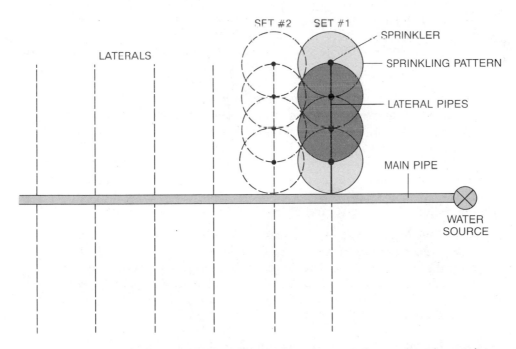

Figure 6-12. In a hand-move system, the lateral is moved from one set to the next along the main line. Note that the sprinklers are spaced so their patterns overlap. A solid-set system is similar, except that all of the laterals are set up at once and remain in place.

intensive and the operating cost is high. Hand-move systems are most suitable where irrigation is used to supplement natural rainfall rather than to replace it.

Solid-Set Irrigation. This uses the same equipment as hand-move setups, except that an entire field is set up at planting. The pipes remain in place until harvest. The larger number of pipes needed to supply all fields increases the cost of the additional initial equipment purchase but almost eliminates additional labor during the growing season.

Solid-set systems are commonly installed for both irrigation and for frost protection (figure 6-13). Strawberries are an example of a crop needing frost protection. A late spring temperature of 28 degrees Fahrenheit will freeze the flowers. The technique of wetting the plants in the event of a frost is successful because as long as there is liquid water *in the process of freezing* on the plant, the temperature at the plant will not go below 31 degrees. When a frost threatens, the grower turns on the system. The plants are kept wet, and even though they become crusted with ice, they do not cool to 28 degrees. The system is turned off the next day after all of the ice has melted. The most popular crops for solid-set irrigation are high-value crops like small fruits, especially those that benefit from frost protection.

Traveling-Gun Irrigation. This uses one very large sprinkler mounted on a trailer that moves across a field (figure 6-14 and figure 14-7). The sprinkler sends out a single large stream of water that reaches a height of 40 feet and covers several acres. The gun is very liable to wind problems. Because it has a very large nozzle, it has also been used to spray liquid manure and other slurries, a subject covered in chapter fourteen.

Center-Pivot Irrigation. This system is anchored to a center pivot point. The watering line is elevated above the crop by towers mounted on wheels (figure 6-15). As the system operates, the line slowly turns around the pivot point. By the time the circle is complete—60 to 120 hours later—as many as 160 acres have been watered. Center-pivot systems have the lowest labor requirement of any irrigation method and so are very popular. Their use has spread rapidly in recent years. The evidence of this can be seen by airline passengers as they fly over large green circles in many dry parts of the world (figure 6-16).

Wheel-Move Irrigation. This consists of a line of sprinklers mounted on wheels at both ends. In operation, the line of sprinklers slowly rolls down the field until it reaches the end of its hose. The pattern of wetting for wheel-move irrigation is rectangular rather than the circular pattern of center pivot. Some growers prefer the wheel-move system because it irrigates all parts of the field and does not leave unirrigated corners as does center-pivot irrigation.

Trickle Irrigation. *Trickle irrigation,* also known as *drip irrigation,* was pioneered in Israel where water conservation is critical. A trickle system is made of 1/2- to 3/4-inch flexible plastic pipes running down a crop row on the ground

Figure 6-13. Solid-set sprinkler irrigation system in operation in a field of strawberries. It is used here both for irrigation and frost control.

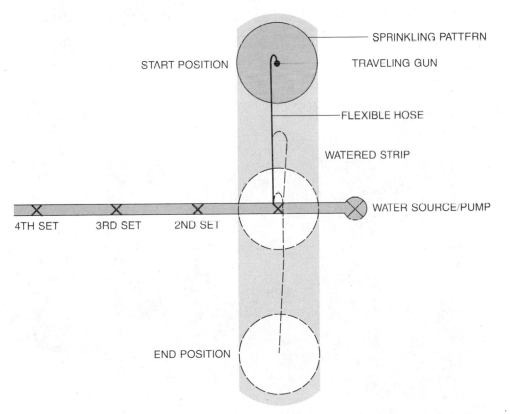

Figure 6-14. In the traveling-gun irrigation system, the gun moves along the path indicated, dragging the flexible hose behind it. When the gun reaches the end of the areas to be irrigated, the grower hitches it to a tractor and moves it to the next set on the main line.

Figure 6-15. Center-pivot irrigation has been responsible for the recent rapid growth of irrigation in some areas. (USDA, Soil Conservation Service)

Figure 6-16. The round half-mile diameter fields resulting from center-pivot irrigation are visible from the air in many parts of the world. Note that this method leaves the corners of square fields unirrigated. (USDA, Soil Conservation Service)

with special "emitters" spaced along the pipe (figure 6-17 and figure 4-11). The emitters daily (or nearly daily) drip water on the soil surface near the plants at carefully controlled rates. Figure 6-18 shows how trickle irrigation wets only the part of the soil occupied by crop roots.

Trickle irrigation has several benefits. The system operates at low water volume and pressure. This means that limited water sources and small pumps and pipes can be used. The system is nearly 100% water efficient compared with an efficiency of 50% to 75% for other methods of irrigation. Generally, less water is needed in trickle irrigation to produce a better yield compared with surface or sprinkler irrigation. For example, a farm near Fresno, California, reported that tomatoes under trickle irrigation produced 13 to 31 more tons per acre than tomatoes under furrow irrigation.

Early trickle systems had two major problems: plugging of the emitters and variation in flow rates between emitters. Plugging occurs when sand, hard water scale, or algae growth lodge in the the tiny holes. Some modern emitters, good filters, and chemical treatments have lessened the problem considerably. Modern emitters and pressure regulators have also made the flow rates more uniform.

USING IRRIGATION

The ideal irrigation brings just the soil in the crop's root zone to the field capacity. Irrigation engineers—and growers who use the systems they design—decide how much water must be added to reach that moisture level. They must also decide how dry the soil should be before irrigating.

Irrigation practices vary in different parts of the country. In general, irrigation begins when 50% to 60% of the *available* water has been used by plants. An exception to this general rule is trickle irrigation, which is designed to maintain field capacity.

Deciding When to Irrigate.   The moisture level of a soil can be judged using a feel test similar to ribbon testing. A soil sample is taken several inches below the surface. The feel of the soil is then compared with the chart in figure 6-19. Irrigation should begin when the soil reaches the "fair" moisture level. Experience helps make the test reliable.

Potentiometers (tensiometers) and resistance blocks work very well for deciding when to irrigate. Resistance blocks work over the widest moisture range, but potentiometers are more accurate within a narrower range. Since potentiometers work best within the fairly moist range between 0.0 and −0.8 bar, they are best suited to a fairly moist irrigation regime. Resistance blocks are needed where irrigation is infrequent. An irrigation system can be automated by adding a controller to the potentiometer. When the soil becomes too dry, the reading from the potentiometer causes the controller to turn on the irrigation automatically.

Figure 6-17. Trickle irrigation on greenhouse roses. This emitter drips water at carefully controlled slow rates. Water losses to percolation and evaporation are almost eliminated.

Some irrigators use a very detailed method of scheduling irrigation that involves tracking consumptive use. One way of doing this is to measure the water lost from a surface of free water, called an *evaporation pan.* This loss can be correlated to evapotranspiration rates by charts. Knowing how much water a soil holds, and how much has been lost, irrigation can be scheduled.

Modern technology is developing ways to schedule irrigation that greatly increase water-use efficiency. These methods employ computers and measure actual water stress in the plants, not the soil moisture conditions. For example, USDA researchers in Colorado recently attached devices to plants to sense moisture stress as a function of the closing of stomata. The devices were connected to a computer that controlled a trickle irrigation system on corn. Compared with regular scheduling, yield dropped from 159 bushels per acre to 146 bushels, but water use was cut from 33 to 16.5 inches—a great increase in water-use efficiency.

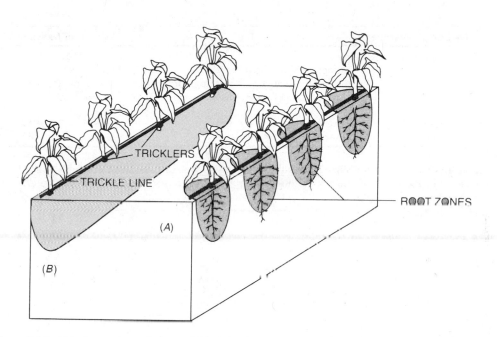

Figure 6 18. This figure shows how a trickle irrigation system waters crop plants. Trickles can be spaced to water individual plants (A), or spaced more closely to water a solid strip (B). Nearly ideal moisture levels are maintained in part of the root zone of the crop plant, while nearby soil remains unwetted.

How Much to Water.   It is important to add the right amount of water when irrigating. Too much water causes excess percolation and runoff, resulting in leaching of nutrients, wasted water, possible pollution, and even erosion. Too little water fails to bring soil to the best moisture level. Two factors affect the amount of water to be applied during irrigation: soil texture and rooting depth.

Figure 6-20 is a simplified table showing how much available water soils of different textures hold at field capacity. An "inch" of water is an amount that would cover the soil surface to a depth of one inch. Water-holding capacity is measured as inches per foot, or the number of inches of water each foot of soil will hold.

Figure 6-21 gives examples of the average rooting depth of most roots for several crops. As an example of how to calculate the amount of water to be applied in one irrigation, let's consider soybeans in a sandy loam soil.

1. A medium coarse soil holds 1.2 inches/foot of water.
2. Soybeans root to a 2-foot depth.
3. Total water available to soybeans = 1.2 inches/foot x 2 feet = 2.4 inches water.
4. Irrigation is turned on when 50% of the available water is gone. Thus, 50% of the water is to be replaced. Total added = 2.4 inches × 0.5 = 1.2 inches per irrigation.

Degree of Moisture	Feel	Amount of Moisture
Dry	Powdery dry	0
Low	Crumbly, will not hold together	<25%
Fair	Somewhat crumbly, will hold together	25%–50%
Good	Forms ball, sticks together slightly with pressure	50%–75%
Excellent	Forms pliable ball	75%–100%
Too wet	Can squeeze out water	Over field capacity

Figure 6-19. Moisture feel chart. Check the soil below the soil surface. Begin irrigation at the "fair" level. The amount of moisture is expressed as a percentage of the remaining available water.

Soil Texture	Available Water per Foot of Soil
Coarse	0.3–1.1 inches
Medium coarse	1.1–1.8 inches
Medium	2.0–2.9 inches
Medium fine	1.8–2.6 inches
Fine	1.2–2.0 inches

Figure 6-20. Available water held in soils of different textures at field capacity. This information is used to design irrigation systems and to calculate how much water is to be applied during irrigation.

Average Rooting Depth (ft)			
1–2	**2–3**	**3–4**	**4–6**
Beans	Beets	Alfalfa	Nuts
Cabbage	Cane berries	Cotton	Tree fruits
Carrots	Corn	Grapes	Shade trees
Cucumbers	Grains		
Lettuce	Melons		
Onions	Peas		
Pasture	Potatoes		
Peanuts	Sweet potato		
Strawberries	Tobacco		
Tomatoes			
Turfgrass			

Figure 6-21. The average rooting depth of most of the roots for a variety of crops. The actual depth may vary depending upon conditions.

This calculation tells a grower that if watering is started when the water is 50% gone, 1.2 inches of water will bring the root zone of soybeans in coarse soil to field capacity.

Saving Water. Saving water is an increasingly important task for the grower. In irrigation systems using a pump, saving water also means saving energy. It also helps avoid water pollution from unused irrigation water flowing into streams or seeping underground. The following points are old and new ways in which irrigation water can be saved:

- Use the most water-efficient system that is practical. Where feasible, trickle irrigation uses the least amount of water.
- In surface systems, land should be leveled carefully. The system should be designed to reuse excess tailwater.
- Make sure that all systems are designed correctly to fit the crops, soil, and terrain. For example, the application rate of a sprinkler system should be no greater than the infiltration rate of the soil.
- Maintain all systems for efficiency. For example, sand in irrigation water wears away at sprinkler nozzles, increasing the nozzle size and the application rate.
- Water should be transported through sealed ditches to avoid seepage, or through pipes, which also stops evaporation.
- All systems should contain devices to measure and control the water flow.
- Use the amount of irrigation water that gives the best return. Using less than the ideal amount may cause some yield loss, but it results in a savings in water. Researchers are developing models for determining the most efficient amounts of irrigation water to be applied.
- Base the scheduling of irrigation on the actual crop needs, as noted previously, not on a time schedule.
- Stay informed of new developments. For example, in surface irrigation, a fairly new method is surge-irrigation. In this method, water is applied very heavily at brief intervals. In this way, more of the water reaches the end of the furrow before it seeps into the soil near the furrow heads.
- Use computers to automate irrigation systems and to make decisions about what crops need to be irrigated when.

WATER QUALITY

Both groundwater and fresh surface water are used for irrigation. The choice depends upon the type of irrigation system used and on the water source that is most practical locally. When obtaining water, the first consideration is the legal availability of the water. Most states have laws controlling access to water, such as water-use permits. Growers using water from federal water projects must also meet federal regulations. The second consideration is the quality of the irrigation water. Water may be contaminated by suspended solids, boron, or soluble salts.

Suspended Solids. Suspended solids are small bits of solid material floating in the water. Groundwater may contain grains of sand or silt. Surface water often has bits of organic matter or small aquatic organisms like algae. Most irrigation systems are not bothered by small amounts of solids, but the tricklers in drip systems can be clogged by suspended solids. All trickle systems should include filters to remove suspended solids. Water with too much material floating in it may not be suitable for trickle irrigation.

Boron. Tiny amounts of boron are needed for plant growth, but slightly larger amounts can be toxic to plants. Some irrigation water has an excessively high boron level, especially for sensitive plants. Most fruits and nuts, such as citrus, apple, and walnut, are sensitive to boron levels. Some crops, such as alfalfa and sugar beet, are relatively tolerant.

Soluble Salts. The most widespread water quality problem is the presence of *soluble salts*. Soluble salts are compounds of sodium, calcium, and magnesium that can dissolve in water. These compounds are found in various levels in rocks, soil, and all water. The problems of soluble salts will be examined in detail in chapter ten, but their effect on irrigation will be surveyed briefly here.

When irrigation water evaporates from the soil surface or is removed by plants, the salts contained by the water will be left in the soil. Over time, irrigated fields may accumulate high levels of salts and become *saline*. The problem is most common in the western United States. In the East, because of high amounts of natural rainfall, enough water percolates through the soil to leach salts out of the root zone.

Soluble salts cause two problems. First, all salts cause an increase in the salt potential of the soil, causing the plant to work harder to absorb water. Second, one of the cations, sodium, tends to destroy soil aggregates, something like puddling. As a result, the soil surface is sealed and crusts form. Relatively salt-tolerant crops include barley, sugar beets, and cotton. Most vegetables, fruits, and alfalfa do not tolerate a saline soil.

Salinity is most severe in arid areas where land is heavily irrigated with water containing fairly high salt levels. Irrigation water can be classified by salt and sodium levels, as shown in figures 6-22 and 6-23. The units in the table are explained in chapter ten. One answer to the problem of salt buildup from irrigation is to use water low in salts. However, as the demand for water increases, irrigators are usually forced to use increasingly salty water. Thus, growers must learn how to manage salty water.

The key to using salty water is to *overirrigate* so excess water leaches salts below the root zone. However, if the soil is impermeable or there is a high water table, which is very common in saline soils, the salts will return to the root zone by capillary rise. The second approach to managing salty water is to install drain tiles to carry the salty water off the field.

Salinity Class/Hazard	Conductivity (Micromhos/cm)	Description
I Low	100–250	Suitable for most crops, little leaching needed
II Medium	250–750	Moderate salt-tolerant crops or moderate leaching needed
III High	750–2,250	Plants with good salt tolerance on drained soil with salinity control
IV Very high	2,250–up	Not suitable for irrigation except for occasional use under high salinity control

Figure 6-22. Irrigation water salinity classes. (USDA, *Agriculture Handbook* #60, 1954)

Sodium Class/Hazard	Sodium Adsorption Ratio	Description
I Low	0–10	Suitable for irrigation except for crops very sensitive to sodium
II Medium	10–18	Suitable for coarse-textured or organic soils with good drainage
III High	18–26	Soil will need treatment for sodium, or water must be treated to remove sodium
IV Very high	26–up	Generally not suitable for irrigation

Figure 6-23. Irrigation water classes for sodium hazard (USDA, *Agriculture Handbook* #60, 1954)

In irrigating saline land, one must balance the salt coming into the field with the salt going out to avoid a buildup. The saltier the irrigation water, the more excess water must be applied. This excess is called the *leaching requirement*. It is the amount of water to be applied in excess of that needed to wet the root zone of the plant.

SUMMARY

Artificial drainage allows a grower to make a productive field out of soil that is too wet to grow crops. In addition, drainage effectively prolongs the growing season by allowing earlier planting and better growth. While naturally wet soils are good candidates for drainage, wetlands are best as a natural resource.

Poorly drained soil is deficient in oxygen. The conditions can be indicated by the presence of standing water, water-loving plants, or subsoil color. Surface drainage carries excess surface water off the field by means of ditches. Subsurface drainage moves excess underground water from the soil through buried drainage lines.

Irrigation is primarily used to supply some or all of the water needs of crop plants. Subsurface irrigation uses capillary rise from a natural or artificial water table to water plants. This method is not widely used in the United States. Surface irrigation is very widely used. Surface irrigation involves flooding a field through border strips or furrows. Sprinkler irrigation sprays water out over the soil surfaces, through several different systems like center pivots and solid set. Trickle irrigation, the newest type of irrigation, drips water on the soil surface near crop plants. It is the most efficient system in terms of water use.

One goal of irrigation is to avoid water stress on the plant. The need for irrigation can be judged by feeling the soil, by using tensiometers or resistance blocks to measure soil moisture levels, or by more exacting budgeting methods. By knowing the soil type and rooting depth of the crop, the grower can calculate how much water is required.

Water quality is a concern to all irrigators. Some water contains too much boron or suspended solids. Soluble salts are the more common problem, especially in the western United States. Growers can manage soil salinity by making sure drainage is adequate and by overirrigating to leach salts out of the root zone of crops.

Irrigation is one of the most important developments in American agriculture. As water supplies dwindle, growers are becoming more concerned about their water use. Over time, irrigation must continue to become more efficient. In addition, more efficient and cost-effective ways must be developed to solve salinity problems.

REVIEW

1. Wet soils shorten the effective growing season (True/False).
2. Many ocean creatures depend on coastal wetlands.
3. No crops grow in wet soils.
4. Somewhat poorly drained soil should be drained for best alfalfa growth.
5. Surface drainage systems are the most expensive.
6. The easiest drainage system to install is concrete tile.
7. Furrow irrigation is an example of a surface system.
8. Hand-move sprinklers are the most labor intensive.
9. Irrigation usually begins when most of the available water is gone.
10. The problem with trickle irrigation is reduced yields.

11. In poorly drained soils, large pores tend to be filled with _____

12. Tile lines are spaced most closely in _____-textured soil.

13. A plant nutrient that can reach toxic level in some irrigation water is _____ .

14. The zone of wet soils above a water table is _____ .

15. The most water efficient irrigation system is _____ .

16. The irrigation system in which a sprinkler line goes in a circle is _____ .

17. The type of irrigation that can be used for frost protection is _____ .

18. _____ irrigation depends on upward capillary movement of water.

19. Land leveling is often needed for _____ irrigation.

20. Evapotranspiration can be estimated from _____ .

21. List three benefits of wetlands.

22. How much water should be applied to corn on a loamy soil?

23. List three benefits of artificial drainage.

24. Which soil texture, coarse, medium, or fine, matches the following?
    a. must be watered most frequently.
    b. more water is applied each watering.
    c. irrigation at the highest application rate.

25. List five ways to improve irrigation efficiency.

26. Why is saline water a problem?

27. Explain several ways to schedule irrigation.

28. Explain why poor drainage harms most plants.

29. How can sprinkler irrigation equipment be used besides watering plants?

30. How would terrain influence the choice of irrigation systems?

SUGGESTED ACTIVITIES

1. Observe the effects of salinity by watering previously established potted tomato plants with saline solutions. Use common table salt in water to create solutions of varying concentrations. Grow some control plants in untreated water, and compare differences in growth over time.

2. Try using different irrigation methods in school or home gardens.

3. Obtain tensiometers or resistance blocks and practice using them to schedule irrigation.

chapter seven
Life in the Soil

Here's real antagonism. At the microscopic level, one fungus (Gliocladium virens) attacks another (Rhizoctonia solani). This picture is magnified three thousand times. The fungus under attack is one of the bad guys: it attacks and rots the roots of some 200 plant species. This attack, indeed called antagonism, is one of the beneficial activities of life in the soil. Life makes soil more than just some pile of sand and clay. This chapter will introduce you to that life.

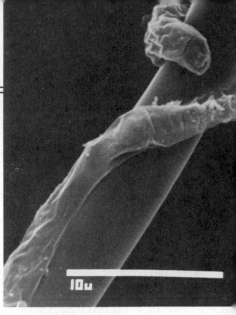

(Courtesy of USDA-ARS)

OBJECTIVES

After completing this chapter, you should be able to:

- define the carbon cycle and explain its importance
- briefly describe soil organisms
- list ways that soil organisms are important
- describe how to promote populations of beneficial soil organisms

TERMS TO KNOW

actinomycetes	hyphae	parasite
aerobic	immobilization	predator
algae	inoculation	primary consumer
anaerobic	microorganisms	primary producer
antagonism	mineralization	rhizosphere
arthropods	mycelium	saprophyte
autotroph	mycorrhizae	secondary consumer
decomposers	nematodes	symbiosis
denitrification	nitrogen fixation	symbiotic nitrogen
fungi	nonsymbiotic nitrogen	fixation
heterotroph	fixation	
humus	organic matter	

We live in a world teeming with life, yet few people know the creatures inhabiting the soil beneath our feet. Every acre of soil is home to two or more tons of living things (figure 7-1). What are these organisms, and why are so many of

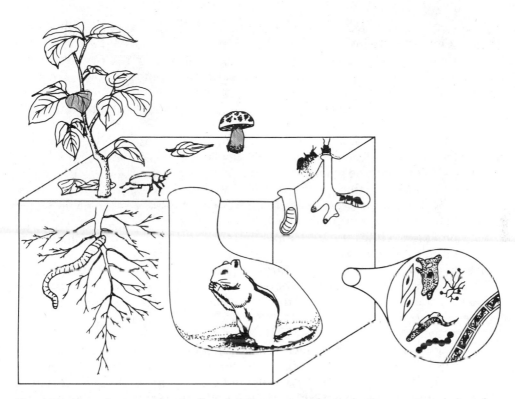

Figure 7-1. The soil teems with microflora, insects, worms, and animals. These creatures help make the soil the way it is. (Adapted from USDA, *Conserving Soil*)

them important to agriculture? It has been suggested that as little as one-fourth of a teaspoon of fertile soil is home to:

- 50 nematodes
- 62,000 algae
- 72,000 amoebae
- 111,000 fungi
- 2,920,000 actinomycetes
- 25,280,000 bacteria

This chapter will describe these organisms and their important functions. Let us first discuss their most critical role—the breakdown of organic matter.

THE CARBON CYCLE

The livelihood of all living creatures depends on photosynthesis. During photosynthesis two events occur that make life possible. First, carbon dioxide in the air is changed to organic carbon, the building block of living tissue. Second, photosynthesis converts sunlight energy to chemical energy stored in simple sugar. One estimate holds that each year world-wide photosynthesis produces more than 150 billion tons of sugar. One might add that, in this day of concern for the "greenhouse effect" and global warming, billions of tons of carbon dioxide are removed in the process. Later, plants and animals "burn" the sugar during respiration to reclaim the energy. Higher animals, like cattle or people, use the energy to keep warm, grow, or be active.

Chlorophyll-containing plants carry on the photosynthesis. Since they manufacture the basic food that animals eat, plants are called *primary producers*. Animals that harvest the plants are called *primary consumers*. Animals of prey are called *secondary consumers*. These levels make up the well-known *food chain*.

If the food chain were only producers and consumers, the chain would collapse and all the world's carbon would be fixed in the bodies of living and dead creatures. This does not happen because certain microorganisms, called *decomposers*, decompose the bodies of dead plants and animals. During decay most of the carbon is changed to carbon dioxide to complete the carbon cycle. Figure 7-2 shows the complete carbon cycle. As the figure shows, some carbon is also recycled by plant and animal respiration, and some carbon enters a geologic cycle.

Decay organisms, most of which live in the soil, break down *organic matter*. This is material of living origin such as leaf litter, dead animals, and crop residues. The material that resists decay becomes *humus*, the complex mix of organic chemicals that gives soil its dark color. An important result of decay is the release of plant nutrients that were tied up in the bodies of plants and animals.

In the complex ecology of the soil, there are four roles of interest to those who study soil:

- *Producers* include mostly plants and a few microorganisms. They produce their own food from inorganic carbon (like carbon dioxide) by photosynthesis or by certain reactions with soil chemicals. The technical term for producers is *autotrophs,* from a Greek term meaning to supply one's own food. All other organisms are *heterotrophs,* meaning they get their food from others.

- *Parasites* feed on plant roots and are often responsible for plant diseases.

- *Predators* prey on other soil life. They help keep parasite populations in check.

- *Saprophytes*, or decomposers, feed on dead organic matter.

Most organisms in the soil need oxygen; these are called *aerobic.* However, a few need no oxygen. These *anaerobic* organisms exist in low oxygen sites in all soils, but are most plentiful in poorly drained soils.

Now let's examine organisms that live in the soil.

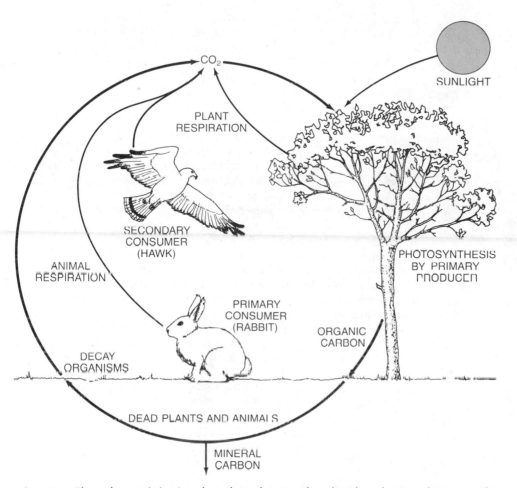

Figure 7-2. The carbon cycle begins when plants change carbon dioxide in the air to plant tissue. The cycle closes when microorganisms that aid in decomposition change organic matter to carbon dioxide.

MICROORGANISMS

Microorganisms, mostly simple plants (flora) or animals (fauna), are so named because they are too small to be seen by the naked eye. They are observed by soil scientists under a microscope or electron microscope. Actually, the simple division of all life into flora and fauna is no longer considered adequate for microorganisms. A number of ways to classify microbes are being proposed. Figure 7-3 briefly describes one classification that is useful for soil organisms.

Many categories of microorganisms inhabit the soil, but this text will consider just four of these categories: bacteria, fungi, actinomycetes, and algae.

Microoorganisms do not fit well into either the plant or animal kingdom, and a number of ways have been suggested to resolve the problem by reclassifying the kingdoms. One system divides living things into five kingdoms and is useful in classifying soil organisms:

- *Plantae.* Plants do not move and have no muscles or nervous system. Most are multicelled and have chlorophyll. Examples in the soil include plant roots and algae.
- *Animalia.* Multicelled creatures with muscles and a nervous system. They move to search for food or remain still but catch food. In the soil, animals include insects, worms, nematodes, and mammals.
- *Fungi.* Single-celled (yeasts) or multicelled (mushrooms, molds) organisms that lack chlorophyll and the type of roots and vascular system typical of plants. Fungi must feed on either decaying organic matter (saprophytic) or on other organisms (parasitic) because they cannot photosynthesize their own food. These organisms previously were classed as primitive plants.
- *Protista.* Single-celled organisms that behave like animals in that they move and capture food. However, they do not "eat," but engulf food. Protista include amoebae and protozoa like the well-known paramecium. While common in the soil, they do not strongly affect soil properties. These organisms previously were classified as animals.
- *Monera.* Mostly single-celled organisms that differ from all other organisms by the fact that their genetic material does not occupy a distinct nucleus in the cell. Important soil monera include bacteria, actinomycetes, and cyanobacteria. Actinomycetes resemble fungi, but they are classified as monera because their genetic material is not in a nucleus.

Figure 7-3. Classification of living things

Bacteria. *Bacteria*, simple one-celled organisms, are the most abundant inhabitants of the soil. Up to 100 million organisms may live in one teaspoon of soil. Common soil bacteria are rod-shaped, though many assume other shapes as well. They are about 1/25,000 inch wide and slightly longer. While they are single-celled, many may cling together to form chains. Bacteria usually grow as small colonies on the surface of soil particles.

Most soil bacteria are saprophytic. They comprise one of the groups most responsible for breaking down organic matter in the soil. A few species are parasites, causing plant diseases such as crown gall (*Agrobacterium tumefasciens*), which causes a tumor-like growth on roots of many plants. A few important species of bacteria are actually primary producers, or autotrophs. These organisms obtain their energy not from the sun but from chemical reactions with certain soil substances.

Fungi. Most *fungi* resemble a mass of tangled threads *(hyphae)* called a *mycelium* (figure 7-4). Fruiting bodies grow from the mycelium. These bodies release spores that may be considered the "seeds" of fungi. In the strictest sense, fungi are not microorganisms, since many grow to become quite large—the common mushroom is a fungus. The mushroom is the fruiting body of a fungus whose hyphae feed on decaying material in soil (figure 7-5). However, much of the fungi in the soil must be examined under a microscope. Up to 450,000 fungi may reside in a teaspoon of soil.

Along with bacteria, fungi act as the main soil decomposers. Fungi can attack matter that resists breakdown by bacteria, partly because hyphae can grow *into* the material. Many fungi are plant parasites, such as the wilt fungus *(Verticillium albo-atrum)* that attacks potatoes, tomatoes, and several other plant species.

A few odd fungi are predators. For instance, certain fungi capture and consume nematodes (a microscopic worm). These fungi trap nematodes either by growing rings that can tighten around the body of a nematode (figure 7-6) or by growing knobs covered with a sticky substance. Once the nematode is trapped, hyphae grow into its body until it is consumed.

Actinomycetes. *Actinomycetes,* also called *mold bacteria,* resemble bacteria because they have a similar cell structure and are often classified as bacteria. They also look like fungi because they grow a threaded network. Like fungi, actinomycetes can work on resistant organic matter. One teaspoon of soil contains about 12 million actinomycetes.

Many actinomycete species produce chemicals that stop the growth of other microorganisms, a phenomenon called *antagonism.* Many useful antibiotics used

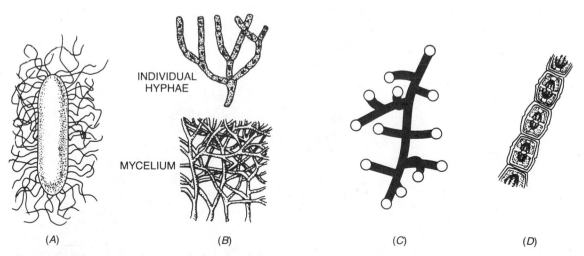

INDIVIDUAL
HYPHAE

MYCELIUM

(A)   (B)   (C)   (D)

Figure 7-4. Soil organisms show a great diversity of shapes. *(A)* A rod-shaped bacteria with "hairs" that "wiggle" to move the bacteria. Many bacteria have no hairs. *(B)* Fungal mycelium, composed of individual hyphae. *(C)* An actinomycete. *(D)* Algae.

(A)

(B)

Figure 7-5. (A) These mushrooms are the fruiting body of a mushroom whose (B) hyphae are underground, rotting the wood poles.

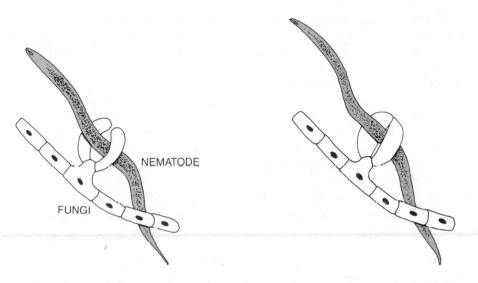

Figure 7-6. A fungus is shown trapping a nematode. Once the nematode enters the ring the fungus constricts, trapping the animal. The nematode is then digested by the fungus.

in modern medicine are derived from actinomycetes. In fact, the characteristic odor of damp soil comes from the most important genus of antibiotic-forming actinomycetes, the *Streptomyces*. In the soil, these natural antibiotics sometimes protect plant roots from attack by disease organisms. All but a few species of actinomycetes are saprophytes. On the other hand, a few actinomycetes produce plant diseases, such as potato scab *(Streptomyces scabies)*.

Algae.   While algae growing in water may be quite large, most soil algae are single-celled. One teaspoon of moist soil is host for about 250,000 algae. Algae are simple chlorophyll-containing plants that live in high-water environments. In the soil, most live in water films. Like higher plants, algae can photosynthesize and so are considered primary producers.

As producers, algae add slightly to the organic matter of soil. Certain algae combine with fungi to form lichens. Lichens growing on rocks release mild acids that dissolve minerals, adding to soil formation (see chapter two).

DISTRIBUTION AND FUNCTIONS OF MICROORGANISMS

Organic matter decay is an important task of soil organisms. However, many organisms perform other tasks that also are important to agriculture. Before examining these tasks in detail, let's look at where organisms live in the soil, because their location affects their function.

Distribution in the Soil.   Most microorganisms need air, water, and food to thrive. These materials are best supplied in the top two feet of soil, especially the

A horizon. Here, organisms will find the greatest amount of organic matter for food, good soil structure, water storage, and will be close to the soil surface for air. Thus, most soil organisms live near the soil surface (figure 7-7), as do most plant roots.

Plant roots "leak" a variety of chemicals into the surrounding soil, including sugars and other organic compounds. In addition, roots slough root caps, bits of bark, and old root hairs. All this material acts as food for microorganisms, which, in response, multiply in great numbers. This area of high biological activity surrounding plant roots is called the *rhizosphere*. The rhizosphere extends several inches from the plant roots.

The effect of the rhizosphere and the preference of microbes for the top layer of soil mean that microbe populations concentrate near plant roots. As a result, the desirable activities of microbes reach their peak near plant roots—to the benefit of plants.

Nutrient Cycling. Nutrients taken up from the soil by plants cannot be used by other plants. In a similar manner, the chemicals in the bodies of living microorganisms, animals, or fresh organic matter cannot be used by plants. The nutrients in living bodies or fresh organic matter are said to be *immobilized*. These nutrients are bound in complex organic forms.

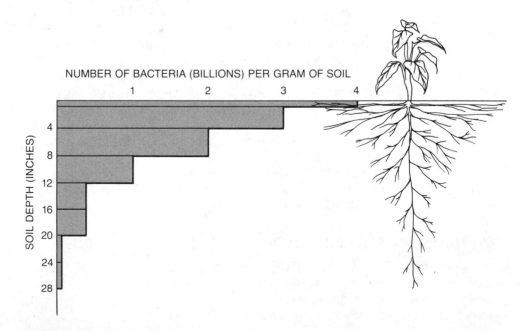

Figure 7-7. The population of bacteria decreases from a maximum at the soil surface to a minimum near the greatest soil depth for a fertile loam. (Drawn from data in Frobisher, Hinsdill, Crabtree & Goodheart, *Fundamentals of Microbiology*, W. B. Saunders Co., 1974)

Unlike animals, plants need nutrients in simple, inorganic forms. Thus, plants cannot use immobilized nutrients until they have been changed to simple, inorganic forms by microbial decomposers. This process is called *mineralization*. Immobilization and mineralization are opposite processes (figure 7-8). The sulfur cycle is an example of these processes (figure 7-9).

Most soil sulfur comes from the weathering of sulfur-containing minerals. Some of it comes from industrial pollution, in the form of sulfur dioxide in acid rain. The sulfur is changed by microorganisms in the soil to sulfate ions, a form of sulfur that can be absorbed by plant roots.

Plants absorb the sulfate to make protein and other compounds, thus immobilizing the sulfur. When leaves fall they decay and soil flora mineralize the sulfur in the leaves to sulfate ions. Some sulfate is taken up again by plants, some is again immobilized in the bodies of organisms, and some leaches away.

The sulfur cycle shows how immobilization and mineralization lead to elements being recycled by plants and microbes. Fully one-third of all the naturally occurring elements on earth are recycled in this way, including many plant nutrients. An essential element of this recycling process is the storage of nutrients for plant use. Many mineralized nutrients easily leach from the soil. This loss is reduced by soil flora, which capture the nutrients for their own use. When they die in the rhizosphere, the nutrients are available again to plants. Therefore, soil organic matter and life can be seen as a means of nutrient storage.

Microorganisms are involved in another important cycle, the nitrogen cycle. The nitrogen cycle will be covered in detail in chapter eleven, but we can look here at the role microbes play in the cycle.

Nitrogen Fixation. Nitrogen comes from nitrogen gas in the atmosphere—approximately 34,500 tons of it over every acre of the earth's surface. Higher plants cannot use even one molecule of this nitrogen gas. However, certain bacteria, blue-green algae, and actinomycetes can use it. They absorb the gas and convert it to ammonia that plants can use. This process is called *nitrogen fixation*.

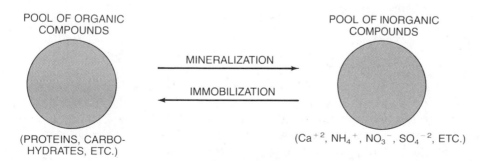

POOL OF ORGANIC
COMPOUNDS

MINERALIZATION

IMMOBILIZATION

POOL OF INORGANIC
COMPOUNDS

(PROTEINS, CARBO-
HYDRATES, ETC.)

$(Ca^{+2}, NH_4^+, NO_3^-, SO_4^{-2}, ETC.)$

Figure 7-8. Nutrients can be viewed as occupying two pools of compounds and can pass back and forth between the pools. When organisms die, their parts create a pool of organic compounds. Decay microbes mineralize these compounds to inorganic forms. These, in turn, are taken up by plants or other organisms, being immobilized back into the organic pool.

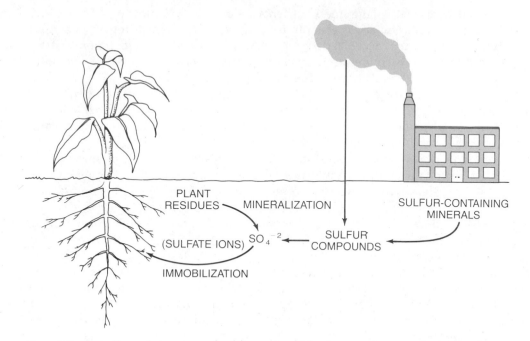

Figure 7-9. The sulfur cycle is an example of the way in which microorganisms recycle plant nutrients.

Some plant roots play host to an important group of nitrogen-fixing bacteria. These bacteria invade the root hairs, which respond by surrounding the bacteria with plant cells to form a lump, or *nodule*, on the roots (figure 7-10 and 7-11). The bacteria get minerals and food from the plant roots. When a bacterium dies, the nitrogen it has fixed can be used by the plant. Thus, this association is useful for both the bacteria and the plant.

Biologists apply the term *symbiosis* to mutually helpful associations between two organisms. The way the bacteria gather nitrogen is thus called *symbiotic nitrogen fixation*. For agriculture, the most important bacteria is of the genus *Rhizobium*. This bacteria grows on legume roots. Alfalfa and soybeans are common agricultural legumes. Up to 300 pounds of nitrogen per acre can be added to the soil yearly by the legume-*Rhizobium* association.

A number of noncrop plants also host nitrogen-fixing bacteria. Some trees, locusts for instance, are also legumes. Many nonlegume plants, such as alders, host nitrogen-fixing actinomycetes. Alders can add between 70 and 150 pounds of nitrogen per acre each year. Such trees add to the nitrogen status of woodlands, and they can be used to good effect in efforts to replant forests. Some have been useful in the reclamation of surface mines, dumps, and other heavily disturbed areas.

A few genera of free-living bacteria (*Clostridium, Azotobacter,* and others) also fix nitrogen. These nonsymbiotic bacteria do not live on plant roots.

Figure 7-10. Nitrogen-fixing nodules on the roots of an alfalfa plant. (Courtesy Dr. J. Burton, The Nitragin Company)

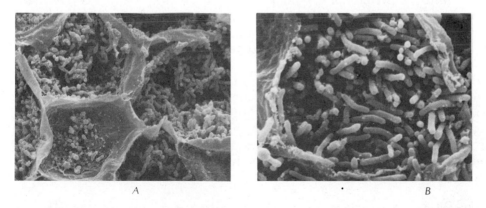

A   *B*

Figure 7-11. *(A)* Root nodules broken open to show the bacteria (magnification 640X). *(B)* Detail of *Rizobium* bacteria (magnification 5000X). The photographs were taken with a scanning electron microscope. (Dr. Carroll Vance, Research Plant Physiologist, Department of Agronomy and Plant Genetics, University of Minnesota)

Nonsymbiotic nitrogen fixation, for the most part, is not considered important to agriculture. Under the best conditions, it adds about 40 pounds of nitrogen yearly to the acre.

Another interesting group of free-living nitrogen fixers is *blue-green algae.* These organisms defy neat classification—some biologists place them as algae, others as an order of bacteria called *Cyanobacteria.* Blue-green algae grow in aquatic environments; thus, they thrive in water films in the soil. However, their small numbers in well-drained soil limit the amount of nitrogen they can add. Blue-green algae can achieve high populations in rice paddies. Traditionally, blue-green algae have been an important source of nitrogen for rice production.

Nitrification and Denitrification.   The nitrogen fixed by soil microbes is, of course, immobilized in the bodies of the microbes or host plants. When these die, they decay to form a pool of organic nitrogen. The nitrogen is mineralized to *ammonium* ions (NH_4^+), which may be absorbed by plants for growth. Most of the ammonium nitrogen, however, is oxidized by a group of bacteria *(Nitrosomonas)* to another form of nitrogen—*nitrite* ions (NO_2^-). Nitrite ions are then quickly oxidized by other bacteria *(Nitrobacter)* to *nitrate* ions (NO_3^-). Nitrite ions are toxic to plants and animals, but they reside in the soil for only a very short time. Nitrates, on the other hand, are the favored form of nitrogen for the growth of many plants.

Some nitrate ions are taken up by plants or other microbes. This action completes a cycle in the soil: from living matter to organic matter to ammonium to nitrites to nitrates and back to living matter. Some of the nitrates, however, are changed by other bacteria to nitrogen gas again. It then escapes back to the atmosphere. This process is called *denitrification.* Since denitrifying bacteria are anaerobic, using nitrates instead of oxygen, the process occurs most rapidly in wet soils. All of these nitrogen transformations make up the biological portion of the *nitrogen cycle* (figure 7-12). Note that the nitrogen cycle is really two nested cycles. In the outer cycle, nitrogen enters the soil by fixation and leaves the soil by denitrification, recycling nitrogen between the earth and air. The inner cycle, just described, recycles nitrogen inside the soil.

Soil Structure.   Soil microbes are important agents of soil aggregation. Fungi and actinomycetes are the most effective of these organisms. First, their thread-like hyphae twine throughout the soil particles, pulling them together to form loose aggregates. Both organisms also produce gummy substances that glue the aggregates together. These substances resist wetting and so peds do not fall apart when they get wet. The improved strength and wetting resistance of these soil aggregates keep soil structure sound during tillage, rainfall, and irrigation.

Mycorrhizae.   Mycorrhizae are fungi that form a symbiotic relationship with plant roots, like the nitrogen-fixing bacteria already mentioned. These fungi infest the roots of plants to obtain food and nutrients. In return, the host plant gains a number of benefits:

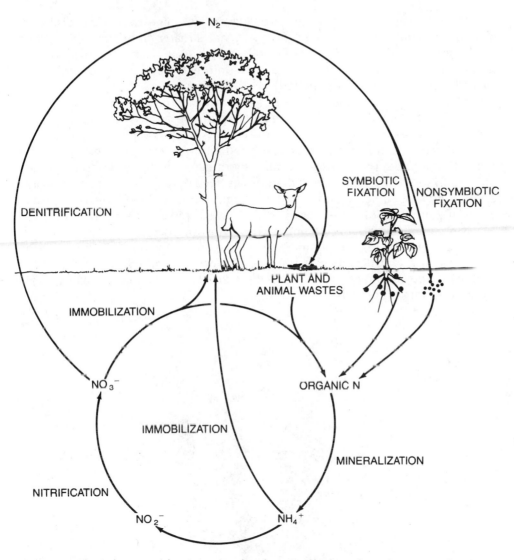

Figure 7-12. This is the core of the nitrogen cycle—the natural biological transformations of nitrogen. See Figure 11-2 for the complete cycle.

- Roots are better able to absorb phosphorus. This is probably the most certain and important benefit. Other gains for the plant may derive from better phosphorus nutrition.
- Roots are better able to absorb water, making plants more drought resistant.
- Roots are better able to absorb nutrients, especially phosphorus, zinc, copper, and others.
- Infected rootlets live longer than uninfected ones.
- Some mycorrhizae protect roots from disease.

Mycorrhizae were first noted growing on jack pine roots. This discovery clarified why jack pines grew best on soil that already grew jack pine. When seedling trees are planted in prairie soils, which usually lack the mycorrhizae fungus, infected seedlings grow better than uninfected ones. We now know that virtually all forest trees are infected.

Mycorrhizae growing on many forest trees create a thick growth, or "mantle," of fungal hyphae on the *outside* of the root. These fungi, called *ectomycorrhizae,* penetrate *between* the outer few cells of the plant root. These infected cells are presently used in the production of tree seedlings grown for replanting woods. The roots are artificially infected in the greenhouse. When planted in the field, these infected seedlings have better survival rates, become established more quickly, and grow faster than uninfected seedlings. This proves especially useful in planting heavily disturbed soils like mine tailings.

Most plants, including common crops, host mycorrhizae that grow *inside* root cells (figure 7-13). These fungi are called *endomycorrhizae.* The association improves crop productivity. For instance, the fungus improves the growth and yield of soybeans. Beans are larger on infected plants, and plant tissues contain larger amounts of four plant nutrients.

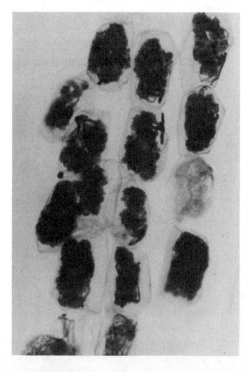

Figure 7-13. Endomycorrhizae colonizing root cells with hyphal strands in some cells. These strands will grow out of the cell into the surrounding soil to act as an extended root system for the plant. (USDA, *Agricultural Research,* April 1984)

Both types of mycorrhizae act as extensions of the root system of the plant. Indeed, mycorrhiza means "fungus root." Hyphal strands extend out of the plant roots into the surrounding soil, greatly increasing the absorbing area of the root system. Because the strands are much finer than plant roots, they can grow into tinier pores than can roots. There they can forage for water, phosphorus, and other nutrients that plants cannot reach. As a result, mycorrhizae greatly enhance plant growth in low-phosphorus soils (figure 7-14).

Much research is being done to find ways to use mycorrhizae more often in plant production. While most plant species host mycorrhizae—estimates range from 65% to 95%—agricultural systems often work against their use. Ectomycorrhizae can be grown artificially in the laboratory. However, endomycorrhizae cannot be grown artificially, making it difficult to produce large quantities of the agriculturally important organisms. High phosphorus levels in soil, a common condition in farm soils because of phosphorus fertilization, suppress the growth of mycorrhizae. Research continues and the practical uses of mycorrhizae continue to expand.

Figure 7-14. The geraniums shown were grown in phosphorus-deficient sand. The plants on top were then inoculated with endomycorrhizae, resulting in an improvement in phosphorus uptake. (USDA, *Agricultural Research*, April 1984)

Breaking Down Chemicals. Fortunately for modern society, there are organisms living in the soil that can break down the chemical products and refuse deposited in the soil. Chapter one mentioned the importance of soil in waste disposal. The cleanup of increasingly frequent oil leaks and spills is also aided by organisms that digest oil.

Farms avoid a buildup of agricultural chemicals in the soil mainly because of microorganisms. While some chemicals leave the soil by leaching or by evaporation, biological decomposition is the most important means of removing chemicals. The ability of a soil to degrade a chemical depends upon the substance. Some pesticides disappear quickly, while others, such as DDT, persist in soil for years.

Interestingly, some microbes have adapted to soil chemicals so well that they have become a new but growing problem. It is increasingly common for an herbicide or soil insecticide to fail altogether because new strains of organisms break them down so fast. Until more is known, researchers suggest that where the problem has occurred, growers should use soil pesticides as little as possible, rotate crops, and rotate chemicals.

How do pesticides and soil flora interact? In general, when growers apply a chemical to the soil, the number of organisms that can feed on it rises dramatically. The population of sensitive flora, in contrast, declines (figure 7-15). The result is an abrupt change in the makeup of the soil's living community. In some cases, products of decomposition are as toxic or more so than the original pesticide. We have yet to fully understand how farming chemicals modify soil populations and what their effect is on soil.

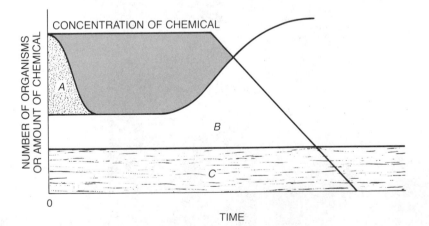

Figure 7-15. Soil microbe populations are affected by the use of a chemical such as 2,4-D. The numbers of sensitive organisms *(A)* decline. Organisms that feed on the chemicals *(B)* multiply rapidly in response to the food source. Unaffected organisms *(C)* remain the same. As organisms feed on the chemical, its concentration in the soil goes down. (Adapted from Cullimore, "Interactions Between Herbicides and Soil Microorganisms," *Residue Reviews*, 35:65, 1971)

Other Functions. Microbes are active in other useful ways. Some of these activities are well-known and understood. The exact impact of others is unknown, but they are interesting activities that are probably helpful to plant growth.

- Organic acids produced during decay help dissolve soil minerals to make some plant nutrients more available to plants. Also, certain autotrophs live by "feeding" on soil minerals, releasing their elements in forms useful to plants.
- Some organisms living in the rhizosphere produce plant hormones and vitamins that can promote plant growth. These microbes need organic matter as a food source.
- Several organisms suppress the growth of harmful parasites, a process called *antagonism* (see photograph front page of chapter). Examples include nematode-trapping fungi and antibiotic-producing actinomycetes. Most such microbes also need organic matter as a food source.

PROMOTING SOIL ORGANISMS

Some growers talk about having a "healthy" soil—a soil with a large population of microorganisms. How is this condition achieved? Let us look at some suggestions for populating the soil with healthy numbers of organisms.

Inoculation. *Inoculation* is purposely infecting soil with useful organisms. As an example, there have been many attempts, not completely successful, to speed up nonsymbiotic nitrogen fixation by soil inoculation. Inoculation with mycorrhizae has been successful, but much remains to be done in this area. Researchers continue to explore ways to infect soil or plants with "friendly" microorganisms.

Inoculation of legume seed, however, has long been an important farming practice (figure 7-16). Innoculants can be purchased and applied to seeds to ensure good nodule growth on the roots of host crops.

Soil Conditions. "Healthy" soils provide a good place for the growth of microorganisms. This list suggests some ways growers can make sure of good soil conditions:

- A constant supply of fresh organic matter is needed as a food source for most organisms. Other factors being adequate, this is the single most important factor for successful microbe populations.
- Good aeration supplies soil flora with the oxygen they need. While some organisms are anaerobic, most of the agriculturally important organisms need oxygen.
- Adequate moisture is important. Many organisms survive periods of drought by going dormant. Most, however, need moist soil to grow and multiply actively. Generally, the best moisture level is near field capacity.

Figure 7-16. Good nodule formation on legumes, such as this alfalfa, is obtained when the plants are inoculated with the right bacteria. (Dr. J. Burton, The Nitragin Company)

- Soil temperature affects soil flora. Few organisms grow actively at less than 41 degrees Fahrenheit, and most grow best between 77°F and 99°F. In cold soils, many beneficial activities slow down, including mineralization of phosphorus and nitrogen and nitrification.
- Most organisms grow best at neutral pH. Acid soils suppress growth, so liming of acid soils is a helpful practice. Potato growers, in contrast, prefer acid soil because it slows the growth of the scab-forming actinomycete. Of the soil organisms, fungi are most acid-tolerant.
- Proper nutrients are as important to microorganisms as they are to plants. For example, legume root nodules grow best in high-phosphorus soil. The number of mycorrhizae, on the other hand, is kept down by a high phosphorus content. Microbes have a high enough need for nitrogen that they can rob the plants of that needed nutrient (see chapter eight).

Controlling Harmful Organisms. Part of making a soil healthy for growing crops is to control harmful organisms like nematodes and parasitic fungi and bacteria. In cases of severe infestation, soil sterilant chemicals can be used. Some of the chemicals may be in the form of gases, such as methyl bromide, that are injected into the soil. Some chemicals are solids, washed into the soil by rain or irrigation.

Sterilizing an entire field is very expensive; therefore, the practice is often reserved for high value crops like strawberries. In addition to the cost, a difficulty of sterilizing soil is that it kills the good along with the bad. The key to controlling harmful organisms is to *prevent* their occurrence and to use other control methods where possible. A few suggestions for alternative methods of control are as follows:

- Practice sanitation. Start with disease-free seeds and plants. Don't drag infected soil into a field on tillage equipment.
- Take advantage of certification programs where they exist, such as the certified seed potato program. The programs certify growers to grow disease-free plants and seeds for other growers.
- Obey quarantines. Quarantines are intended to prevent the transportation of parasitic nematodes or other diseases on plants into uninfected areas.
- Control soil pH. For instance, potato scab is not a serious problem in acid soils. Some wilts are more of a problem in heavily limed soil.
- Crop rotation can help suppress diseases that are fairly limited in their hosts. For example, one parasitic nematode (soybean cyst nematode) feeds on soybean roots. During years in which corn is grown in rotation with soybeans, the nematode population declines because of a lack of a host. However, some harmful organisms are able to survive for years living as saprophytes.
- Incorporating large amounts of organic matter *may* help control parasites by promoting the growth of decay organisms. These organisms compete with the parasites and many, like some actinomycetes, "fight off" the harmful creatures.
- A fascinating though not widely used trick is the planting of certain species of marigolds to control nematodes. Apparently marigolds leak a chemical into their rhizosphere that repels nematodes.

Soil sterilization is a must in the greenhouse, unless media are used that contain no soil. This practice is discussed in detail in chapter sixteen. Even in a greenhouse, sanitation remains one of the best defenses against soil-borne diseases.

SOIL ANIMALS

Many animals, from tiny nematodes to larger animals like badgers, make their home in the soil. Animals, or fauna, affect cultivated soil less than microorganisms. Undisturbed soil, which provides a better animal habitat, can be heavily changed by soil animals. Let us look at some soil animals, starting with nematodes.

Nematodes. *Nematodes* are microscopic eel-like worms. Many inhabit the soil or plant roots. They can swim short distances through water films on soil particles. Those that infest plants puncture the roots with their needle-like mouthparts (figures 7-17, 7-18). A teaspoon of soil may contain up to 200 nematodes.

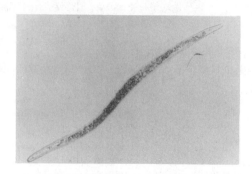

Figure 7-17. This parasitic nematode is the root lesion nematode *(Pratylenchus penetrans).* This adult female is eel-like and about 0.7 mm long. (Dr. D. H. MacDonald, University of Minnesota)

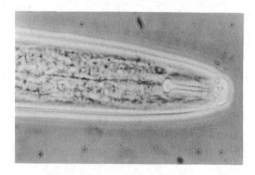

Figure 7-18. In the anterior ("head") portion, one can see the bulb-like muscle that controls the stylet, or needle, that punctures plant cells for feeding. (Dr. D. H. MacDonald, University of Minnesota)

Farmers are most concerned with parasitic types of nematodes. These animals infest plant roots, sapping plants of their strength and reducing yields (figure 7-19). The tiny puncture wounds also provide entry for other fungal or bacterial diseases. For that reason, nematode feeding is often related to infections by other soil-borne disease.

Most nematodes are actually saprophytic, feeding on decaying organic matter. A few prey on other soil fauna, including other nematodes. In fact, one of the exciting advances in insect control is the harnessing of insect-attacking nematodes. For instance, work has been done with a soil inhabiting nematode that attacks corn rootworm.

Arthropods. Mites, millipedes, centipedes, billbugs, and insects are the most common soil *arthropods.* Arthropods are easily recognized because they have jointed legs and a hard outside skeleton. Many arthropods, like some mites, millipedes, and insects, feed on decaying organic matter. Others, like centipedes and some mites and insects, feed on other soil fauna.

A number of arthropods feed on plant roots. June beetle larvae are important plant pests. These white grubs, as they are called, often injure crops and turf areas. Grubs are fat white larvae, about one inch long, with a black head and three pairs of legs just behind the head. Larvae live in the soil and feed on plant roots.

Ants and some termite species alter soil more than other insects by their tunneling behavior (figure 7-20). Tunneling mixes the soil, a subject to be considered later. Ant and termite burrows also aid soil aeration.

Earthworms. Earthworms feed on intact organic matter and excrete it as digested matter called "casts." In doing so, the nutrients become more available to plants. Earthworm burrows aerate the soil and both burrows and casts improve soil structure. Species of worms that feed on surface debris dramatically improve moisture infiltration. Earthworms cannot, however, *by themselves*, improve a problem soil.

Earthworms develop best in moist fine loams with a good supply of fresh organic matter and neutral pH. Between 200 to 1,000 pounds of earthworms may occupy an acre of soil. Earthworm populations may be viewed as a visible sign of soil health, because conditions in which earthworms thrive are also good for other soil flora and fauna and plant roots.

Figure 7-19. Field damage by the Soybean Cyst nematode. Notice the stunted plants. (Courtesy of *Crops and Soils Magazine,* American Society of Agronomy)

All of the farming practices that promote soil organic matter levels and structure are good for earthworms. These practices include crop rotation, minimum tillage, and additions of organic matter. Liming acid soils also helps. In the late fall, earthworms migrate deep into the soil for protection from the cold. On soils left bare over winter, the soil may freeze so rapidly and deep as to kill many of them. Leaving the soil covered with mulch or crop residues protects earthworms from winter kill.

Lawns present an interesting problem. Earthworms help prevent thatch buildup (a layer of dead grass that weakens turf) by feeding on it. Earthworms also help counteract the compaction common to lawns. Homeowners may object however, to the piles of casts some species leave in the lawn. Earthworms may also attract moles, whose burrows can severely damage a lawn. Where earthworms present a problem, using a soil insecticide will usually destroy them.

Mammals. Mammals affect soil by burrowing. The greatest number of burrowing mammals are rodents, such as gophers, woodchucks, and prairie dogs. Their population is highest in undisturbed soils such as pastures, forests, and prairie.

Figure 7-20. As an example of soil mixing, ants have almost covered this low-growing evergreen in a nursery with soil dragged up from below.

Figure 7-21. Prairie dogs mixed the soil of the western prairies, renewing the soil. Most prairie dogs are now confined to parks and zoos.

Rodent digging, like that of ants and earthworms, alters soil by mixing the layers. The mixing counteracts, to some degree, the natural soil aging process in which clay particles and nutrients leach into the B horizon. Mixing has the effect of rejuvenating the soil.

The native prairie of the western United States, before its use as agricultural land, was an example of the soil-mixing activities of burrowing animals. An important part of the animal community was the prairie dog town. Prairie dog burrows extend some five feet into the ground. In digging the burrows, prairie dogs carried a lot of subsoil to the surface and piled it on the topsoil (figure 7-21). As a result, the subsoil was mixed with the topsoil. The soils of these areas are surprisingly high in clay.

SUMMARY

More than two tons of living creatures can inhabit an acre of soil. The most important role of microorganisms is the decay of organic matter. In this process, nutrients are returned to the soil, and the level of humus is improved. An even more essential result of decomposition is the return of carbon from plants to the atmosphere.

The most numerous soil flora are bacteria. Most bacteria decay organic matter, though a few cause plant diseases. Important bacteria include the symbiotic nitrogen fixers *Rhizobia*, nonsymbiotic nitrogen fixers, and other bacteria involved in the nitrogen cycle. Both fungi and actinomycetes are excellent decay organisms and help preserve soil structure. Some actinomycetes fix nitrogen and some produce antibiotics. Mycorrhizae, which help plants absorb water and nutrients, are fungi. Algae add organic matter to the soil by being primary producers.

Microorganisms need proper soil conditions to grow and multiply. The most basic requirements are a constant supply of fresh organic matter, good soil aeration, and enough moisture. The growth of organisms is further influenced by pH, nutrient levels, and warmth.

Common soil animals include nematodes, arthropods, and earthworms. Some nematodes are helpful, but growers are concerned primarily about the types that cause plant disease. Earthworms feed on fresh organic matter, making the nutrients more available to plants. They also improve soil permeability and structure. The burrowing of worms, ants, and larger soil mammals mixes the soil layers. This slows the "aging" of a mature soil and helps keep it fertile.

REVIEW

1. Plants are heterotrophs (True/False). F
2. Plants are aerobic. T
3. Saprophytic organisms feed on dead organic matter. T
4. Most bacteria are parasitic. F
5. Microorganisms are uniformly distributed through the soil profile. F
6. Nitrogen in the soil undergoes a series of microbe-caused changes. T
7. The most important role of mycorrhizae is the improvement of phosphorus uptake. T
8. Most pesticides kill all soil organisms. F
9. Nematodes can dispose a plant to infection by other pathogens. T
10. Earthworms are good for the health of a home lawn. T
11. The organisms that grow a threadlike network into decaying organic matter are called __Fungi__, and the threads are called __mycellium__.
12. The organisms that resemble both bacteria and fungi are __Actinomycetes__.
13. Algae differ from other microorganisms in that they can __photosynthesize__.
14. __Mycorrhizae__ fungi help plant roots absorb water and nutrients.
15. __Nematodes__ attack roots, creating tiny holes.

16. Growers can increase nitrogen fixation of legumes by _Innoculation_
    with bacteria of the genus _~~Clostridium~~_. _Rhizobia_
17. The food source for most soil microorganisms is _organic matter_.
18. Besides food, most soil organisms need ___pH___ and
    _____ to grow and multiply.
19. An arthropod that mixes the soil is _____ .
20. A mutually helpful joining of two organisms is called _Symbiosis_

21. Where are microbes most numerous in the soil? _Near plant roots_
22. Name two important symbiotic relationships in the soil.
23. Name three element cycles involving microbes that are described in
    this chapter.
24. Where does nitrogen come from?
25. List three kinds of nitrogen-fixing organisms.
26. Many fertilizers make a soil more acid. How could such fertilizers affect
    earthworms?
27. Describe the nitrogen cycle.
28. Describe good soil conditions for microorganisms.
29. Most burrowing animals are annoying to growers. However, are they
    helpful to a soil?
30. Distinguish and give examples of autotrophic and heterotrophic, and
    aerobic and anaerobic organisms.

SUGGESTED ACTIVITIES

1. Isolate nematodes from a soil sample and examine them under a
   microscope. Test kits for this purpose are available.
2. Culture soil microorganisms on an agar medium and observe them under a
   microscope.
3. Grow several soybeans or other legume plants in sterilized soil. Inoculate
   one group with the correct *Rhizobium* but not the other. Compare the
   plants' growth. When the experiment is over, carefully wash soil off the
   roots and note the difference between the two root systems.
4. Survey a pasture for soil disturbance created by insects, worms, or
   mammals.
5. Follow the adage about lifting a rock to see what crawls out from under it.
   Can you identify the arthropods and other animals?

chapter eight
Organic Matter

This strange place is a peat bog in Minnesota—the remains of a glacial lake filled in with organic matter. Unusual plants grow here, like cranberries, sphagnum moss, and plants that eat insects. This soil is almost entirely organic matter. However, even mineral soils can be strongly affected by their organic matter content. This chapter will discuss soil organic matter—and the kind of soils that develop in bogs like this.

(Courtesy of Howard Hobbs, Minnesota Geological Survey)

OBJECTIVES

After completing this chapter, you should be able to:

- explain what organic matter is and how it forms
- describe what organic matter does in the soil
- list several ways to maintain soil organic matter
- discuss the problem of nitrogen tie-up
- define organic soil, listing uses and problems

TERMS TO KNOW

carbon:nitrogen ratio	humus	organic matter
colloids	lignins	organic soils
compost	muck	oxidation reactions
cover crop	nitrogen-depression	peat
green manure	period	subsidence
humic acid		

Early settlers in America had to clear the woodlands of the eastern colonies to create their farms but many of those farms were later abandoned and have since returned to forest (figure 8-1). By the middle of the nineteenth century, pioneering farmers were turning over the prairie sod of the Midwest. Farming in the Midwest continued to expand until the prairie retreated to a few preserved areas.

Why could soils of the grassland support long-term agriculture while some eastern woodland soils could not? One difference is the high organic matter content of the prairie soils. We'll examine organic matter carefully in this chapter. A good starting point is a description of the nature of soil organic matter.

Figure 8-1. Some of the woodland soils of eastern United States were unable to support long-term agriculture and are now devoted to forest products such as lumber and maple syrup. This sugar maple tree is being tapped for the sap that is used in making maple syrup. (USDA, Soil Conservation Service)

THE NATURE OF ORGANIC MATTER

Organic matter is that portion of the soil that includes animal and plant remains at various stages of decay. In forests, it comes from fallen leaves, dead tree trunks, and dead forest animals. In prairies, much of the organic matter comes from grass roots and tops. In farmland, crop residues add to the organic matter.

Chemical Makeup of Organic Matter. Organic matter consists of complex, carbon-containing compounds. Carbon atoms, unlike other elements, naturally form long chains. These long chains provide a framework upon which can be "hung" other elements like hydrogen, oxygen, nitrogen, and sulfur, to make a wide array of organic compounds. Of the many carbon-containing compounds, the most important are carbohydrates, lignins, and proteins.

Carbohydrates are long chains of simple sugars, each "link" in the whole molecule being a sugar molecule. Sugars are short carbon chains of five or six carbons with many oxygen atoms attached. *Starches*, familiar as an important food humans get from plants, are one common form of carbohydrate. Much of the tissue of grasses, tree trunks, and other plants is *cellulose*, another form of carbohydrate. Cellulose forms long fibers in plant tissue. Carbohydrates are important foods for soil microflora, which rapidly break them down to carbon dioxide and water. Most plant tissue is starch and cellulose.

Lignins make up 10% to 30% of plant tissue. Lignins make plants rigid by cementing together cellulose fibers. Lignins are complex molecules that resist decay. Corn has a high lignin content, which is why cornstalks remain visible in soil longer than other crop residues. Lignin accounts for much of the soil humus, so it is an important substance.

Protein is a long chain of simpler nitrogen-containing compounds called amino acids. Amino acids are also short carbon chains, with some nitrogen atoms and sometimes sulfur atoms attached. Residues from decayed protein become part of humus, supplying most of its nitrogen.

Decomposition of Organic Matter. Decay of organic matter happens in two basic stages. In the first stage, soil flora quickly digest organic materials, releasing carbon dioxide and water. Easily decayed compounds like carbohydrates are first consumed. Long carbon chains are split into shorter chains, attached atoms split off, and simpler compounds are produced. Some of these simple compounds react further to become part of the complex molecules that are humus. More resistant materials like lignin stay, but they are changed slightly to form humus. During the second stage, humus itself decays very slowly.

Decay organisms, like other living creatures, use oxygen. Soil flora combine oxygen with organic compounds in a process like respiration. Reactions in which chemicals combine with oxygen, called *oxidation reactions*, give energy. Thus, soil organisms use the energy stored in organic matter during decomposition. We can show decay in two steps, as shown below. Note that the carbon and hydrogen in the organic matter have been oxidized, or combined with oxygen:

$$\text{Organic matter} + O_2 \rightarrow CO_2 + H_2O + \text{humus}$$
$$\text{Humus} + O_2 \rightarrow CO_2 + H_2O$$

The first reaction proceeds rapidly and, under good conditions, requires weeks or months. The second reaction is much slower. Well-drained soil loses about 1% to 3% of its humus each year to oxidation. As one might expect from the reactions, low oxygen conditions limit organic matter decay, so organic matter tends to accumulate in poorly drained soils.

Humus is a collection of complex compounds that would be difficult to describe here. It does contain many different elements, including many plant nutrients. About 50% of humus is carbon, 5% is nitrogen, and about 0.5% is phosphorus. Humus is dark in color and is made of very tiny particles of clay size.

Factors Affecting Organic Matter. Four major factors directly affect the amount of organic matter in the soil. These factors are vegetation, climate, soil texture, and tillage.

Prairies generate the most organic matter. Root masses in a humid tallgrass prairie sum between 5.8 to 7.6 tons per acre. In a North Dakota mixed prairie, growth each year generates about 1.4 tons of shoots and about 4 tons of roots. Note that in native grasslands, most of the growth is in the soil, where natural turnover of roots enriches soil organic matter.

In contrast, forests generate more organic matter as litter on the soil surface. The litter decays into a thin organic layer, the O horizon, on the surface. Insects, worms, and other animals mix the material into the top few inches of soil, making a shallow humus-rich A horizon. Further, the needles of conifers contain chemicals that discourage many soil organisms, so conifer forests have even less organic matter than other woods.

Grass tops also die back each year, while trees do not. This means most of a grass plant returns to the soil each year. The differing growth of grasses and trees causes the following differences in prairies and woodlands and their soils (see figure 8-2):

- There is about twice as much organic matter in grassland soil as in an otherwise similar woodland soil.
- Organic matter extends deeper into the prairie soil, since grass roots can decay deep in the soil while organic matter in forest soils mostly comes from the decay of surface litter.
- Most of the organic matter of the prairie is in the soil. In forests, most of the organic matter resides in standing trees.

Since soils in arid climates support very little vegetation, they are lower in organic matter than either prairie or forest soils. Arid soils, unlike other soils, may gain organic matter under cultivation if irrigation is used. The gain results from the greater amount of green matter growing under cultivation than is provided by native plants.

Temperature and rainfall are key climatic factors that affect soil organic matter. The more rainfall, the greater the total amount of vegetation. Thus, soils in high rainfall areas tend to develop more organic matter than those in drier sites.

High average temperatures also promote plant growth. However, organic matter decays more rapidly at higher temperatures. Soils in warmer climates, receiving the same rainfall, tend to contain less organic matter than those in cooler climates. As a simple guide, organic matter is generated faster than decay when temperatures are below 77°F. (25°C.). In these cooler soils, organic matter can accumulate.

Fine-textured or clayey soils tend to have more organic matter than coarse soils like sand. Finer soils grow a large supply of plant materials because they hold water and nutrients well. Because coarse soils are better aerated than fine-textured soils, they have a better supply of oxygen and, as a result, organic matter decay is more rapid in sandy or coarse soils. Fine-textured soils also tend to contain more organic matter because clay protects humus from further decay by a process whose description is beyond the scope of this text.

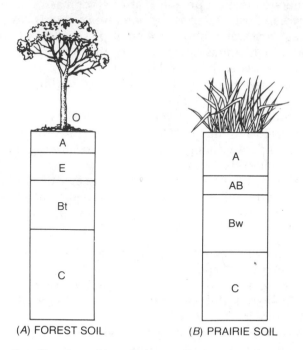

(A) FOREST SOIL   (B) PRAIRIE SOIL

Figure 8-2. Typical soil profiles of a prairie and a forest soil are compared in these examples. (A) The forest soil has a thin O horizon of leaf litter over a thin A horizon. The A horizon is thin because little of the litter mixes deeply into the soil. (B) The prairie soil has a deep A horizon because grass roots decay deep in the soil.

Virgin soils almost always lose organic matter when they begin to be farmed. Organic matter levels drop rapidly at first, but eventually the loss of humus slows and a new balance is reached. The loss occurs partly because erosion washes away some humus along with topsoil. Cropping usually returns less organic matter to the soil to replace the loss than does native vegetation. Interestingly, cultivated crops reverse the root: shoot ratio noted above for grasslands. That is, crops produce far more mass above ground than roots; after harvest, less root mass is left to contribute organic matter to the soil. Tillage also stirs oxygen into the soil and raises its average temperature. In one study in the Great Plains, 42% of soil carbon was lost after 36 years of cultivation.

FUNCTIONS OF ORGANIC MATTER

Organic matter improves the soil conditions of all mineral soils for several different reasons. Organic matter helps sandy soils, for instance, by increasing their water and nutrient-holding capacity. It improves clay soils by loosening them and improving their tilth. This section describes these and other functions of organic matter in the soil.

Nutrient and Water Storage.   Organic matter stores many of the nutrients used by plants. This storage occurs in two different ways. The first method of storage results from the size of the humus particles. Like clay particles, humus particles are extremely small with a relatively large surface area. Particles of this size are called *colloids*. Water and nutrients cling to the large surface area of the colloids. In addition, the colloids—clay and humus—attract some nutrients the same way a magnet attracts iron filings.

Organic matter also stores nutrients as a part of its own chemical makeup. Nutrient compounds are released for plant use as the organic matter decays. Humus contains most of the soil's supply of boron and molybdenum, about 60% of its phosphorus, and 80% of the soil's sulfur. Most of the nitrogen in the soil is stored in the organic matter. Thus, organic matter acts as one of the major reservoirs of soil nutrients.

Both fresh organic matter and humus absorb water like a sponge, holding about six times its own weight in water. This is extremely important in naturally dry and sandy soils. In fact, the water and nutrient-holding capacity of organic matter is its major benefit in sandy soils.

Nutrient Availability.   Humus not only stores nutrients, but it also makes several nutrients more available for plant use. As organic matter decays, it releases mild acids, called *humic acids*, which dissolve soil minerals, freeing them for plant use. Phosphorus in the soil tends to form compounds that do not dissolve in water. These forms cannot move in the soil, nor can plant roots absorb them. Humic acid acts on these compounds, making soluble phosphorus that moves in the soil in a form that plants can use.

Some nutrients, such as the metals iron and zinc, react with other soil chemicals to form insoluble compounds. Certain humus molecules form a ring around the metal atom in a process called *chelation* (key-lay-shun). These chelates protect the metal atoms from being locked in the soil. In this way, organic matter helps keep iron, zinc, and some other nutrients water soluble and available to plants. Copper, on the other hand, is so tightly bound to humus that it is least available in high organic matter soils.

Soil Aggregation.   As mentioned earlier, organic matter causes soil particles to clump together to form soil aggregates. The organic carbon in humus aids aggregation. More importantly, the gummy substances produced by soil organisms during the decay of fresh organic matter also bind the soil clumps. Better aggregation improves soil tilth and permeability. The soil is easier to work, better aerated, and absorbs water more readily. This is probably the most important way that heavy clay soils respond to organic matter.

Preventing Erosion.   Soils kept supplied with organic matter have an improved structure that greatly improves water infiltration. Because water infiltrates soils high in organic matter quickly during rainstorms, less water runs off—water that can remove soil from the field. Thus, organic matter makes soil less susceptible to erosion. Data used in the Universal Soil Loss Equation (a tool for predicting erosion rates, described in chapter eighteen) indicate that increasing a soil's organic matter from 1% to 3% can reduce erosion by one-third to one-fifth. An equivalent loss of organic matter would increase erosion.

Undesirable Effects.   Two undesirable but temporary effects can occur during the decay of fresh organic matter. The first effect is that nitrogen is tied-up in the bodies of microbes during the decay process. The nitrogen is immobilized and is not available for use by plants. A second effect is that certain plant residues are toxic to other plants. The remains of some plants release chemicals during decay that harm the growth of other plants. Dead quackgrass roots, for instance, until completely decomposed, may slow the growth of crop plants.

MAINTAINING SOIL ORGANIC MATTER

For most conventional large farming operations, it is impractical and uneconomical to try to raise organic matter levels significantly. However, unconventional systems like those employed by Amish farmers and organic growers do show much higher organic matter levels than their neighbors. It is therefore possible, and should be a goal of all growers to maintain organic matter at the highest practical level. Equally important is the *frequent* addition of fresh organic matter. It is *new* organic matter that provides most nourishment for soil microbes and releases nutrients rapidly. Decomposition of fresh organic matter supplies the greatest amount of gum for improving soil structure. These gummy substances decompose and disappear if not renewed.

Crop Residues.   Simply leaving crop residues in the soil is an easy way to provide organic matter (figure 8-3). With the exception of root crops, such as carrots, plant roots automatically stay in the field after harvest. The aboveground parts of crops for which only the seed is harvested are also usually left in the field. Nationally, growers harvest about one-third of crop residue for feed, animal bedding, or fuel. While there may be good economic reasons for harvesting or burning crop debris, the practice results in a loss of organic materials for the soil.

Growers can increase the quantity of crop residues being returned to the soil by proper fertilization. A well-fertilized crop produces a greater bulk of vegetation, both roots and tops, resulting in more organic matter for the soil.

Green Manuring.   Some crops are planted to be turned into the soil rather than for harvest. This practice is called *green manuring*. Green manuring has its most marked effects in sandy soil. The nutrients contained in the green manure are slowly released for crop growth during the growing season as the green manure rots. Many nursery growers depend on green manuring, because the trees and shrubs they grow leave almost no organic residues. In general, a green manure does not add enough organic matter to justify taking land out of production of

Figure 8-3. Crop residues left in the field add organic matter to the soil. Cornstalks, high in lignin, decay slowly.

paying crops. For most farmers, leaving crop residues is a more practical way of increasing organic matter.

Two types of plants can be grown as green manures. Legumes such as clover or vetch are useful because, in addition to the organic matter they leave behind, the nitrogen they fix is important. However, legume seed is rather costly for green manuring. To get the most bulk of organic matter at the least cost, grasses like oats or rye may be used. Sudangrass, a tropical grass that grows to six feet, develops the greatest amount of green matter (figure 8-4). Sometimes, grasses and legumes are combined.

Green manures can also control weed growth and erosion on land not protected by a cash crop. When grown for this purpose, the green manure is serving as a *cover crop*—plants grown to cover and protect the soil surface. For example, some growers on sandy soil may need a planting of winter rye in late summer to keep the soil covered through winter. The rye is plowed under before the following season's crop is planted.

Figure 8-4. This farmer grew sudangrass as a green-manure crop to satisfy the requirements of the Payment-in-Kind program of 1983. The grass controlled erosion, smothered weeds, and contributed organic matter to the soil.

Crop Rotation.   Economic factors often cause growers to avoid rotating crops, because it means growing less profitable crops some years. However, rotation, where feasible, improves soil humus. Studies have shown that continuous cropping of row crops like corn causes the greatest decline in soil organic matter. Grains cause smaller loss, while the meadow or hay legumes (e.g., alfalfa and clover) actually increase organic matter levels. Thus, a crop rotation of row crops, small grains, and legume hay is better than the continuous cultivation of row crops.

Organic Matter Additions.   The sources of organic matter described so far are all grown in the field. Many growers or gardeners can use other organic materials, including animal manures, organic wastes, and sewage sludge. The availability of these materials depends on the kind of farming operation and the presence of local organic waste producers.

Manure is an important source of both organic matter and plant nutrients. For the farming operation that both grows crops and feeds animals, manuring recycles nutrients and organic carbon on the farm. Most manure is cycled in this way. The cycle may be broken, however, if the feeding of animals and the growing of feed crops are separated, as in some large feed-lot operations. The handling of manure and sewage sludge is covered in chapter fourteen, where their nutrient content is discussed.

A number of industries generate organic wastes that may be locally useful. Forestry byproducts, such as sawdust or woodchips, may be available. Meat packing operations and canneries also produce organic wastes. One operation in the author's home state composts sawmill wastes with turkey manure to sell to local growers.

Homeowners usually have leaves, grass clippings, or other sources of organic matter in their gardens. Some gardeners compost the leaves, clippings, and even table scraps for their own use (figure 8-5). Garden centers usually sell bagged composted manures and peat moss, both useful for the home garden.

Mulches. Home gardeners often mulch their gardens by spreading straw, sawdust, woodchips, or other materials several inches deep on the ground. As the organic matter decays during the growing season, it enriches the humus content of the top few inches of soil. Besides adding organic matter, mulches have other benefits:

- Thick mulches smother weeds, especially annuals. Many of the more aggressive perennial weeds can grow through normal mulches.
- Mulched soil absorbs water much more readily than bare soil, improving soil-water content and reducing erosion.
- Mulches limit water evaporation from the soil surface, also improving water content of the soil.
- Organic mulches reduce the range of soil temperatures. For example, during a hot, sunny day, mulched soil remains cooler than bare soil, and it also stays warmer on cool nights.

The financial returns on typical farm crops like corn do not justify mulching. Conservation tillage, however, does leave a mulch of crop residue on the soil surface. Many growers mulch high-value crops like berries (figure 8-6) or nursery stock. Mulches almost always make striking improvements in the yields of blueberries, strawberries, raspberries, and tree fruits.

Increasing Soil Water. Any of the methods of water conservation or irrigation increases the bulk of organic matter growing in the field. The more green matter, the greater the amount of organic matter returned to the soil.

Maximum Cropping. As much as possible, keep the soil covered with crops. The soils will be cooler, and more organic matter is produced. Where possible, double-cropping, or the production of two crops a year, is desirable. Avoid fallowing as much as possible.

Figure 8-5. A handful of composted leaves

Conservation Tillage. Conservation tillage preserves organic matter in the soil primarily because less of the topsoil, which is high in organic matter, is lost to erosion. Another reason is that the soil is tilled less often, resulting in slower decay as compared with standard tillage. In addition, more crop residue stays on the soil surface where it decays more slowly than when it is buried in the soil.

No-till systems involve no tillage at all—seeds are simply planted right through residues left from the previous crop. Since the soil is not tilled, oxygen is not stirred into the soil and decay is less rapid than in other tillage systems. In some areas of the country, no-till allows two crops to be grown in one season that could not otherwise be grown. This is a result of the fact that no time is spent preparing a seedbed for the second crop. Two crops mean that double the amount of organic material can be grown on the soil. For these reasons soils worked for some time by no-till systems tend to have a high organic matter content in the top layer of soil.

Figure 8-6. The mulch applied to these strawberries conserves moisture and smothers weeds. It also keeps the fruit clean.

NITROGEN TIE-UP AND COMPOSTING

Soil flora (microorganisms) need both carbon and nitrogen in their diet to grow and multiply. When fresh organic matter is added to the soil—whether it be crop residues, green manure, or mulch—the number of organisms rises because of the new food. These organisms may compete with crop plants for nitrogen, causing a slowing of crop growth.

The organic matter of greatest concern contains a lot of carbon compared with nitrogen. This can be measured by the *carbon-nitrogen ratio* (C:N ratio) of the material. For instance, the C:N ratio of well-rotted manure is about 20:1, meaning that there are 20 parts of carbon for each part of nitrogen. Figure 8-7 shows the C:N ratio for soil and several common organic materials. Matter with a low C:N ratio is nitrogen-rich; material with a high C:N ratio is nitrogen-poor.

Let us see what happens when a large amount of fresh, nitrogen-poor material begins to decay in the soil. In response to the new food source, the population of decay organisms rises rapidly (figure 8-8). Bacteria themselves have a C:N ratio in the range of 4:1 to 5:1, and thus they need to incorporate a lot of nitrogen into their bodies. The flora feed on both the carbon and nitrogen in the material, but with high C:N materials, the nitrogen is quickly used up. To make up the difference, the flora draw nitrogen from the soil. In the initial stages of decay, then, soil nitrogen is rapidly tied-up or immobilized.

During the period when nitrogen is being immobilized, there is a temporary loss of free nitrogen. Crops growing on the soil will be short of nitrogen. Crop growth may slow, and crops may exhibit nitrogen-shortage symptoms. This period of decay is the *nitrogen depression period*.

After a time, most of the food is used up and the process reverses. The microorganism population declines and, when microorganisms die, the nitrogen stored in their bodies is released to the soil. In other words, nitrogen is being mineralized. When decay is complete, the net gain in soil nitrogen can be measured—the original nitrogen plus that in the new organic matter.

Nitrogen tie-up can be viewed in terms of the balance between nitrogen immobilization (which makes nutrients unavailable) and mineralization (which makes nutrients available). Both processes occur at the same time but not at the same rate. The balance depends on the stage of decay and how much nitrogen flora must pull out of the soil. Materials with a C:N ratio of more than 30:1 favor immobilization; those with a ratio of less than 20:1 favor mineralization. At rates between 20:1 and 30:1, the two processes balance. Ask yourself what happens when you add alfalfa hay to soil, and what happens when you add sawdust to soil.

Phosphorus, sulfur, and some other nutrients undergo a similar process when organic matter is added to soil. Nitrogen tie-up, however, produces the most noticeable effect.

Nitrogen tie-up can be avoided in several ways. One way is to plant crops after the nitrogen depression period is over—when the decay of previous crop residues is mostly complete. Another way is to fertilize with enough nitrogen to provide for the needs of the microorganisms. For instance, if sawdust is being used to amend soil, some additional nitrogen fertilizer can be added as well. Gardeners, and some growers, often lower the C:N ratio of organic material, like sawdust, before putting it in the soil. This is done by composting.

Composting. Gardeners *compost* organic material by storing it in a pile while providing proper conditions for decay. During decay of the compost, the carbon:nitrogen ratio narrows until it reaches about 15:1. The compost can then be added to soil without fear of nitrogen tie-up.

Material	C:N Ratio
Soil humus	10
Garden soil	12–15
Young alfalfa	12
Compost	15–20
Rotted manure	20
Clover residue	23
Corn stover	60
Straw or leaves	60
Sawdust	400

Figure 8-7. Common organic materials exhibit a wide range of carbon:nitrogen (C:N) ratios. The lower the ratio, the richer the material is in nitrogen.

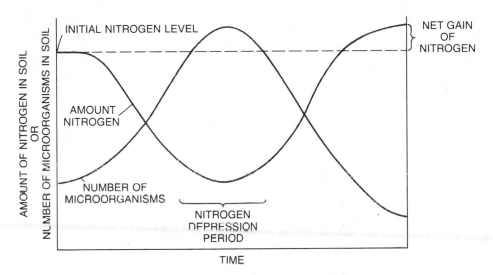

Figure 8-8. When a nitrogen-poor material begins to decay, the rising number of microorganisms in the soil robs the soil of nitrogen. During the nitrogen depression period, nitrogen is tied-up faster than it is released. As decay is completed, the number of organisms declines and nitrogen is released.

Common composting materials include leaves, grass clippings, wood chips, and even household table scraps. Foods that may attract rodents or dogs, like meats and fats, should not be composted. Materials with a high C:N ratio decompose slowly, so many composters add nitrogen fertilizer to speed up the process. Others include nitrogen-rich materials like manure in the pile. They may also add other fertilizers or minerals to enrich the compost.

Figure 8-9 shows the composition of a common home compost pile. The gardener makes a bin out of snow fencing or chicken wire. Different materials are added in layers, and the pile is finished off with a depression in the top to capture water. Periodically, the pile may be turned by hand to aerate it and water is added. Turning the pile takes materials from the outside of the pile to the center, where most decay occurs. Stirring also adds oxygen to the pile to speed up the decay rate. The compost should be ready in a couple of months.

Large-scale composting is practiced by a few growers. The most common large-scale use is in the nursery industry. Many nurseries grow stock in containers, and they compost the shredded bark or other nitrogen-poor materials used in potting mixes. In large composting operations, the materials are usually piled in large windrows and periodically turned by machine. See chapter fourteen for more details on commercial composting.

ORGANIC SOILS

So far, we have discussed the organic matter of soils whose traits are set by their mineral particles. These mineral soils contain only a small percentage of

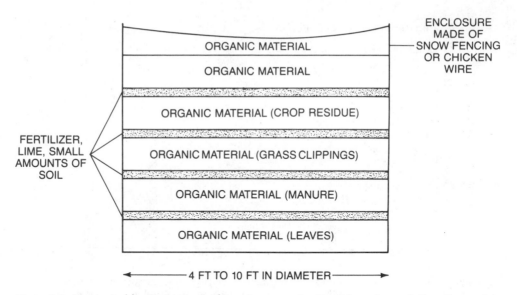

ENCLOSURE MADE OF SNOW FENCING OR CHICKEN WIRE

ORGANIC MATERIAL

ORGANIC MATERIAL

ORGANIC MATERIAL (CROP RESIDUE)

ORGANIC MATERIAL (GRASS CLIPPINGS)

ORGANIC MATERIAL (MANURE)

ORGANIC MATERIAL (LEAVES)

FERTILIZER, LIME, SMALL AMOUNTS OF SOIL

◄─── 4 FT TO 10 FT IN DIAMETER ───►

Figure 8-9. In a typical home compost pile, various organic materials are layered. The pile may be turned several times to promote decay. The addition of manure, nitrogen fertilizer, lime, or other nutrients speeds up decay and enriches the compost.

organic matter. Soils containing more than 20% to 30% organic matter are called *organic soils*. These soils are much different than mineral soils. In organic soils, the soil traits are set by the organic matter. Approximately 1 out of every 200 acres of American soil is organic. The five states with the most organic soil, in order, are Alaska, Minnesota, Michigan, Florida, and Wisconsin.

Organic soils form in marshes, bogs, and swamps. As aquatic vegetation, such as reeds or cattails, dies each season, it sinks to the bottom. Lacking the air needed for decay and oxidation, the material builds up on the floor of the bog. Eventually the bog will be completely filled in by the organic deposits. Deposits of partially decayed plant remains may reach depths ranging from a foot to as much as 80 feet (figure 8-10). Figure 8-11 shows a common organic soil profile.

Organic soils in which the plant remains are only slightly decayed are called *peat*. If the deposit consists primarily of fully decayed materials—usually because it has been exposed at some time—it is called *muck.* Squeezing a handful of fresh, wet soil can often tell them apart. Water that may be brown, but not muddy, squeezes out of a handful of peat. Muddy water runs out of muck. Further, peat contains plant remains that are at least partially identifiable; muck does not.

The nature of organic soils varies not only by the amount of decay, but by the types of plants the peat derives from. Sphagnum peat forms from sphagnum moss; it is very acid. Hypnum peat contains hypnum moss; compared to sphagnum moss it contains more lime and nitrogen and is less acid. Reed-

Figure 8-10. This organic soil profile shows a 12- to 18-inch deposit of peat on a mineral soil base. The vegetation is typical of a bog filling itself in. Note the sphagnum moss near the ruler. Sphagnum moss is a common plant of northern bogs. Sphagnum peat is the type used in most horticulture. (USDA, Soil Conservation Service)

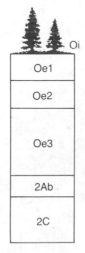

Figure 8-11. A profile of a typical organic soil formed in what is now a black spruce bog. The growth of different types of vegetation at different periods resulted in the several layers of organic material shown. The lower horizons are old horizons buried by vegetation.

sedge peat forms in marshes full of reeds and sedges; it generally is less acid and more decomposed than the other types.

Organic soils are very light, porous, and loose. A cubic foot of peat can weigh as little as a twentieth of a mineral soil. More decomposed types, like muck, weigh more, as do soils that have been cultivated. The heaviest organic soils may weigh about a third that of mineral soils.

These soils can soak up great amounts of water. Sphagnum peat can hold ten to twenty times its own weight in water, reed-sedge about five times, and cultivated mucks about twice. Peats are often added to soil to improve their water-holding-capacity. However, because they also increase the wilting point, the *available* water-holding-capacity may not improve.

Before fertilization, many organic soils are quite low in plant nutrients. The three major elements, nitrogen, phosphorus, and potassium are low, especially potassium. However, once fertilized, organic soils have a high ability to hold nutrients.

Organic soils are excellent for the production of certain vegetables, including onions, celery, lettuce, carrots, and others. Warm-season vegetables like tomatoes and melons don't perform as well. Mint, hay, and turfgrass seed are also favorite crops for peats. Peats are especially valued for sod production

Figure 8-12. Organic soils are favored by sod producers. Turf growth is lush and the sod is lightweight because of the low bulk density of peat. (USDA, Soil Conservation Service)

(figure 8-12), because of easy harvest and light weight. About 700,000 acres of organic soils are planted to these specialty crops, for a gross value of about one billion dollars annually.

Unfortunately, peat presents some interesting challenges. After a bog is cleared of brush and drained, exposure to air speeds decay. This means that the soil begins to disappear, changing to carbon dioxide. Added to compaction and wind erosion, the land sinks, a process called *subsidence.* The warmer the climate, the more rapid the loss. In some parts of Florida, soils may loose as much as two and a half inches of depth annually.

In addition, organic soils are flammable, and peat fires are notoriously hard to put out. Being loose and light, wind carries off organic particles easily. Since peats are at low elevation, they are often frost pockets, making it difficult to grow tender crops. Lastly, the black soil can get so hot during warm sunny days as to damage young seedlings.

The following practices can help avoid these problems:

- Design drainage systems that keep the water table as high as possible. This will reduce subsidence by decay.
- Install sprinkler systems. They can be used to control frost, cool the soil on hot days, wet soil to reduce wind erosion, or even to drench peat fires.
- Use wind erosion control techniques (see chapter eighteen). However, avoid tall windbreaks that reduce air movement. That can increase the chance of frost and increase the soil temperature.

Horticultural Peat. Peat can be harvested for two uses: fuel and horticulture. The Irish have long dug peat for fuel, and it has been investigated here. Because of fear of ecological damage, no large-scale energy use is yet found in the United States.

Horticultural peat, however, is widely dug. In 1978 some 800,000 short tons were harvested, for a value of about thirteen million dollars. Peat is used to amend problem soils, as a mulch, in soils for container plants, and as shipping material for nursery stock.

To harvest peat, a bog is cleared and drained. The soil is then plowed and disced or otherwise stirred to promote drying of the surface layer. The dried peat can then be picked up and stockpiled and bagged for sale (figures 8-13 and 8-14).

Sphagnum peat is most often used in greenhouses for soil mixes. Because it is so acid, it makes an ideal growth medium for acid-loving plants like azaleas. It is also fine for other uses. Hypnum peat is also good for most uses, and better for plants that dislike high acid. Reed-sedge peats work best for soil conditioning, and are not recommended for potting mixes.

One word of warning about horticultural peat. Once it dries, it resists being wetted again. Used as a mulch, one must be careful that it does not repel water. When used in a potting mix, make sure it is damp before use!

Figure 8-13. A peat field in Minnesota. Harvested peat is lying in piles, awaiting processing. (Courtesy of Howard Hobbs, Minnesota Geological Survey)

Figure 8-14. A peat harvester. This one operates like a giant vacuum cleaner. (Courtesy of Howard Hobbs, Minnesota Geological Survey)

SUMMARY

Organic matter is the remains of plant and animal material. It is made up of such compounds as carbohydrates, lignins, and protein. Microorganisms decay organic matter into carbon dioxide and the more resistant residue, humus. During the decay process, microbes can tie up soil nitrogen.

The amount of organic matter in soil depends on vegetation, climate, soil texture, and tillage. The highest organic matter mineral soils are usually virgin prairie soils formed under fairly cool, moist conditions. Forest soils and those of warm climates are lower in organic matter.

Organic matter and humus store many soil nutrients. They also improve soil structure, loosen clay soils, help prevent erosion, and improve the water and nutrient-holding capacity of coarse or sandy soils. Organic matter comes from crop residues, animal and green manures, compost, and other organic materials. Proper fertilization, conservation tillage, and crop rotation help preserve organic matter.

Organic soils, widely used for vegetable and sod production, form under the low-oxygen conditions of a swamp or bog. Peat is light, porous, and holds water well. Once the swamp or bog is drained, however, peat slowly disappears by oxidation and wind erosion.

REVIEW

1. Forest soils tend to have more organic matter than prairie soils. (True/False)
2. Organic matter may actually increase in some soils under irrigation.
3. Most soils lose organic matter under cultivation.
4. Loss of organic matter makes a soil more erodable.
5. Double-cropping is not a good idea because it wears out the soil.
6. Alfalfa hay causes severe nitrogen tie-up if mixed into the soil.
7. Composting can prevent nitrogen tie-up.
8. Organic soils are naturally rich in nutrients.
9. Low-oxygen conditions under water are responsible for organic soils.
10. Of the peats available for sale, sphagnum peat is best for acid-loving plants.
11. Of the three major components of fresh organic matter, _____ decomposes most rapidly.
12. Organic matter disappears from the soil because it oxidizes to _____ and _____ .
13. The residue from decay is called _____ .
14. Organic soils contain more than _____ % organic matter.
15. The amount of nitrogen tie-up by a material depends on its _____ ratio.

16. Probably the most available source of organic matter for farmers is _____ .

17. The natural loss of organic soils under cultivation is called _____ .

18. The chemical that adds to plant rigidity and resists decay in the soil is _____ .

19. The most decomposed organic soil is _____ .

20. The peat most often used in potting mixes is _____ .

21. Name three crops commonly grown on organic soils.

22. List three ways to avoid nitrogen tie-up.

23. List five practices a grower can use to maintain soil organic matter.

24. Other things being equal, which would you expect to have the most organic matter, a clay loam or a sandy loam?

25. Name something mentioned in the text that should not be added to a compost pile.

26. Some have thought that soil should be "given a rest" by sometimes not growing crops on it. Why is this not normally a good idea?

27. Why does high fertility help maintain organic matter in the soil?

28. Describe the conditions that create organic soils.

29. Explain the difference, in terms of organic matter creation, between prairies, forests, and cultivated fields.

30. Explain how organic matter improves both coarse and fine textured soil.

SUGGESTED ACTIVITIES

1. Observe the effects of nitrogen tie-up by growing corn in pots using two different soil mixes. Grow one group of plants in a normal but unsterilized soil mix. Grow another group in a mix that is half fresh sawdust. Note differences in crop appearance or growth.

2. Prepare a miniature compost. Tightly pack a jar half full of grass clippings, and sprinkle a few drops of water on the clippings. Place the cap on loosely, then put the jar in a warm dark place. Observe what happens every two days for six weeks. You can also try this using different materials. See if adding a little fertilizer or soil makes a difference.

3. Make an actual compost pile. Note that as the compost decays, the pile gets smaller. Why?

chapter nine

Soil Fertility

Here is a photograph of a project in a college botany student lab. These two sunflower plants are being grown hydroponically; that is, in solutions of water and fertilizer, rather than in soil. The plant on the right needs nitrogen, so is small and pale. How exactly do plant scientists know what elements plants really need for growth? By conducting hydroponic studies such as this. This chapter will tell you some of what they have found out about plant nutrients and soil fertility.

OBJECTIVES

After completing this chapter, you should be able to:

- name and classify the essential elements
- list four sources of nutrients in the soil
- describe soil colloids
- define cation exchange capacity and related terms
- describe how plants absorb nutrients
- explain other soil fertility factors

TERMS TO KNOW

adsorption	isomorphous substitution	primary nutrient
cation exchange	luxury consumption	root interception
cation exchange capacity	macronutrient	secondary nutrient
colloid	mass action	sesquioxide
diffusion	mass flow	silicate clay
essential element	micelle	soil fertility
exchangeable base	micronutrient	soil solution
expanding clays	non-exchangeable ions	trace element
hydroponics	percent base saturation	

Soil fertility is simply defined as the ability of soil to supply nutrients for plant growth. The soil is a storehouse of plant nutrients. The nutrients are stored in many forms, some very available to plants, some less so. The concept of soil fertility includes not only the quantity of nutrients a soil contains, but how well nutrients are protected from leaching, how available the nutrients are, and how easily roots can function. A beginning point for discussing soil fertility is to define the term *plant nutrient*.

PLANT NUTRIENTS

What exactly is a plant nutrient? Plant nutrients are the *essential elements* needed for plant growth. Plants absorb at least 90 different elements. Some of these are needed for plant growth. Some are not needed by plants, but, like cobalt, are needed by the animals that eat plants. Many elements are not needed by either plants or animals, and some, like lead, are even toxic. Thus, plants contain many elements that are not needed for growth. Which elements are essential? The most commonly accepted rules (though not the only ones) for determining if an element is essential are as follows:

1. A lack of the element stops a plant from completing growth or reproduction.
2. The element is directly involved in plant nutrition, not merely "taking up space" in plant tissues.
3. A shortage of the element can be corrected only by supplying that element.

Based on these rules, 16 essential elements are identified by most scientists (figure 9-1). (Several others play a role in the nutrition of some plants, but cannot yet be considered true essential elements for all plants. These are identified as "other" in figure 9-1.) Of the 16, 3 account for 95% of all plant needs: carbon, oxygen, and hydrogen. These three elements are obtained from air and water. The other 13 mineral nutrients are obtained from the soil. It is these 13 elements that will be covered in this chapter.

Plants use 6 of the 13 mineral elements in large amounts. These elements are labeled *macronutrients,* "macro" meaning large. The six nutrients are nitrogen, phosphorus, potassium, calcium, magnesium, and sulfur. In the following relationship, the six nutrients are listed in decreasing order from the greatest amount used by plants (nitrogen) to the smallest amount (sulfur). The ">" symbol means greater than, and "≥" means greater than or equal to. Refer to figure 9-1 for the meanings of the chemical symbols.

$$N \geq K > Ca > Mg \geq P > S$$

Soils are less likely to be deficient in calcium, magnesium, and sulfur than the other three nutrients (nitrogen, phosphorus, and potassium). Since most soils (but not all) supply enough calcium, magnesium, and sulfur, soil scientists call them the *secondary macronutrients,* or simply, secondary nutrients. The *primary macronutrients,* sometimes called fertilizer elements, are not usually available in large enough amounts for best growth. The three primary nutrients—nitrogen, phosphorus, and potassium—are most likely to be added to soil by fertilization. Note that the division into primary and secondary macronutrients is not based on the relative amounts used by plants, but on their importance as fertilizers.

The six macronutrients, except for potassium, are part of the materials that make up the bulk of the plant. Protein, for instance, includes both nitrogen and sulfur. Living tissue also contains very small amounts of certain important chemicals that control life processes, like the vitamins that people take. The other seven essential elements form part of these key materials in plants.

The other seven nutrients, also listed in figure 9-1, are labeled *micronutrients* or *trace elements,* because they are used in such small amounts. Iron, for example, plays a role in the process that forms chlorophyll. Only a small amount of iron is needed, but too little iron means that chlorophyll fails to form (figure 9-2). Without chlorophyll, photosynthesis slows down, resulting in a decline in plant growth. The term "micronutrient" does not, however, mean the elements are unimportant (figure 9-3). Plants will not grow normally without enough of these trace elements.

The 13 "micro" and "macro" elements are furnished by the soil. Plants absorb these elements in a specific way. Their roots take in the nutrients in certain ionic forms. Knowing this process helps us understand how best to fertilize plants for growth.

Name	Symbol	Ionic Form	Ion Name
Carbon	C	—	—
Hydrogen	H	H^+—(not used by plants in this form)	
Oxygen	O	—	—
Primary Macronutrients			
Nitrogen	N	NO_3^-, NH_4^+	Nitrate, ammonium
Phosphorus	P	HPO_4^{-2}, $H_2PO_4^-$	Orthophosphates
Potassium	K	K^+	—
Secondary Macronutrients			
Calcium	Ca	Ca^{+2}	—
Magnesium	Mg	Mg^{+2}	—
Sulfur	S	SO_4^{-2}	Sulfate
Micronutrients			
Boron	B	BO_4^{-2}	Borate
Copper	Cu	Cu^{+2}	—
Chlorine	Cl	Cl^-	Chloride
Iron	Fe	Fe^{+2}, Fe^{+3}	Ferrous, ferric
Manganese	Mn	Mn^{+2}	Manganous
Molybdenum	Mo	MoO_4^{-2}	Molybdate
Zinc	Zn	Zn^{+2}	
Others			
Nickel*	Ni	Ni^{+2}	
Sodium	Na	Na^+	
Silicon	Si	SiO_3^{-2}	
Cobalt	Co	Co^{+2}	

*Nickel was tentatively identified as an essential element in 1984.

Figure 9-1. Essential elements and their ionic forms. Nickel was tentatively identified as an essential element in 1984.

Nutrient Ions. Ions, as explained in appendix 1, are charged atoms or molecules. The charge, either positive (a cation) or negative (an anion), results when there is a different number of electrons than protons. One way ions form in soil is when compounds dissolve in water. For example, when the soluble fertilizer potassium nitrate dissolves, the molecule breaks into two ions:

$$KNO_3 \xrightarrow{\text{solution}} K^+ + NO_3^-$$

The concept of nutrients as ions is important, as will become apparent later in this chapter. Plant roots *absorb* nutrient ions, and soil particles *adsorb* them. *Absorb* means to take in something, like a sponge absorbing water. *Adsorb* means to attract a thin layer of molecules to a surface, where they stick. Figure 9-1 lists each nutrient in the ionic form(s) most commonly absorbed by plants. Any special name for the ion is also listed.

Figure 9-2. One healthy pin oak leaf is shown with one that exhibits the symptoms of a lack of iron. Iron is a trace element involved in chlorophyll formation. In the iron-deficient plant *(foreground)*, little of the dark green chlorophyll forms, resulting in a yellow leaf.

Figure 9-3. Micronutrient deficiencies. Plant on the right lacks micronutrients, while the one on the left enjoys complete nutrition. Sometimes micronutrients are called "minor elements"—but obviously a shortage is not minor in its effect.

SOURCES OF ELEMENTS IN SOIL

Nutrient elements are present in the soil in four forms, as shown in figure 9-4. Together, these four sources perform two functions: to store nutrients and to make them available to plants. We can compare soil to a bank and nutrients to the money deposited there. Customers keep money in a bank in several ways. Money in checking accounts can be obtained simply by writing checks, but a person uses up this money quickly. Thus, checking accounts are a short-term form of money storage, and the cash is readily available. One can also buy savings certificates but can only cash them after some time has passed. Thus, they are a long-term form of money storage, and the cash is not readily available.

Soil Minerals. These minerals are the source of all soil-supplied nutrients except nitrogen. Soil minerals are the longest-term storage. Weathering frees the elements slowly over time, dissolving the minerals into ions. Figure 2-4, for instance, lists the nutrients contained in several rocks.

Organic Matter. This supplies large amounts of several elements. Organic matter is the main way of storing the nutrient anions listed in figure 9-1. Organic matter is an intermediate form of storage, since elements are freed for plant use by decay. Some of the nutrients in fresh organic matter are released fairly quickly; however, those in humus are released more slowly.

Adsorbed Nutrients. These nutrients are held in the soil because they are attracted to clay and humus particles. Adsorption occurs because clay and humus particles are negatively charged. Many plant nutrients are positively charged and so stick to the soil particles. Adsorbed nutrients are held fairly tightly by particles, but most are considered to be available to plants.

Dissolved Ions. Dissolved ions are the most readily available form of nutrients. The mixture of ions and soil water is termed the *soil solution*. Plants absorb ions

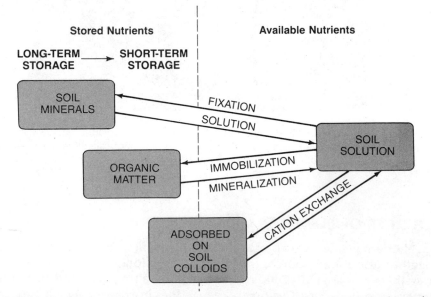

Figure 9-4. There are four sources of plant nutrients in the soil. The sources of nutrients that are least available to plants are shown on the left. The source of nutrients in their most available form is shown on the right. The elements can change forms. When crops remove nutrients, some stored nutrients go into solution and become available to plants.

directly from the soil solution. However, these nutrients can be used up rapidly and can be leached away by percolating water.

As shown in figure 9-4, the nutrients can change form, so that some are available while others are being stored for later use. As plants withdraw nutrients from the soil solution, elements being held in reserve can become available to plants. While nutrients are held in reserve, they are protected from leaching or other losses.

SOIL COLLOIDS

As shown in figure 9-4, adsorption serves as a source of stored nutrients when plants take nutrients out of the soil solution. Nutrients are adsorbed on certain soil particles called *colloids*. A colloid is a very tiny particle, a few thousandths of a millimeter in size. Soil colloids bear a slight negative charge, for reasons described later. This negative charge is important, because it attracts nutrient cations. The soil contains three types of colloids: silicate clays, oxide clays, and humus.

Silicate Clays. Clay minerals are not simply pieces of silt or sand broken into tinier particles. A clay particle is a tiny crystal of mineral formed in the soil from the weathered products of minerals like feldspar or mica. Their formation could be shown as:

$$\text{primary mineral} \xrightarrow[\text{solution}]{\text{weathering}} \text{ions in soil} \xrightarrow{\text{crystalization}} \text{clay material}$$

A particle of silicate clay, called a *micelle,* is a flat, platelike crystal made of many layers. Each layer, in turn, is made of two or three sheets. The sheets are mainly composed of three elements: silicon (Si), oxygen (O), and aluminum (Al). These three elements combine to form several kinds of sheets, which can combine to form several kinds of clays.

In the soil, silicon combines with oxygen to form the *silica sheet.* The basic unit of the silica sheet is the silica tetrahedron: a silicon atom surrounded by four oxygen atoms. This forms the shape of a four-sided pyramid or tetrahedron (figure 9-5). Many tetrahedra join together by sharing oxygen atoms to form the silica sheet.

The second important sheet in silicate clays is the *alumina sheet.* The basic unit of the alumina sheet is the alumina octahedron (figure 9-5). Here an aluminum atom is surrounded by six hydroxyl groups (OH^-) to form an octahedron or eight-sided figure. Octahedra join through the hydroxyl groups to form the alumina sheet.

These sheets can stack atop one another in several ways to form a complete clay crystal. The simplest stacking is to join a single alumina sheet to one silica sheet, forming what is called a 1:1 layer (figure 9-6). The alumina sheet sheds some of its hydroxyl groups by sharing oxygens at the "tips" of the attached silica

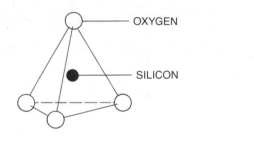

(A) **Silica Tetrahedron**

(B) **Alumina Octahedron**

Figure 9-5. *(A)* In the silica tetrahedron, each silicon atom is located in the center of four oxygen atoms. *(B)* For the alumina octahedron, each aluminum atom is centered among six hydroxyl groups.

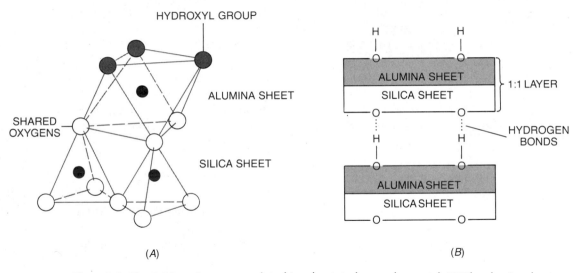

(A)

(B)

Figure 9-6. The 1:1 layer is one way of stacking sheets to form a clay crystal. *(A)* The alumina sheet bonds to the silica sheet by sharing oxygen atoms at the "tip" of each silicon tetrahedron. *(B)* Hydrogen bonding between layers permits several layers to bond together to form a complete micelle.

sheet. Note the clear order of the 1:1 layer: hydroxyl groups at the top and oxygen atoms at the bottom. Two layers can now join by hydrogen bonding—the hydroxyl groups of one layer bond to the oxygen atoms of another. Hydrogen bonds hold the layers in the clay crystal together tightly. The clay mineral composed of bonded 1:1 layers is *kaolinite*.

A second way to stack sheets is to sandwich an alumina sheet between two silica sheets (figure 9-7). Here the alumina octahedra replace all but two of their hydroxyl groups by sharing oxygen atoms with the silica sheets. This is a 2:1 structure. No hydroxyl groups are exposed at the surface, so layers are not

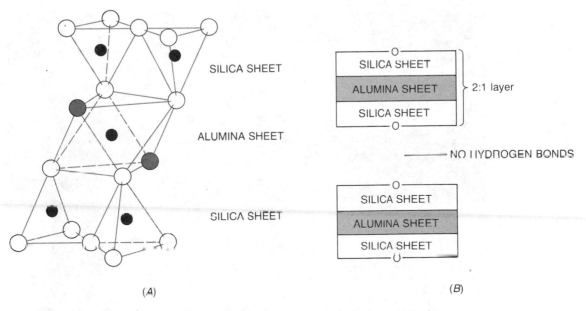

Figure 9-7. Another way of stacking sheets to form a clay structure is the 2:1 layer. *(A)* In this structure, the sheets are bonded together by sharing oxygen atoms. *(B)* Hydroxyl groups are not exposed, so the layers are not held together by hydrogen bonds. The layers of many 2:1 clays are less firmly bonded together and thus can open up.

cemented by hydrogen bonds. The bonds that hold the layers together are much looser than hydrogen bonds, so most 2:1 clays can "open up."

Types of Silicate Clay.   Several types of clays result from the ways in which 1:1 or 2:1 layers bond together. Some clays are highly charged and hold cations well, others are not. Some clays are sticky, some are plastic, and some swell when wet. Note that some of these traits are the soil consistence factors listed in chapter three.

Two important traits of silicate clays depend on how easily the layers can be separated. If they loosen easily, then water can enter the micelle between the layers, and the particle will swell when wetted and shrink when dried (figure 9-8). Such clays are called *expanding clays*. If the layers separate, more surface area is exposed for the adsorption of cations. Thus, clays with loosely bound layers usually hold more nutrients. Figure 9-9 sketches the structure of several types of clays. Figure 9-10 summarizes some of the characteristics of these clays.

Mica Clays.   Mica clays are 2:1 clays resulting from the weathering of mica minerals. The layers of one mica clay, *illite*, are firmly bound by bridges of potassium ions. If some of the potassium leaches away, then the layers can open up slightly, so illite is slightly expanding. If all the potassium is lost, a new clay

Figure 9-8. Mudcracks in expanding clay. Swelling and shrinking as the soil wets and dries forms these cracks. (Courtesy of Howard Hobbs, Minnesota Geological Survey)

forms, *vermiculite*. Vermiculite layers are loosely bound by magnesium ions that are surrounded by six water molecules. Vermiculite expands greatly when wetted.

Smectite Clays.   Smectites result either from the weathering of feldspar or from the advanced weathering of vermiculite. They are 2:1 clays, sticky, and highly expanding. Water fills the space between the layers, so the layers are very loosely held. The bonding force is too slight to hold a large particle together. This means that smectite clays are formed of very small particles. The best known smectite is *montmorillinite*.

Chlorite Clays.   Chlorite layers are tightly bound by a fourth clay sheet. Chlorite is often termed a 2:1:1 clay. The fourth sheet is either another alumina sheet or a sheet of magnesium-oxygen octahedra. This sheet binds the 2:1 layers together fairly tightly.

Kaolinite Clays.   These are 1:1 clays. Hydrogen bonds bind the layers tightly, so water cannot get between the layers. Thus, kaolinites swell least of all the clays. They present the smallest surface area for the adsorption of soil cations. The strong bonds allow particle sizes as large as silt. Kaolinite is highly plastic and is used for making pottery.

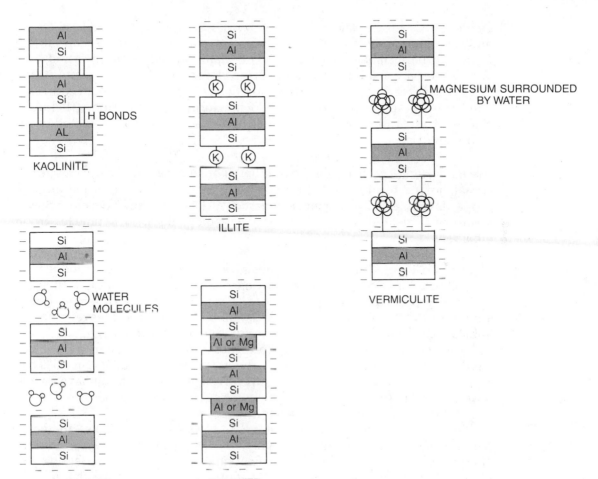

Figure 9-9. Each drawing illustrates the micelle structure of a different type of silicate clay. The dashes indicate where negative charges, or cation exchange sites, can exist.

Clay	Wet Consistency	Relative Swelling When Wetted	Interlayer Bonding
Smectites (2:1)	Very sticky	High	Very weak
Vermiculite (2:1)	Sticky	High	Moderate
Illite (2:1)	Slightly sticky	Low	Strong, by K ions
Chlorite (2:1:1)	Nonsticky	None	Strong, by fourth sheet
Kaolinite (1:1)	Plastic, slightly sticky	Very low	Strong H bonds
Sesquioxides	Nonsticky	None	—

Figure 9-10. Characteristics of important soil clays

Oxide Clays. Oxide clays, called *sesquioxides,* are tiny particles of iron (Fe_2O_3) and aluminum oxides ($Al(OH)_3$). Oxides are common to old soils in warm, humid, tropical climates. Long periods of weathering leach out silica and some alumina, leaving behind the oxides. Oxide clays tend to aggregate into strong, coated, sand-sized peds, so that the soil behaves much like sand. Oxide clays can form stacked sheets, but they do not form the crystalline structure of silicate clays. Oxide clays do not swell, are not sticky, and have limited power to hold nutrients.

Humus. Humus particles are the residues of organic matter decay. They are not crystalline and form irregular, round shapes. They have none of the physical properties of clays, like stickiness or plasticity. However, they have more power to adsorb nutrients than clays. Humus is not stable; it decays over time to carbon dioxide.

Charged Colloids. Colloids carry a negative charge that attracts cations from the soil solution. Clay particles gain a negative charge in two ways. First, some hydroxyl groups on the broken end of a clay micelle lose their hydrogen ion (figure 9-11). The hydrogen ion is simply a proton, so this leaves a charge imbalance. The remaining oxygen therefore has a negative charge.

The second process is *isomorphous substitution.* One cation can replace another cation of similar size in a clay sheet. For instance, aluminum (Al^{+3}) can replace a silicon atom (Si^{+4}) in a silica layer. While the cation fits, it has a *smaller positive charge.* This leaves a "spot" in the crystal that is short of positive charges. These spots acquire a negative charge.

Because a micelle contains a number of these negative "spots," the entire particle maintains a negative charge. The minus charge attracts positively charged cations, so the micelle is surrounded by a "cloud" of cations. The minus and plus charges balance, so the entire unit has a net zero charge.

Clays differ in the number of sites for negative charges, so their ability to retain cations also differs. The types of clay shown in figure 9-7 differ as follows:

- Kaolinite has little minus charge, since almost no isomorphic substitution occurs. The only exchange sites are from hydroxyl groups on the broken ends of a clay micelle.
- Smectites have many negative sites because magnesium ion (Mg^{+2}) substitute for some aluminum (Al^{+4}) in the alumina sheets. In addition, cations can adsorb on sites between the 2:1 layers.
- Vermiculite has an even greater negative charge, because about one in four silicon atoms are replaced by aluminum. Because it is a swelling clay, some cations can be retained between the 2:1 layers.
- Illite has the same substitutions as vermiculite. However, since the layers are held together by potassium, few cations can be held between the layers. Thus, illite holds fewer cations than vermiculite.

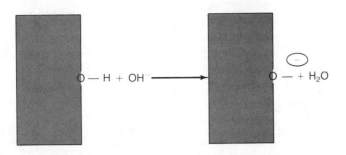

Figure 9-11. The loss of a hydrogen ion (proton) from a hydroxyl group on the surface of a clay micelle leaves the oxygen with a negative charge that can attract cations from the soil solution.

- Sesquioxides have very little negative charge, except for a few hydroxyl groups on the surface.
- Humus has numerous sites. Many organic compounds found in humus have hydroxyl groups as part of their structure. These groups can lose a hydrogen to form negative spots over much of the humus particle surface.

CATION EXCHANGE

The negative charge of soil colloids plays a key role in the way nutrients behave in the soil. Since the tiny particle bears a negative charge, it attracts positively charged ions and repels negatively charged ones. The action follows the rule that "opposites attract, likes repel." This is much the way magnets work: like poles repel and opposite poles attract (figure 9-12).

To explain this further, let's see how it works with a clay particle. Clay particles are plate-like in shape, and negatively charged. Figure 9-13 shows a colloid surrounded by soil solution. The negative charges on the colloid, or *cation exchange sites,* attract cations from the soil solution. The force of attraction holds the cations on the clay particle, in a manner similar to iron filings adhering to a magnet. This is called *adsorption.*

As figure 9-13 shows, cations can move on and off the particles. When one ion leaves, it is usually replaced by some other cation. We call the replacement of one ion for another *cation exchange.* Cations that can be replaced on exchange sites are said to be *exchangeable,* such as exchangeable potassium.

The ability of a soil to hold nutrients is directly related to the number of cations it can attract to soil colloids. This value is determined by the *amount of clay,* the *type of clay,* and the *amount of humus.* The number of sites is measured by the *cation exchange capacity* (CEC), which is expressed in milligram equivalents per 100 grams of soil (mg/100 g). An equivalent weight is the weight of the number of atoms of an element that is the same as the number of atoms in one gram of hydrogen.

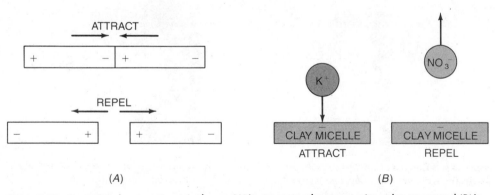

Figure 9-12. Opposite charges attract as shown *(A)* in magnets where opposite poles attract and *(B)* in the soil where negatively charged clay particles attract positively charged cations.

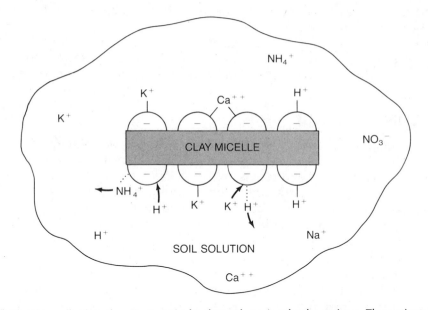

Figure 9-13. The colloid surface is negatively charged, so it adsorbs cations. The cations are "exchangeable" in that if one cation leaves, another one takes its place. The greater the change of the colloid, the higher the "cation exchange capacity" of the soil. (*Note:* The semicircles used to represent the exchange "sites" are for illustration only.

A milliequivalent is 1/1000 of an equivalent. A milliequivalent is used rather than the simple weight of adsorbed cations, because it cancels differences in the weight and charge of the cations on the sites. For instance, one milligram of calcium occupies fewer exchange sites than one milligram of magnesium, simply because it weighs more. Using milliequivalents per 100 grams of soil ensures that the CEC values are the same whatever the cations on the exchange sites.

Cation exchange units can be thought of as "fertility units." The number of units can be used to compare the fertility of soils. In other words, as the number of units increases, the CEC increases and the soil becomes more fertile.

Figure 9-14 lists the cation exchange capacity values of several clays, humus, and several soil textures. Note that humus has a much higher CEC than clay. However, clay usually adds more cation exchange capacity to a soil than does humus, simply because there is so much more clay than humus in most soils. Organic soils, which are mostly organic matter, are an exception. Sandy soils may also gain a large portion of their exchange capacity from humus. Since sandy soils are very low in clay, the contribution from humus may outweigh that of clay.

Cation Behavior at the Exchange Sites. The negatively charged surface of a clay micelle strongly attracts cations. The cations cluster most densely near the micelle surface, neutralizing the negative charge. The cations can move on the micelle and trade places or exchange with cations in solution. Several factors control the selection of cations that leave the micelle or become adsorbed. Two important factors are (1) the relative bonding strength of each cation and (2) the number of each type of cation.

If two cations are present in the soil in equal numbers, then the one that bonds most tightly to exchange sites will tend to be the one found on the micelle. The most strongly adsorbed cation is hydrogen, followed in decreasing order by:

$$H^+ \geq Al^{+3} > Ca^{+2} \geq Mg^{+2} > K^+ \geq NH_4^+ > Na^+$$

Colloid	Cation Exchange Capacity (mEq/100 g soil)
Humus	100–300
Vermiculite	80–150
Montmorillinite	60–100
Illite	25–40
Kaolinite	3–15
Sesquioxides	9–3
Soil Texture *(Temperate Climate Soils)*	
Clay loam	30
Silt loam	27
Loam	24
Sandy loam	17
Loamy Sand	9

Figure 9-14. Cation exchange capacity of several colloids and soils. The values for soil textures are provided only to show differences in the capacity. An individual soil may vary greatly from these values.

Assume that a soil has equal numbers of calcium and sodium (Na) ions. Calcium will tend to take over the exchange sites because it adsorbs more strongly on the micelle. Sodium will tend to leach out of the soil solution.

The second controlling factor is *mass action*. Mass action means that the greater the number of an ion in the soil, the more exchange sites it will occupy. As an example, in high lime (calcium carbonate) soils, most exchange sites are occupied by calcium.

Consider the treatment of high sodium soils with gypsum (calcium sulfate). In a high sodium soil, many exchange sites are taken up by sodium (more than 15%, as a rule). When gypsum is added, calcium displaces sodium on the exchange sites. The displaced sodium enters the soil solution and is leached away by heavy watering. Calcium replaces the sodium on the exchange sites by means of mass action (there are many calcium ions in solution) and because calcium is adsorbed more strongly than sodium.

Cations that are weakly held, in direct contact with the soil solution, are exchanged fairly easily. These are termed *exchangeable* cations. Some are held very tightly against the colloid, or may be trapped between layers of a clay micelle. These do not normally pass into the soil solution easily, and are said to be *nonexchangeable*. Even these may be given up slowly if the surrounding solution becomes very low in those ions. This may be thought of as long-term storage.

Anion Storage.   A number of nutrients exist in the soil as negatively charged ions, or anions (refer to figure 9-1). The negative charge means that an anion is repelled from a cation exchange site. The most important form of storage for many anions like sulfates and borates is organic matter. There is, however, an anion exchange process that can store small amounts of some anions.

Anion exchange sites are the opposite of the cation exchange sites where hydrogen is lost from a hydroxyl group. At the anion exchange site, an extra hydrogen joins the hydroxyl group to produce a net positive charge (figure 9-15). The positive charge can then attract anions. For most soils, anion exchange capacities are quite low. Typical values are a few tenths of a milliequivalent per 100 grams of soil.

Applications of the Cation Exchange Capacity.   How growers use soil is strongly influenced by the cation exchange capacity. High CEC soils, measuring between 11 and 50 units, usually contain a lot of clay. Low CEC soils, measuring below 11 units, usually have a high sand content. Sticky soils are high in the types of clay having the highest cation exchange capacity. Thus, the cation exchange capacity is reflected in physical properties of soil, such as texture and consistence. A number of other differences among soils are related to the CEC.

Cation exchange capacity is one of the factors that determines how much herbicide should be spread on the soil. Colloids adsorb pesticides as well as nutrients. Therefore, clay and humus tend to tie up many chemicals. As a result, most soil-applied pesticides have a high and low application rate printed on the

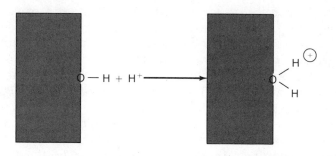

Figure 9-15. At the anion exchange site, the hydroxyl group on the surface of the particle picks up a hydrogen ion (proton) The site now has a net positive charge that can attract anions from the soil solution.

label. The high rate is recommended for fine-textured soils to compensate for the amount that will be tied up in the soil. The low rate is recommended for coarse-textured soils. Application rates also often depend upon the humus content of the soil.

The amount of lime to be applied to a soil to change the soil pH is also a function of the CEC. In the process of liming soil, calcium displaces hydrogen on the cation exchange sites. The more exchange sites in the soil, the more lime is needed. Therefore, growers apply much more lime to correct the acidity of fine-textured soils than they do to correct coarse-textured soils.

Fertilization practices often depend on the cation exchange capacity. For example, soils with a high CEC store cationic nutrients well. These nutrients tend to leach out of low CEC soils. As a result, growers with finer soils can apply an entire season's fertilizer at one time and be sure the soil will hold it. Growers with coarse soils, on the other hand, may divide their fertilization into two applications during the season to avoid losing fertilizer through leaching.

In another example, the concept of cation exchange suggests that it is easier to improve the cation exchange capacity of a sandy soil by improving the organic matter content than by adding clay. For example, in temperate humid climates, most clays are kaolinite (CEC = 3–15 mEq) and illite (CEC = 25–40 mEq). In a soil composed of these clays, each percentage of clay in the soil adds between 0.03 to nearly 0.5 mEq to the cation exchange capacity of the soil. Each percentage of humus, on the other hand, adds between one and three full mEq. Thus, far less organic matter is required, compared with clay, to raise the exchange capacity of the soil.

Percent Base Saturation.   Soil fertility is influenced not only by the cation exchange capacity (how many cations it can store), but also by how much of the CEC is actually filled with plant nutrients. Each exchange site can be filled by members of either of two groups of cations. One group consists of hydrogen and aluminum, which are not plant nutrients at all. Their primary contribution is to acidify the soil. The other cations are called *exchangeable bases* and include elements such as calcium, magnesium, potassium, and sodium. Except for sodium, the bases are plant nutrients.

The percentage of the cation exchange sites filled with exchangeable bases is called the *percent base saturation.* It tells us how much of the soil's "potential fertility," the CEC, holds exchangeable bases. For example, if the total CEC of a soil is 10 milliequivalents per 100 grams of soil and bases occupy 6 of the 10, then the base saturation percentage is 60%. Most crops grow best at a base saturation of 80% or more. These crops require a good supply of nutrients. Some trees that grow on infertile soils do well at a base saturation of around 50%.

Plants need nutrients in balance for balanced nutrition. If one cation dominates the soil, it tends to replace other cations on the exchange sites by means of mass action, subjecting them to leaching. This causes an imbalance of nutrients. Figure 9-16 lists the optimum range of base saturation of macronutrient cations for most crops.

NUTRIENT UPTAKE

This chapter has already noted two factors that affect soil fertility: (1) the amount of storage capacity of a soil (cation exchange capacity), and (2) how much of that storage actually contains nutrients (percent base saturation). A third fertility factor is how easily roots take up nutrients. How do plants take up nutrients from the soil?

Plants absorb nutrients in the form of the ions listed in figure 9-1. While roots absorb nutrients from the soil solution, the elements do not "soak in" with the soil water. In fact, roots may have a concentration of some nutrient hundreds of times that of the soil solution. For nutrients to passively "soak-in" against such a gradient would be like water running up hill. Roots actively transport nutrient ions through root cell membranes. The transport is an active process that uses energy. Since roots produce their energy by respiration, conditions that limit respiration, like a waterlogged soil, also limit nutrient uptake. In addition, the active transport of ions across a cell membrane allows some selection—the root can take up some elements more than others.

The soil solution surrounds roots growing through soil pores. Root hairs get ions directly from the soil solution through their own form of cation and anion exchange. If a cation is removed from solution, the root gives up a hydrogen ion (H^+) to replace it in the soil solution (figure 9-17). If an anion is absorbed from solution, the root gives up an anion to replace it. The exchange maintains the electrical balance in the root and in the soil.

Note that plants take exchangeable bases from solution and replace them by releasing hydrogen ions. Because hydrogen forms stronger bonds on the exchange sites, the hydrogen ions will eventually replace another cation on the exchange sites (figure 9-17). This exchange renews the number of nutrient cations in the soil solution, allowing plants to continue to draw nutrients from the soil. Over time, however, the exchange increases the number of hydrogen ions bonded to the exchange sites and lowers the percent base saturation.

Cation	Percent Saturation
Calcium	60–70
Magnesium	10–20
Potassium	2–5
Hydrogen	10–15
Others (iron, etc)	2–4

Figure 9-16. Optimum range of percentage of saturation of several elements for most crops

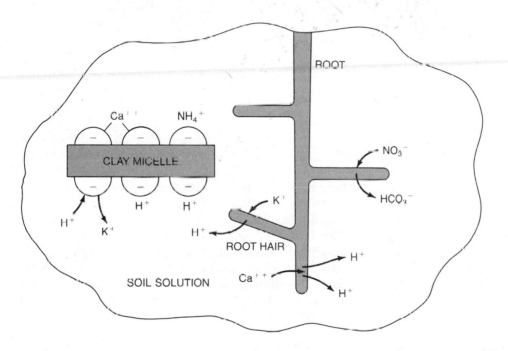

Figure 9-17. Root hairs absorb nutrients from the soil solution. To maintain electrical balance, the plant replaces an absorbed ion by releasing another ion to the soil solution. A loss of bases in the soil solution and a buildup of hydrogen and ions causes hydrogen/base exchange on soil colloids.

When growers fertilize and lime soil, they are reversing the loss of nutrient cations. If, for example, potassium fertilizer is supplied in the form of potassium chloride, potassium will replace some other cations, including hydrogen, by mass action (figure 9-18).

Figure 9-17 raises a question about nutrient uptake. Near the root hair, nutrients should be depleted. How does the root continue to obtain nutrients? Remember, plant roots only absorb nutrients which are in solution at the root surface. Roots continue to grow through the soil mass to find new supplies

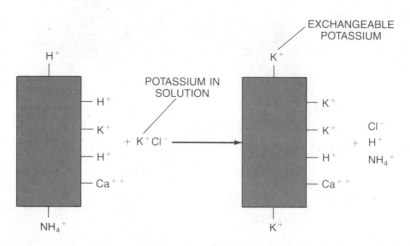

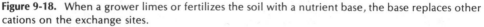

Figure 9-18. When a grower limes or fertilizes the soil with a nutrient base, the base replaces other cations on the exchange sites.

of nutrients. This important means of obtaining nutrients is called *root interception*. Roots actually come into close contact with very little of the soil mass, however, so other means of obtaining nutrients are also needed.

Two other processes help root hairs draw on nutrients by bringing ions to the root. The first process is called *mass flow*. In mass flow, ions are simply carried along with water moving toward the root by capillary flow. Review figure 4-13 if needed to understand how this works.

A second process, known as *diffusion*, also moves ions through the soil. Consider calcium as an example. Near the root there is less calcium because the root has been removing it. In response, calcium ions move toward the root *through* soil water (not along with it) to make a new balance. The movement of ions is from areas of greater concentration to areas of lesser concentration. Both mass flow and diffusion move nutrients short distances if the soil is moist. Both processes depend on a suitable film of water around soil particles.

Factors Affecting Uptake.   Several factors affect how plants take up nutrients. Plant use of nutrients is affected by features of the soil, such as oxygen supply, water supply, and soil temperature. Root distribution in the soil is another factor.

Nutrient uptake uses energy, which is produced by respiration (refer to figure 1-4). Since respiration uses oxygen, conditions that limit oxygen supply also limit nutrient uptake. Both poor drainage and soil compaction slow the movement of oxygen into the soil. As a result, these conditions also limit the ability of plants to absorb nutrients. This is another reason for draining wet soils and for avoiding compaction.

Dry soils lower nutrient uptake because the lack of water impedes nutrient flow toward the root hairs by mass flow and diffusion. Phosphorus, for example, moves in the soil only by diffusion. This is a very slow and short-range process that depends on soil moisture. Phosphorus uptake is sharply reduced in dry soil.

Soil temperature also affects nutrient uptake. The rates of all chemical reactions, including those in soil and plants, depend on temperature. Respiration rates go down in cold soil, so the plant has less energy to take up nutrients. Root growth is also slowed in cold soil, limiting root interception of nutrients. Mineralization of organic matter declines as well, so fewer organic nutrients are made available. For these reasons, cold soil hinders nutrient uptake. Phosphorus and iron deficiencies are common, for example, in spring when soils are cold and wet. Farmers often place an extra amount of phosphorus fertilizer beside planted seeds to overcome deficiencies. Greenhouse growers must be aware of the effect of cold soil, because irrigation water may be cold enough during the winter to harm plants.

An increase in the amount of nutrient ions in the soil improves absorption. This factor is obviously one reason growers fertilize their crops. When some elements, like potassium, are present in very high amounts, plants even take up more than they can use. This condition is called *luxury consumption*. However, since the excess is stored in plant cells, it may be used later if something happens to slow uptake by the roots.

Uptake is most rapid, of course, where roots are most numerous. We know roots grow best where air, water, and nutrients are in good supply. Drainage, compaction, and fertilization influence how well roots grow. So does the depth of the soil or the natural root depth of a particular crop. Plants with deeply growing roots need less fertilization than plants with shallow root systems. Soils which have restricted zones—those with high water tables, plow pans, or bedrock—can cause shallow root systems.

Figure 9-19 summarizes the factors affecting soil fertility and nutrient uptake.

Raises Fertility	Lowers Fertility
High clay content	High sand content
High humus content	Loss of organic matter
Good structure	Compaction
Warm soil	Cold soil
Deep soil	Shallow soil
Moist soil	Dry or wet soil
Good drainage	Excess irrigation or drainage
Fertilization	Erosion
Desirable microbes	Root damaging pests
Near neutral pH	pH too acid or alkaline

Figure 9-19. Factors affecting a soil's ability to supply plant nutrients

SUMMARY

For normal growth most plants require 16 essential elements. Carbon, oxygen, and hydrogen come from air and water; the other 13 elements are absorbed by plants from the soil. Crops take up primary and secondary nutrients in large amounts; trace elements are needed in small amounts.

Four sources of nutrients work together to both store and release nutrients to plants. These sources are soil minerals, organic matter, the soil solution, and adsorption by clay and humus. The ability of colloids to adsorb nutrients is based on their large surface area and on the negatively charged sites that are part of the colloids' structure. This ability is measured by the cation exchange capacity. Percent base saturation is the percentage of the exchange capacity filled with exchangeable bases.

Plants absorb nutrients by transporting ions into cells of the root. Uptake uses energy. To find new nutrient supplies, roots grow through the soil. In addition, nutrients flow either with or through soil water toward roots. Extreme soil conditions, including soil that is too dry, too wet, too cold, or badly compacted, impair the ability of roots to absorb nutrients.

Nutrient uptake is aided by a deep, well-drained soil. Several farming practices can improve soil fertility. Artificial drainage helps if a soil is poorly drained. Avoiding compaction or subsoiling already compacted soils is useful. Organic matter additions improve the cation exchange capacity and provide nutrients. Organic matter also keeps soil loose to enable sufficient oxygen to be supplied to roots. Proper fertilization also improves soil fertility.

REVIEW

1. Plants take up nutrients less readily from a cold soil (True/False).
2. Other things being equal, high CEC soils are more fertile than low CEC soils.
3. The most readily available nutrients are stored in organic matter.
4. Nutrients are absorbed as ions at the root surface.
5. Soil colloids consist of clay and sand.
6. Clays are structured like stacks of plates.
7. Clay particles are negatively charged.
8. Nutrient diffusion occurs rapidly in dry soils.
9. Adding humus to a sandy soil will add more CEC than a similar amount of clay.
10. Aluminum is needed by all plants.
11. _____ is the mineral nutrient used in the greatest amount by plants.

12. The primary macronutrients are _____, _____, and _____ .
13. The secondary macronutrients are _____, _____, and _____ .
14. Because of CEC, the ionic form of nitrogen best held by the soil is _____ (see figure 9-1).
15. Cation exchange capacity is measured by the units _____ .
16. Fertilization with a _____ macronutrient is most likely to cause a growth response.
17. _____ clays shrink and swell when dried and wetted.
18. Application rates of herbicides are highest in _____ (high/low) CEC soils.
19. Exchangeable ions like calcium are called exchangeable _____ .
20. When nutrients move through soil by moving with soil water, it is called _____ .
21. List the micronutrients.
22. Name three factors affecting the CEC of the soil.
23. List four soil conditions that inhibit uptake.
24. What are the three ways that roots and nutrients come into contact?
25. Define and give examples of soil fertility.
26. Explain the basic structure of a clay particle.
27. Why does montmorillinite expand but not kaolinite?
28. What are ions and what is a solution?
29. Can you explain why tropical soils are often infertile?
30. Explain the concept of percent base saturation.

SUGGESTED ACTIVITIES

1. Grow three groups of potted corn plants under three different soil moisture conditions: grow one group in dry soil, one in moist soil, and one in fairly wet soil. Using a tissue testing kit, measure the nutrient content of the leaves of each treatment. The tests should be done before the roots have completely filled up the pots.
2. Can you devise a way to test the effect of cold soil on nutrient uptake?
3. Using molecular model kits, construct models of silica tetrahedra and alumina octahedra. Then try to construct a portion of a clay layer.
4. Examine granules of horticultural vermiculite. Note that it is made of expanded mica sheets, and that it can soak up water between the sheets. Physically and chemically, the granule resembles the structure of a vermiculite micelle.

5. Gentian violet is a positive dye, and eosin is a negative dye. Prepare a water solution of each dye. Pour each solution into a pot of soil until water begins to drain from the bottom. Collect the drainage water. Which dye passed through the soil? Why? Try this on both a very sandy soil and finer-textured soil to see if there is a difference.

6. This little test can demonstrate solution and chemical fixation of nutrients. Mix a little silver nitrate (used in photo processing) in distilled, deionized water. It will dissolve, like nutrients in solid form dissolving in soil water. Also mix a bit of table salt into some distilled water. Now mix the two solutions together. The white material that results is silver chloride, which is not soluble in water. In the soil, chemical reactions can occur that have the same effect.

chapter ten
Soil pH and Salinity

This poor tree—a pin oak—is trying to grow on a college campus in Minnesota. Why are the leaves pale yellow, and some branches bare? Because the tree lacks iron. But the soil here doesn't lack iron. Minnesota, once home to a huge iron-mining industry, has plenty of it in its soils. Read this chapter to discover why this tree is short of iron.

OBJECTIVES

After completing this chapter, you should be able to:

- describe soil pH and describe its development
- describe how pH affects plant growth
- tell how to lime or acidify soil
- describe saline and sodic soils
- describe methods to treat saline and sodic soils

TERMS TO KNOW

acid soil	calcium carbonate	saline soil
agricultural lime	equivalent	saline-sodic soil
alkaline soil	chemical guarantee	SMP buffer test
basic soil	dolomite	sodic soil
buffering	fluid lime	sodium adsorption ratio
burned lime	hydrated lime	soil reaction
calcareous soils	marl	soluble salts
calcitic limestone	physical guarantee	total neutralizing power

"Sweet" and "sour" are old terms used to describe soil quality, which remind us of a farmer raising a handful of soil to the lips to taste sweetness or sourness. "Sweet" and "sour" are simple terms for soil reaction, or soil pH. What is soil reaction and how does it affect crop growth?

SOIL pH

The term *soil reaction* describes the acidity or alkalinity of a soil. Growers are concerned about soil reaction because it strongly affects plant growth. Reaction is measured by the pH scale. Figure 10-1 shows the pH scale, which gives sample pH values for common substances. The scale runs from a pH of 0 to a pH of 14.0. Readings between 0 and 7.0 are said to be *acid*. A pH of 1.0 is extremely acid and a pH of 6.0 is slightly acid. Examples of acid materials include foods like vinegar, tomato juice, and lemon juice. These acid foods have a sour taste.

Readings between 7 and 14 are *alkaline* or *basic*. The larger the number, the stronger the base. Soap is slightly basic. Household ammonia, with a pH of 11, is strongly basic. Bases, or alkaline substances, taste bitter.

The midpoint of the scale at pH 7.0 is the neutral point, which is neither acid nor base. Pure water, which has a neutral pH, can be a model in our discussion of pH. A few water molecules break up to form a cation and an anion as shown in reaction *(a)*:

(a)
$$H_2O \rightleftharpoons H^+ + OH^-$$

water   hydrogen   hydroxyl
               ion   ion

Soil Reaction	pH Scale	Common Solution
	0	
	1	Hydrochloric acid
	2	Lemons / Vinegar
	3	
Lowest pH for most mineral soils	4	Tomatoes
	5	Boric acid
Strongly acid		
Moderately acid	6	
Slightly acid		Milk
Very slightly acid	7	NEUTRAL
Very slightly alkaline		
Slightly alkaline	8	Sea water
Moderately alkaline		Bicarbonate of soda
Strongly alkaline	9	
	10	Milk of magnesia
Highest pH for most mineral soils	11	Ammonia
	12	
	13	Lye
	14	

Figure 10-1. The pH scale runs from 0 (most acid) to 14.0 (most alkaline). The pH values for a number of common substances are shown. Note that soil pH range extends from approximately 3.5 to 10.5.

The cation in the reaction is the hydrogen ion (H^+). It makes a solution acid. The anion is the hydroxyl ion (OH^-). It makes a solution basic. In pure water, the number of hydrogen ions equals the number of hydroxyl ions to maintain a balance. Thus, pure water is neither acid nor base. However, substances dissolved in water may change the balance, causing one form of ion to outnumber the other.

The pH scale indicates how acidic or basic a solution is by giving the number of hydrogen ions. The pH scale is a special scale for expressing hydrogen ion concentration. The smaller the number on the pH scale, the stronger the acidity of a substance. Each pH point multiplies acidity by a factor of 10. For instance, a pH of 5.0 is 10 times more acid than pH 6.0 and 100 times more acid than pH 7.0.

The important consideration is the balance between the hydrogen and hydroxyl ions. Soil that has far more hydrogen ions than hydroxyl ions is very acid. With only a few more hydrogen ions, it is slightly acid. On the basic or alkaline side of the scale, the reverse is true. Figure 10-2 shows this relationship by comparing the relative amounts of the two ions at different pH values.

DEVELOPMENT OF SOIL pH

Soil does not reach the pH limits shown in figure 10-1—the most acid soil has about a pH 3.5 and the most basic soil is pH 10.5. These are extreme values. Growers more commonly find that soil ranges between pH values of 5.0 and 8.0.

Soil pH results from the interaction of soil minerals, ions in solution, and cation exchange. Different reactions govern at different pH ranges. In the simplest terms, high pH is caused by the reaction of water and the bases calcium, magnesium, and sodium to form hydroxyl ions. Low pH is caused by the percolation of mildly acid water, which results in the replacement of exchangeable bases by hydrogen ions. To understand the full range of soil pH, it is easiest to start with alkaline soil.

Very basic soils (pH greater than 8.0) are more than 100% base-saturated. That is, all exchange sites are filled with bases, and the soil contains particles of mineral carbonates (CO_3) like calcium carbonate ($CaCO_3$). The pH of very alkaline soils results from the reactions of carbonates with water to form hydroxyl ions, according to reactions (b) and (c):

(b) $$CaCO_3 + 2H_2O \rightarrow Ca^{+2} + H_2CO_3 + 2OH^-$$

(c) $$Na_2CO_3 + 2H_2O \rightarrow 2Na^+ + H_2CO_3 + 2OH^-$$

This reaction with water is called *hydrolysis* ("hydro" meaning water). The hydrolysis of calcium carbonate (reaction b) results in a pH range of about 8.0-8.5. Soils in this range, which are 100% base-saturated and contain enough free calcium carbonate, are called *calcareous* soils. They result from the weathering of calcareous parent materials like limestone. If the sodium saturation exceeds 15%, then the hydrolysis of sodium produces lye (sodium hydroxide, $NaOH$), which can raise the pH to 10.0 (reaction c). Such high-sodium soils are termed *sodic*.

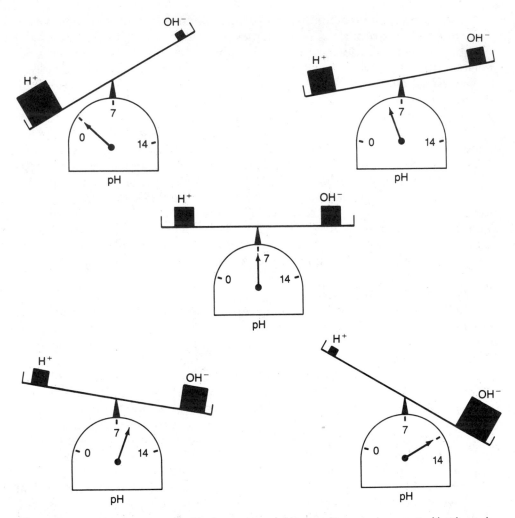

Figure 10-2. The relative amounts of hydrogen ions and hydroxyl ions are represented by the scale readings for several pH values. The total amount of the combined hydrogen and hydroxyl ions remains the same, but the percentage of each changes with the pH value. The pH measures only the amount of the hydrogen ion. In the acid range, hydrogen ions outnumber hydroxyl ions. The opposite is true in the basic range. At a pH of 7.0, they equal each other.

In humid climates, further weathering and leaching remove the excess basic minerals. When these minerals reach a low level, the soil ceases to be calcareous. However, the remaining minerals can still ensure that the soil is 100% base-saturated. This occurs at a pH of about 8.0, though it varies for different soils. At this point, pH begins to be controlled by the hydrolysis of exchangeable bases (rather than free carbonates, as above), as in reaction (d):

(d)　　　　$$\boxed{\text{micelle}}-Ca^{+2} + 2H_2O \rightarrow \boxed{\text{micelle}}\Big\langle{}^{H}_{H} + Ca^{+2} + 2OH^-$$

This reaction maintains a pH level between about 7.0 and 8.0, depending on soil type. Note that reaction *(d)* causes a removal of calcium from the micelle, replaced by hydrogen. As this reaction proceeds, and rainfall leaches out the calcium, base saturation goes below 100%. The same is true of soils of noncalcareous parent materials. Exchangeable hydrogen, as long as it is adsorbed, does not acidify the soil. When it does enter the soil solution by cation exchange, however, it makes the soil more acid, reaction *(e)*:

(e)　　　　$$\boxed{\text{micelle}}-H' \leftrightarrows H^+ + \boxed{\text{micelle}}-$$

The actual pH of soil with a base saturation below 100% depends on the balance between hydroxyl ion production by base hydrolysis (reaction *d*) and hydrogen ion production by hydrogen exchange (reaction *e*). This balance controls pH in slightly acid to slightly alkaline soils.

When pH declines to about 6.0, aluminum begins to leave the structure of silicate clays. Aluminum ions react in several steps with water to form hydrogen ions and aluminum hydroxide compounds. The reactions are summarized in reactions *(f)* and *(g)*:

(f)　　　　$$\boxed{\text{micelle}}-Al^{+3} \rightarrow Al^{+3} + \boxed{\text{micelle}}-$$
(g)　　　　$$Al^{+3} + 2H_2O \rightarrow Al(OH)_2{}^{+1} + 2H^+$$

Aluminum hydrolysis can lower soil pH to about 4.0. This is the most acidic the majority of upland soils become. Figure 10-3 summarizes the pH ranges and associated conditions.

Causes of Acidity.　Relatively young soils, those not exposed to long periods of weathering and leaching, share the pH of their parent materials. Acidic parent materials include granite, sandstone, and shale. These materials are common in New England, the Great Lakes, and the Appalachian states. The soils of many states, including many in the Great Plains, developed from calcareous parent materials like limestone. These soils tend to be neutral to alkaline.

The pH of most soils is controlled by the percolation (or lack of percolation) of acidic water. This percolating water leaches away bases and replaces them on the exchange sites with hydrogen and aluminum ions, reaction *(h)*:

(h)　　　　$$\boxed{\text{micelle}}-Ca^{+2} + 2H^+ \rightarrow \boxed{\text{micelle}}\Big\langle{}^{H^+}_{H^+} + Ca^{+2}$$

Figure 10-4 portrays this reaction graphically. This type of percolation occurs in humid climates, where precipitation exceeds evapotranspiration. In a humid climate, the net movement of water over the course of a year is downward, thus percolating water can leach out bases. In semi-arid or arid zones, the net water movement is upward, since water is being pulled out of the root zone by evaporation or transpiration. With little or no percolation, soils of dry regions tend not to become acidic. They may even become quite alkaline from calcium or sodium being carried upward into the root zone by capillary movement. Figure 10-5 shows that leaching has the greatest effect on the soils of the eastern half of the country and the Pacific Northwest.

A number of processes produce the hydrogen ions that make soil more acidic. Some processes occur naturally, and others result from human activities. A

pH Range	Determining Reaction	Saturation
8.5–10.0	Na_2CO_3 hydrolysis	100% base saturation, sodium saturation more than 15% (sodic soil)
8.0–8.5	$CaCO_3$ hydrolysis	100% base saturation (calcareous soil)
7.0–8.0	Exchangeable Ca hydrolysis	100% base saturation (noncalcarous soil)
6.0–7.0	Hydrogen exchange	Base saturation below 100%, some hydrogen saturation
4.0–6.0	Aluminum hydrolysis	Low base saturation, may be high Al saturation

Figure 10-3. Reactions that determine pH ranges

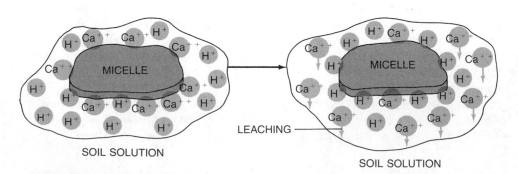

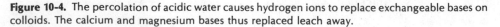

Figure 10-4. The percolation of acidic water causes hydrogen ions to replace exchangeable bases on colloids. The calcium and magnesium bases thus replaced leach away.

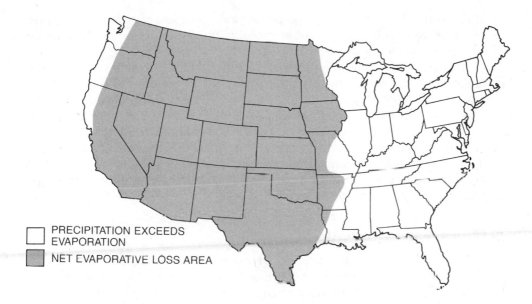

PRECIPITATION EXCEEDS EVAPORATION

NET EVAPORATIVE LOSS AREA

Figure 10-5. Soil acidity is greatest where the average annual precipitation is greater than the average annual evapotranspiration, In these areas (not shaded), there is a potential for leaching. (USDA, Second National Water Assessment, 1975, Appendix C-1, Nationwide Analysis)

major natural process that contributes to soil acidity is the respiration of plant roots and other soil organisms. During respiration, organisms give up carbon dioxide, which reacts with water to produce carbonic acid (reaction *i*). Carbonic acid, in turn, breaks down to release hydrogen ions:

(i) $$CO_2 + H_2O \rightarrow H_2CO_3 \rightarrow HCO_3^- + H^+$$

Thus, plant growth and organic-matter decay both produce hydrogen ions. Crop plants acidify soils in two additional ways. First, when roots take up cation nutrients like potassium, they "give back" an equivalent number of hydrogen ions. Second, growers take calcium and magnesium with each crop harvested. For example, every ton of alfalfa hay is a loss from the soil of 30 pounds of calcium and 8 pounds of magnesium. This removal of magnesium and calcium during harvest speeds up the acidification of soil.

Nitrification also contributes hydrogen ions to the soil. When nitrifying bacteria oxidize an ammonium ion (NH_4^+), two hydrogen ions result:

(j) $$NH_4^+ + 2O_2 \rightarrow NO_3^- + H_2O + 2H^+$$

This reaction is most important because of the increasing use of ammonium fertilizers, otherwise called acid-forming fertilizers. For example, until recently most growers relied on neutral superphosphates to supply phosphorus to crops. During the last twenty years, acidic ammonium phosphates have replaced much of the superphosphate in the fertilizer trade.

Controversy over the effects of acid rain on lakes has largely ignored the effects of acid precipitation on crop growth, including its contribution to soil acidity. Acid rain forms when nitrous oxide gas (NO_2) and sulfur dioxide gas (SO_2) react with water in the air to create sulfuric (H_2SO_4) and nitric acids (HNO_3). As a result, rain, snow, and even fog can be quite acid. Eventually the acids land on soil or in lakes and lower the pH. The oxides are formed as byproducts of burning fossil fuels in automobiles and power plants and other industries. Figure 10-6 shows the pH of rainfall in various parts of the United States and the areas in the United States that generate the sources of acid rain.

EFFECT OF pH ON PLANTS

Each crop grows best in a specific pH range. The pH ranges for a selection of crops are shown in figure 10-7. Most plants growing on mineral soils do well at a pH range of 6.0–7.0, for reasons described below. For organic soils, most crops prefer a pH of 5.5 to 6.0. An exception is a group of acid-loving plants, which includes mostly woody plants like blueberry and azaleas and many evergreens. Alfalfa is one of a few crops that prefer a slightly basic soil.

The actual number of hydrogen or hydroxyl ions does not seem to be the main factor in plant growth. Rather, several soil conditions related to pH are more important to plants. These include (1) the effect of pH on nutrient availability, (2)

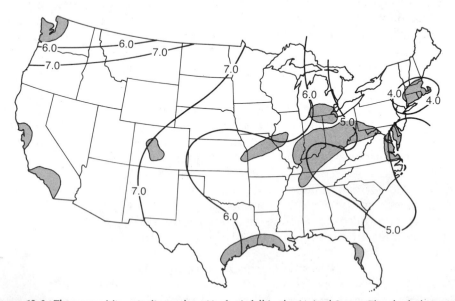

Figure 10-6. The curved lines indicate the pH of rainfall in the United States. The shaded areas are major sources of acid pollution. (Adapted from United States Water Resources Council, 1978)

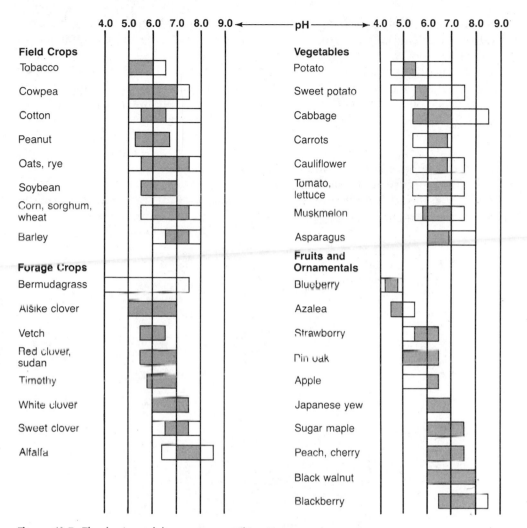

Figure 10-7. The horizontal bars represent the pH range of several crops for mineral soils of temperate regions. The shaded area of each bar shows the preferred pH. The unshaded area of the bar is the pH level the crop will tolerate.

the buildup of toxic levels of aluminum or other metals, and (3) effects on soil microbes. Which of these three factors has the greatest effect on limiting crop growth varies from soil to soil and from crop to crop.

Effect of pH on Nutrient Availability.   Many soil elements change form as a result of reactions in the soil. Plants may or may not be able to use elements in their changed forms. The reactions are controlled by pH. A good example is the element phosphorus. When soil pH falls below 5.8, phosphorus reacts with iron to produce an insoluble iron compound. Above pH 6.0, the reaction tends to reverse to free phosphorus. At a high pH value, phosphorus reacts in the same way with calcium. Therefore, phosphorus is most available to plants between a pH of 6.0 and 7.0.

Figure 10-8 shows the availability of nutrients at different pH levels. Note that the major nutrients and molybdenum are most available in near-neutral soil. The other trace elements are more available in acid soil. Note that pH in the range of 6.0-7.0 is a good average level for all nutrients. This range is also the best pH range for most crops.

Soils that cannot supply enough of a nutrient may actually contain the nutrient element, but the nutrient is tied-up because of acid or alkaline soil. Looking at figure 10-8, you can see why the pin oak at the beginning of this chapter suffered from a shortage of iron—because the soil was too alkaline. Many nutrient shortages are often best solved by changing the soil pH.

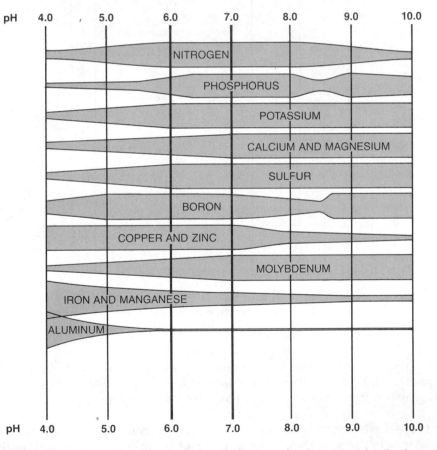

Figure 10-8. The soil pH affects the nutrient and aluminum levels in mineral soils of temperate climates. The thicker the bar, the more available the nutrient. The flares at the acid end of the bars for iron and manganese and aluminum show their solubility in acid soil. (Adapted from Truog, USDA *Yearbook of Agriculture* 1943-1947)

pH and Element Toxicity.   Below a pH of about 5.5, aluminum begins to leave the structure of silicate clays. High levels of soluble aluminum in the soil are injurious to plants, so aluminum toxicity limits growth on strongly acid soil. Aluminum toxicity also increases water stress during dry periods, because poor root growth limits the plant's ability to take up water. Both iron and manganese can have similar effects in some acid soils. Figure 10-8 shows that these three elements become soluble below pH 5.5.

Aluminum toxicity occurs primarily in mineral soils of temperate climates. Aluminum problems seldom appear in organic soils because these soils have little aluminum. In fact, plants tolerate acid organic soils better than acid mineral soils, mainly because of the low aluminum level.

pH and Soil Organisms.   Soil organisms grow best in near-neutral soil. In general, acid soil inhibits the growth of most organisms, including many microbes and earthworms. Thus, acid soil slows many important activities carried on by soil microbes, including nitrogen fixation, nitrification, and organic-matter decay. Liming acid soils helps these organisms do their tasks.

LIMING SOIL

The simplest way to ensure proper pH is to choose a crop that matches the present soil pH. Indeed, matching the crop and pH may be the only answer in some cases. For instance, growers may find it impractical to lower the pH of calcareous soils or to raise the pH of organic soils. In many of these situations, it is preferable to raise a crop that tolerates the soil pH as it is.

The other approach is to change the soil pH to match the needs of the crop. Most common field crops grow best in slightly acid soil. However, leaching of exchangeable bases, acid fertilizers, and other factors slowly may make the soil more acid than is best for good growth. The first crop to suffer from too much acidity is alfalfa, followed by other legumes, and finally by other field crops. *Liming* is practiced by growers to counteract soil acidity.

Benefits of Liming.   Liming acid soils has long been an important agricultural practice. Many growers, however, continue to neglect liming. One reason is that, unlike their response to fertilizers, crops do not respond immediately to liming. Interestingly, one of the most important benefits of liming is to improve crop response to fertilizers by improving the uptake of primary nutrients. Liming makes possible the best crop yields from the dollars invested in fertilizers. For instance, much phosphorus fertilizer applied to acid soils will be wasted since it is tied-up in the soil. Liming acid soils also removes aluminum toxicity and promotes the activities of such desirable organisms as the Rhizobia bacteria that fix nitrogen for legumes. Since calcium is itself a plant nutrient, lime is also a fertilizer, especially for high-calcium crops like alfalfa. Certain limes also supply magnesium, which is important to many acid sandy soils.

Liming Materials. The term *agricultural lime* is applied to ground limestone or other products made from limestone. All of these materials contain calcium. When lime is mixed into soil, it neutralizes excess acidity. Common liming materials include calcitic limestone, dolomitic limestone, burned lime, and hydrated lime.

Calcitic limestone, or "regular" limestone, is nearly pure calcite or calcium carbonate ($CaCO_3$). It forms on the sea floor from the bodies of some sea life and from calcium dropping out of solution in seawater. Limestone deposits are widespread in the United States. The deposits are mined and the limestone rock is ground into agricultural lime (figures 10-9 and 2-5).

Dolomitic limestone or, more simply, *dolomite,* is similar to ground limestone. The rock is a mixture of calcium carbonate and magnesium carbonate ($CaCO_3$ and $MgCo_3$). Liming with dolomitic lime helps the calcium/magnesium balance in soil. Dolomite is especially helpful in sandy soils, because they often lack sufficient magnesium. Magnesium has the same effect on soil pH as calcium.

Figure 10-9. Limestone deposits are mined and the rock is ground into agricultural lime. Here, a spreading truck is being loaded in a limestone quarry.

Burned lime, or quicklime, is made by heating limestone. Heating drives off carbon dioxide resulting in the lighter calcium oxide:

(k) $$CaCO_3 \rightarrow CaO + CO_2 \text{ (gas)}$$

Because calcium oxide is lighter (has a lower molecular weight), a smaller weight of it has the same effect as a larger weight of ground limestone. Burned lime also reacts more quickly in the soil. However, the material costs more and is hard to handle. Burned lime is caustic and may cake during storage. Burned lime can be used where fast action is needed, but it is not usually recommended.

Hydrated lime, sometimes called slaked lime, is produced by adding water to burned lime, forming hydrated lime, or calcium hydroxide:

(l) $$CaO + H_2O \rightarrow Ca(OH)_2$$

Like burned lime, hydrated lime is unpleasant and hard to handle, but it is fast-acting. Hydrated lime is used more often than burned lime. Because of processing steps, it is more expensive than regular ground lime, but it may be used where speed of reaction is needed.

Growers may find other useful materials, depending on what is produced locally:

- *Marl* is a soft, mushy fresh-water deposit in swamps that receive alkaline runoff water from nearby land. Although marl is difficult to harvest and spread, it may be useful where locally mined.
- *Ground seashells,* a byproduct of shellfish industries, may be used in areas where those industries thrive.
- *Slag,* left over from iron smelting or other industries, is used in some areas. It may contain some trace elements, often boron.
- *Refuse-lime* is a byproduct of a number of industries. Some industries that use lime in the manufacturing process may offer this waste product to local growers.
- *Wood ashes* can be used by gardeners who burn wood. The ash of hardwoods such as basswood and oak contains more calcium than softwood ash. Coal and coke ash has no neutralizing effect.

It should be noted that gypsum (CaSO4) does not lower soil pH, so cannot be used as an ag lime. The sulfur in gypsum, as will be noted later in this chapter, increases acidity, so would counteract the effect of the calcium.

How Lime Works. Lime neutralizes soil in two ways. First, calcium replaces hydrogen and aluminum ions on exchange sites by mass action. In doing so, liming raises the percent base saturation. Second, lime converts hydrogen to water. Let's look at a couple of examples to see how this works.

The simplest reaction is that of hydrated lime. Figure 10-10 shows the steps involved in neutralizing soil acidity. As hydrated lime dissolves, it releases calcium and hydroxyl ions. Calcium replaces hydrogen and aluminum on the exchange sites, releasing those cations to the soil solution. Aluminum ions undergo complete hydrolysis to form an insoluble aluminum hydroxide, with the release of more hydrogen ions. All the hydrogen ions react with the hydroxyl ions from the lime, forming water.

Calcite and dolomite act in a similar fashion, with a couple of additional steps. Hydrogen ions resulting from the other steps react with the carbonate to form carbonic acid, which quickly decomposes to carbon dioxide and water. This last step is the same as the cause of pop-fizzing—when you open the bottle, carbonic acid in the pop breaks down to carbon dioxide, which bubbles off as fizz. Figure 10-11 shows this process.

Other liming materials undergo similar reactions. The important thing to remember is that calcium (or magnesium) replaces hydrogen and aluminum on cation exchange sites and hydrogen ions are changed to water. The speed of this overall process varies according to the type of material. Hydrated lime dissolves quickly in the soil and reacts quickly. Ground limestone, on the other hand, dissolves more slowly and takes more steps to neutralize acid.

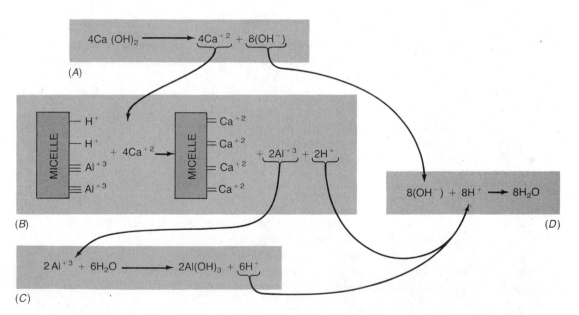

Figure 10-10. The addition of hydrated lime to the soil neutralizes soil acidity as a result of a number of reactions. When calcium hydroxide dissolves, (A), calcium replaces aluminum and hydrogen on cation exchange sites (B). Aluminum is tied up by reacting with water to form insoluble aluminum hydroxide (C). All of the hydrogen ions released by these reactions combine with the hydroxyl ions from the lime to form water (D). *Note:* The numbers in the equations are used only to balance the reactions in the example.

How Much Lime to Apply. Four factors tell the grower how much lime is required: the present pH, the desired pH, the cation exchange capacity of the soil, and the liming material to be used.

By testing the present pH, and knowing the correct pH for a given crop (figure 10-7), a grower or soil testing laboratory can determine how much the pH has to change. For example, if alfalfa grows well at a pH of 6.5, and the present pH is 5.0, then the pH must be raised one and a half points. The pH of the soil solution can be measured with methods described in chapter twelve. However, pH by itself does not tell how much lime to apply, because it measures only the hydrogen ions in solution, not the potential acidity (hydrogen and aluminum) adsorbed on the colloids. Hydrogen ions in solution can be termed *active acidity,* while adsorbed hydrogen and aluminum is *reserve acidity.* While some soil scientists consider these terms out of date, they are useful concepts for this discussion.

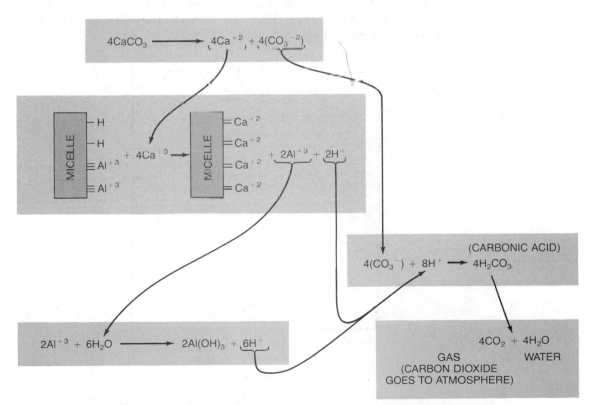

Figure 10-11. The addition of ground limestone to the soil neutralizes soil acidity according to the reactions shown. The cation exchange reaction series is the same as the one in figure 10-10. Hydrogen ions react with the CO_3^- anions to form unstable carbonic acid. Carbonic acid immediately breaks down to carbon dioxide gas and water.

Effect of Cation Exchange Capacity on Liming. Picture the soil, (figure 10-12) storing hydrogen and aluminum in a large bin (reserve acidity adsorbed on soil colloids) attached to a small one (active acidity in solution). pH measures only the active acidity in the small bin. If we add enough lime to neutralize those hydrogen ions, they are quickly replaced from the large bin. Thus the soil resists a pH change—which is called *buffering.* Enough lime must be added to draw down both bins before the soil will become less acid.

The size of the large bin depends on the cation exchange capacity (CEC)— the larger the CEC, the more hydrogen a soil can hold, and the more lime it will need. For instance, a soil with a CEC of 20 will need twice as much lime as a soil at the same pH with a CEC of 10.

The buffering capacity of a soil depends on the amount of clay in the soil, the type of clay, and the amount of humus. The amount of clay can be estimated simply by knowing the textural class of a soil. Figure 10-13 lists suggestions for how much lime to apply to soils of various textures. The type of clay modifies the effects of texture. For the following clays, the lime requirement is the highest for vermiculite and then decreases for each succeeding type of clay: vermiculite, montmorillinite, illite, kaolinite, and sesquioxides. The buffering capacity is also increased by the amount of organic matter in the soil.

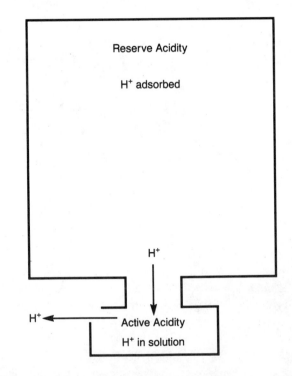

Figure 10-12. pH buffering in the soil. As hydrogen ions are removed from the soil solution by liming, they are replaced by ions held on clay and humus.

Textural Class	Change 4.5 to 5.5	Change 5.5 to 6.5
Sand, loamy sand	25	30
Sandy loam	45	55
Loam	60	85
Silt loam	80	105
Clay loam	100	120
Muck	200	225

Figure 10-13. The values given in the table indicate the amount of ground limestone needed to raise the pH of a 7-inch soil layer in pounds per 1000 square feet. These values apply to soils of different textures in northern and central states. (USDA, Kellogg, *Soils: 1957 Handbook of Agriculture*)

The lime requirement of an acid soil depends on both pH and buffering capacity. The total lime requirement can be measured directly by means of a *SMP buffer test* or other similar test. The test measures the reaction of a soil to a pH 7.5 buffer solution. The test does not measure actual pH; it is a guide to the amount of lime needed to correct pH. Here is an example of how it works.

The pH of a soil sample is first measured to see if liming is needed. This is often the case for a pH below 6.5. Let's say a soil sample has a pH of 5.5. The laboratory technician adds the pH 7.5 buffer solution to the soil sample and measures the pH of the new mixture as 6.2. This means the acidity of the soil lowered the pH of the buffer solution from 7.5 to 6.2. The pH of 6.2 is called the "buffer index." Looking at a table of buffer indexes (figure 10-14), the technician reads how much lime is needed to raise soil pH the correct number of points. In this example, one table suggests applying four tons of lime per acre of mineral soil to bring the pH to 6.5–7.0.

Neutralizing Power of Lime. The SMP buffer test suggests the lime needs of a soil based on an "average" calcitic limestone. However, different lime products have different capacities to neutralize acidity. This capacity is called the *total neutralizing power* or *calcium carbonate equivalent*. Neutralizing power is based on comparing an agricultural lime with pure calcium carbonate, or calcite. Two factors affect the comparison: the chemical nature of the lime and the purity of the lime.

Calcium carbonate has a molecular weight of about 100. Other chemicals weigh more or less, and so have a greater or lesser neutralizing power. For instance, the molecular weight of pure hydrated lime (calcium hydroxide) is 74, less than calcite. Thus, it would take a smaller weight of hydrated lime to obtain the same amount of calcium. Put another way, hydrated lime has a greater neutralizing power. The calcium carbonate equivalent of hydrated lime is simply 100/74, or 135. This means 100 units of hydrated lime has the same neutralizing power of 135 units of calcite.

Buffer pH	(Tons of Lime/Acre)	
	Mineral Soil	Organic Soil
7.0	0	0
6.8	1.0	0
6.6	2.0	0
6.4	3.0	1.0
6.2	4.0	2.5
6.0	5.5	4.0
5.8	6.5	5.0
5.6	8.0	6.0

Figure 10-14. Buffer index tables, such as this simplified version, can be used to measure the amount of lime needed to bring soil pH between 6.5 and 7.0. The table applies to a 6 2/3-inch plow depth and a 90% pure calcitic limestone. (Courtesy of A & L Agriculture Laboratories, Inc. Adapted from "Soil and Plant Analysis.")

The second influence on neutralizing power is purity. For example, calcitic limestone is *mostly* calcite. However, it also contains other materials, like silt, that have no effect on acidity. A ground limestone that is 90% pure is only 90% as active as pure calcite. Since calcite has a neutralizing power of 100, the power of the limestone would be 90.

Figure 10-15 gives sample neutralizing values of several agricultural limes. If the lime requirement from an SMP buffer test was based on 90% pure limestone, and a grower plans to use a different form of lime, a conversion must be made. The following problem shows how much dolomitic lime with a neutralizing power of 90 is needed to replace three tons per acre of the calcitic lime:

$$\text{rate dolomite} = \text{rate lime} \times \frac{\text{neutralization value lime}}{\text{neutralization value dolomite}}$$

$$\text{rate dolomite} = 3 \text{ tons/acre} \times \frac{85}{90} = 2.8 \text{ tons/acre}$$

Most states regulate the purity of agricultural lime to protect the customer. The laws set the *chemical guarantees* for the neutralizing power of lime products used and/or produced in the state. For instance, the average requirement for ground limestone is a calcium carbonate equivalent of about 85.

Lime Fineness. The fineness of the grind affects how rapidly lime acts. The finer the grind, the more rapidly it can neutralize acidity. Lime producers measure the grind by passing the lime through a screen of so many squares per square inch. Figure 10-16 shows the relative efficiency of different grinds.

While finely ground lime acts most rapidly, it is also used up rapidly. A fine powder is also costly and hard to spread evenly. Most labs suggest a medium grind that contains enough "fines" (grains small enough to be dusty) for fast action. All of such a grind would pass an 8-mesh screen, and 25% to 50% would pass a 100-mesh screen.

Form of Lime	Percent Purity	Neutralizing Value
Pure Substances		
Calcium carbonate	100	100
Magnesium carbonate	100	119
Hydrated lime	100	172
Burned lime	100	178
Commonly Available Forms		
Calcitic limestone	85	85
Dolomitic limestone	85	88
Hydrated lime	85	115
Burned lime	85	151
Marl	—	50–70
Basic slag	—	60–90
Wood ashes	—	45–80
Ground seashells	85	85

Figure 10-15. Neutralizing values are given for the major sources of lime. The first four values are for pure chemicals; the remaining values are averages for commonly available products.

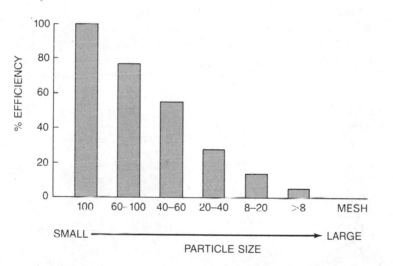
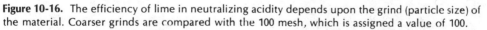

Figure 10-16. The efficiency of lime in neutralizing acidity depends upon the grind (particle size) of the material. Coarser grinds are compared with the 100 mesh, which is assigned a value of 100.

Most states regulate the grind of agricultural lime as well as the purity. The laws specify the *physical guarantee* of agricultural limes. Laws differ from state to state. A sample law, for instance, might specify that all of a lime must pass through a 16-mesh screen and that 35% must pass through a 100-mesh screen.

Lime Application.   The best results are obtained from liming when there is close contact between the grains of lime and the soil. To achieve this, lime should be spread evenly over the field and then mixed well into the soil. Lime-spreading trucks do a good job of spreading the material (figure 10-17). The trucks have a V-shaped bed with a spinning disc mounted on the rear. Lime drops out of the truck bed at a controlled rate onto the spinning disc. The disc flings the lime out in a fan-shaped pattern (figure 10-18).

While most lime is spread in a dry form, some lime is finely ground and mixed with water or a fertilizer solution and sprayed on the field. *Fluid lime* acts more quickly than regular lime, and so is useful where fast action is needed. Fluid lime remains active for a shorter time, so another application must be repeated sooner. Fluid lime is becoming popular, partially because it can be applied in a fertilizer solution (see chapter fourteen), combining two operations and saving time. Fluid lime is also popular with growers with short-term use of land, such as a rented field.

After lime is spread, plowing and/or discing mixes the lime into the soil. In established pasture or other situations where plowing is not possible, the lime is simply spread evenly on the soil surface. If it is not mixed into the soil, the lime will slowly move into the soil.

Growers may lime at any time that is convenient. To avoid compaction, however, it is best not to drive the trucks on wet soil. Lime should not be applied with certain forms of nitrogen fertilizer because it can cause nitrogen losses (see chapter fourteen). The most important consideration is reaction time. Lime should be spread several months before seeding legumes, since legumes have the highest need for lime. It takes a few months for the lime to break down in the soil. If faster action is needed, a grower can use fluid lime, hydrated lime, or more finely ground lime grades.

(A)   (B)

Figure 10-17. (A) A lime-spreading truck has a V-shaped bed with (B) spinning discs at the rear for spreading the lime.

Figure 10-18. Spreading lime

ACIDIFYING SOIL

Crop growth can be inhibited by alkaline soils, common to the areas of the country that are shaded in figure 10-5. Here, excess lime or sodium keeps soil pH high, as shown earlier in this chapter (reactions b, c, and d). Overlimed soils may also become alkaline.

As shown in figure 10-8, zinc, manganese, iron, and other trace elements are tied up in basic soils. In addition, free molybdenum can reach toxic levels. Where soil is strongly alkaline, many crops will grow poorly. Even near-neutral soils present a problem to acid-loving plants like azalea and pin oak. Such plants appear to have a very high iron requirement. Landscapers often modify planting beds with acidic peat moss prior to planting. However, the pH lowering effects are temporary (the peat decays), and peat amendments are not suitable for large areas.

For longer lasting pH reduction, and for larger areas, sulfur is preferred. Once applied and mixed into the soil (as described earlier for lime), the bacteria Thiobacillus alters the sulfur to sulfuric acid:

$$2S + 3O_2 + 2H_2O \rightarrow 2H_2SO_4 + \text{energy}$$

The sulfuric acid releases hydrogen ions, and the soil becomes more acid.

Sulfur is available in granular and powdered forms. The powdered form acts most rapidly, but is more difficult to handle. Granular sulfur, while slower acting, spreads much more easily with application equipment. Figure 10-19 suggests some sulfur application rates.

A number of other chemicals also acidify the soil. These include iron sulfate, $Fe_2(SO)_4)_3$, and aluminum sulfate, $Al_2(SO_4)_3$. These materials are less powerful than sulfur and usually more expensive. They should also be used with caution, because toxic levels of aluminum or iron may build up from repeated use. In addition, when increasing pH is a problem, strongly acid-forming fertilizer may be selected.

Calcareous soils may be very difficult to acidify, because there is such a large reserve of lime that must be leached out. The bin example in figure 10-12 pictures this, except one would relable "reserve acidity" with "reserve alkalinity." Here the free lime buffers soil from pH changes.

Where pH reduction is impractical, fertilization and crop selection are required. Deficiencies may be temporarily corrected by fertilizing with the proper nutrients. Crops should also be selected for tolerance to high pH. In alkaline soils, for instance, alfalfa would outperform soybeans. Varieties of the same species will vary in tolerance, so one soybean variety may grow where another would not.

SOIL SALINITY

In the humid regions of the United States, acidity is a common problem for growers because percolation leaches calcium, magnesium, and sodium from the soil. Growers in the more arid parts of the nation often have a different but related problem—an accumulation of *soluble salts* of these same bases. A salt is a chemcial that results from the reaction of an acid with a base, such as the reaction of hydrochloric acid with lye to form common table salt:

(n) $$HCl + NaOH \rightarrow H_2O + NaCl$$

$$\text{acid} \qquad \text{base} \qquad \text{water} \qquad \text{salt}$$

To Lower pH by This Amount	Ground Sulfur Pints/100 ft²		Pounds/Acre	
	Sand	Loam	Sand	Loam
0.5	2/3	2	360	1,100
1.0	1 1/3	4	725	2,200
1.5	2	5½	2,000	3,000
2.0	2½	8	1,350	4,400
2.5	3	10	1,650	5,400

Figure 10-19. To lower the pH of an 8-inch soil layer, for soils of two textures, the amount of ground sulfur required is given in pints per 100 square feet and in pounds per acre. (Adapted from USDA, Kellogg, *Soil: 1957 Handbook of Agriculture.*)

A soluble salt is defined as a salt that is as soluble or more soluble in water than gypsum (calcium sulfate, $CaSO_4$). The soluble salts of greatest concern in the soil are sulfates (SO_4^{-2}), bicarbonates (HCO^-), and clorides (Cl^-) of the bases calcium, magnesium, and sodium. These salts may come from parent materials or irrigation with salty water. Prominent locations for this problem in the United States include the San Joaquin Valley of California, the lower Rio Grande Valley of Texas, and such western and southwest states as Arizona, New Mexico, Utah, and others. Salinity problems affect about 25% of the irrigated lands of the United States.

Soil scientists define three types of problem soils based on the types of soluble salts: saline, sodic, and saline-sodic. Figure 10-20 summarizes these three.

Saline Soils. *Saline soils* have high levels of soluble salts except sodium. Soil salinity can be easily measured by passing an electrical current through a solution extracted from a soil sample. The greater the salt content, the more electricity will pass. The amount of electrical flow is called *electrical conductivity* and Is measured by the unit millimhos per centimeter (mmhos/cm). This unit of measure is presently being replaced by the siemen per meter, which equals 10 mmhos/cm.

A saline soil is defined as a soil with an electrical conductivity of four or more millimhos per centimeter. However, salinity levels as low as two mmhos/cm can injure very sensitive crops. Most of the salts are chlorides or sulfates. Less than half of the cations are sodium, and little sodium is adsorbed on soil colloids. Soil pH is 8.5 or less. A white crust may be seen on the soil surface, due to salts migrating to the surface by capillary rise.

The main effect of salinity is to make it more difficult for plants to absorb water from the soil. In nonsaline soils, only the attraction of water for soil particles (matric potential) contributes to the total water potential. In a saline soil, the water also is attracted to ions in solution (salt potential), so more of the water is unavailable to plants (figure 10-21).

Soils can be classified for use based on salinity. Figure 10-22 shows the classification system. Figure 10-23 classifies common crops according to their salt tolerance.

Salted Soil Class	Conductivity (mmhos/cm)	Exchangeable Sodium(%)	Sodium Adsorption Ratio	Soil pH	Soil Structure
Saline	> 4.0	< 15	< 13	< 8.5	Normal
Sodic	< 4.0	> 15	> 13	> 8.5	Poor
Saline-sodic	> 4.0	> 15	> 13	< 8.5	Normal

Figure 10-20. Characteristics of salted soil

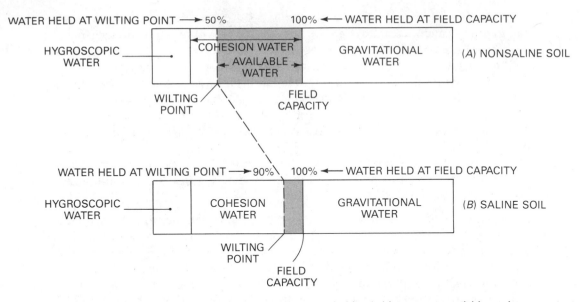

Figure 10-21. (A) In a nonsaline soil, about half the water held at field capacity is available to plants. The remaining water is adsorbed by soil particles. (B) In saline soils, as little as 10% of the water may be available because it is attracted to salt ions.

Class	Salinity (mmhos/cm)	Crop Response
Nonsaline	0–2	Salinity effects unimportant
Slightly saline	2–4	Yields of sensitive crops lowered
Moderately saline	4–8	Yields of many crops lowered
Strongly saline	8–16	Only tolerant crops yield well
Very strongly saline	More than 16	Only most tolerant crops yield well

Figure 10-22. Crop response to soil salinity

Sodic Soils. *Sodic soils* are low in the kinds of salts found in saline soils, but they are high in sodium. The exchangeable sodium percentage (or sodium saturation) is 15 or more, and pH is in the range 8.5 to 10.0. Measuring the exchangeable sodium percentage is time consuming, so sodium is measured by the *sodium adsorption ratio* (SAR). The SAR compares the concentration of sodium ions with the concentration of calcium and magnesium ions according to the formula:

$$SAR = \frac{Na^+}{\sqrt{Mg^{+2} + Ca^{+2}/2}}$$

Using this measurement, a sodic soil has an SAR greater than or equal to 13.

Type of Crop	Tolerant	Medium	Sensitive
Field Crops	Barley Sugar beet Cotton	Corn Soybean Sorghum Wheat Rice	Beans Flax Broadbean
Forage Crops	Bermuda grass Wheatgrass Tall fescue	Alfalfa Orchard grass Birdsfoot trefoil Perennial rye	Clovers
Vegetables	Beets Asparagus	Spinach Tomato Broccoli Cabbage Potato Sweet corn	Sweet potatoes Lettuce Bell pepper Onion Carrot Beans Celery
Fruits	Date palm	Grape Fig Olive	All others

Figure 10-23. Tolerance of crops to soil salinity

Sodic soil has a number of effects on plant growth. The importance of these effects varies according to soil and crop.

- Sodium reacts with water, reaction (c), to form lye. The resulting high pH, which is more than 8.5, can limit the growth of many crops.
- For most crops, the main effect of sodium is the destruction of soil structure. (figure 10-24). When sodium ions saturate cation exchange sites, the colloids separate and disperse soil aggregates. Tiny soil particles lodge in the soil pores, sealing the soil surface. As a result, the soil sheds water so that wet "slick spots" form. Tilth also suffers and crusts hard enough to stop seed germination may form. In many cases, sodic soils also show a poorly drained columnar subsoil structure. The effect of sodium is most extreme on fine-textured soils and least extreme on coarse soils.
- Crop plants may take up enough sodium to injure plant tissues. Crops vary in their tolerance to sodium. For the most sensitive crops, like citrus fruits, the nutritional effects of sodium are more important than its effects on structure. For sodium-tolerant crops, poor growth results mainly from soil conditions.

Figure 10-25 lists the sodium tolerance of selected crops.

Saline-Sodic Soils. *Saline-sodic soils* contain both high levels of soluble salts and sodium. The electrical conductivity is greater than 4.0 millimhos per centimeter, the SAR is greater than 13, and the pH is less than 8.5. The physical structure of

Figure 10-24. A sodic soil. Sodium has dispersed soil aggregates, destroying structure. Vegetation is sparse. (Courtesy of USDA-ARS)

these soils is normal. However, after periods of heavy rain or irrigation with low-salt water, soluble calcium and magnesium may leach out of the soil, leaving behind the sodium salts. The soil may then become sodic, with poor physical structure and drainage.

Reclaiming Salted Soils. The first step in reclamation of salted soils is to decide if the project is practical and will pay for itself. The basic step to reclaiming the soil is to leach out salts, so there must be a source of acceptable water. Very fine-textured soils may not allow sufficient drainage. If a decision is made to reclaim the soil, the next step is to ensure good drainage to allow salted water to leave the soil profile.

Many salted soils have problems with drainage, including a high water table, hardpans, or fine soil texture. Subsoiling can help break up hardpans. Soils with high water tables must be drained to a depth of five or six feet so salty water can be removed from the root zone and carried off the field. Once proper drainage has been installed, the next steps depend on the type of problem.

Sensitive Sodium Percentage (Exchangeable [ESP] = 2–20)	Moderately Tolerant (ESP = 20–40)	Tolerant (ESP = 40–60)	Most Tolerant (ESP Above 60)
Deciduous fruit	Clovers	Wheat	Crested wheatgrass
Nuts	Oats	Cotton	Tall wheatgrass
Citrus fruit	Tall fescue	Alfalfa	Rhodesgrass
Avocado	Rice	Barley	
Bean	Dallisgrass	Tomato	
		Beets	

Figure 10-25. Various crops have different degrees of tolerance to exchangeable sodium. Damage to the most sensitive crops is due to sodium toxicity. Damage to the tolerant crops is due to poor soil conditions. (Adapted from USDA Agriculture Information Bulletin No. 216, 1960.)

Saline soils are most easily reclaimed. Growers flood the soil surface so that percolation leaches salts out of the soil profile. High quality water works best, but larger amounts of fairly saline water will also work. Treatment water should, however, be low in sodium. Ponding is one way to apply leaching water. In ponding, heavy equipment constructs low earthen dikes to divide the affected land into ponds, which are then flooded. The field may be ponded several times, allowing time for drainage between floodings.

The reclamation of saline soils has been improved by the use of organic mulches. The mulches reduce evaporation of water from the soil surface, increasing the net movement of water downward. In addition, organic matter keeps the soil loose and maintains structure to improve drainage.

Sodic soils cannot usually be reclaimed simply by leaching, because the sealed soil surface inhibits drainage. It is usually necessary to first remove the sodium. This is done most often by treating the soil with gypsum. Granular gypsum may be spread on the soil surface, or finely ground gypsum may be applied through an irrigation system. When gypsum enters the soil, it dissolves and calcium replaces sodium on the cation exchange sites. Sodium sulfate leaches out of the soil:

(o) $\boxed{\text{micelle}} \Big\langle {}^{Na^+}_{Na^+} + CaSO_4 \rightarrow \boxed{\text{micelle}}-Ca^{+2} + Na_2SO_4$
(leaches)

Gypsum is the least expensive amendment, but other chemicals may be used as well. If the soil contains some lime ($CaCO_3$), finely ground sulfur will add calcium indirectly. The sulfur is converted to sulfuric acid by bacteria, reaction (m). The hydrogen ions from the sulfuric acid can replace sodium on the exchange sites. More importantly, the acid reacts with soil lime to make gypsum:

(p) $CaCO_3 + H_2SO_4 \rightarrow CaSO_4 + H_2O + CO_2$ (gas)

The conversion of sulfur to sulfuric acid takes some time, so sulfur treatment is relatively slow. Sulfuric acid can be added directly for faster action. This is more expensive and also more dangerous since sulfuric acid is highly caustic. One USDA research project achieved a similar effect on sodic soils by planting a sorghum-sudan grass hybrid. Its roots released large amounts of carbon dioxide, which reacted with soil water to form carbonic acid (reaction *i*). The acid dissolved soil lime, freeing calcium which displaced sodium, as in reaction (*o*).

Once calcium replaces sodium on cation exchange sites, the soil slowly begins to aggregate. As the soil surface begins to improve, some growers plant salt-tolerant crops like barley. The plant roots and tops, if disced into the soil, help rebuild the soil structure.

Saline-sodic soils also must be treated to remove sodium. If they are simply leached with low-salt water, calcium and magnesium salts are removed but sodium remains in the soil, forming a sodic soil. Thus, gypsum treatments are useful. In the initial stages of reclamation, some growers leach these soils with fairly saline water. The calcium and magnesium salts in the water replace some of the sodium on the soil colloids, preventing destruction of soil structure.

Managing Salted Soils.   Saline soils, especially irrigated land in arid climates, may be managed to reduce salt problems. One answer, of course, is to grow salt-tolerant crops. This step, however, does not really solve the problem, but it causes a shift over time to increasingly salt-tolerant crops. A number of practices can be used to help reduce salt problems, as follows:

- Prepare a field properly for irrigation. Proper leveling avoids low spots that collect salts. A grower may also install drainage during field preparation.
- If possible, use high quality irrigation water. Figures 6-22 and 6-23 list the irrigation water classes.
- Keep the soil moist. Water dilutes soil salts, lowering the effect of the salt potential. Salts tend to be most damaging in dry soil, when the salts are concentrated and both the salt and matric potentials are high.
- Overirrigate enough to leach salts out of crop root zones.
- To maintain soil structure, return as much organic matter to the soil as is practical, including manures, crop residues, or green manures.
- Avoid overfertilization. Most fertilizers are salts and can compound salinity problems. Chapter thirteen lists the salinity of common fertilizers.
- Maintain a good soil testing program to monitor salinity and avoid overfertilization.
- In furrow-irrigated fields, crops may be planted on the shoulders of the ridges, because salts tend to concentrate on the top of the ridge (figure 10-26).
- Trickle irrigation tends to reduce salt stress because it keeps the soil uniformly moist, and tends to move salts out of the root zone of the crop plants and into the soil between plants and rows (figure 10-27).

Salted Water Disposal.   One difficulty with methods for reclaiming and managing salted soils is that they do not eliminate soluble salts but move them to

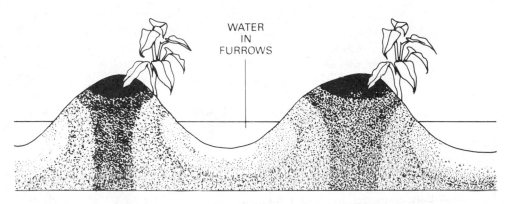

Figure 10-26. In furrow-irrigated fields, planting should be done on the shoulders of the furrows. The top of the furrows has the highest salt content, which inhibits seed germination.

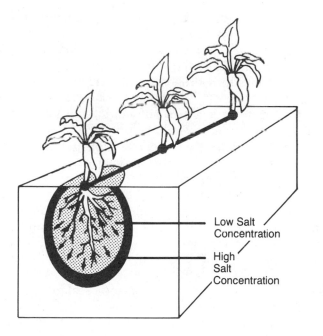

Figure 10-27. Trickle irrigation can reduce salt stress. As water moves away from the trickler by capillary action, it carries dissolved salts with it, out of the root zone.

another place. Salty drainage water reappears in rivers downstream of affected farms, making that water even more salty. For instance, over one 75 mile stretch of the Rio Grande River, salt concentration increases by a factor of five. Solutions can be expensive. At the time this text was being written, a 1.3 billion dollar system was being proposed for carrying salty drainage water out of the San Joaquin Valley.

Individual growers do have some options to help reduce saline discharges from their fields. These suggestions stress reducing water use and retaining salt safely in the field:

- Improve water delivery systems to reduce seepage and evaporation from canals.
- Use techniques to improve irrigation efficiency, like surge irrigation and careful budgeting to reduce percolation and tailwater losses.
- Where feasible, adopt trickle irrigation.
- Practice minimum leaching to carry salts below the root zone but not into the drainage system.
- Reuse salty water on salt tolerant crops like barley or sugarbeets.

SUMMARY

Soil pH depends on the balance of hydrogen and hydroxyl ions in the soil solution. Alkaline soil, with a pH between 7.0 and 10.0, results from the reaction of calcium and sodium with water to form hydroxyl ions. Acid soil, with a pH between 4.0 and 7.0, results from the leaching of these bases by mildly acidic water and from the release of hydrogen ions by aluminum hydrolysis.

Acid soils affect plant growth by lowering the availability of phosphorus and other nutrients, freeing toxic levels of aluminum, and inhibiting helpful soil organisms. Alkaline soils render several micronutrients unavailable, and create many problems associated with salted soils. Most plants grow best between pH 6.0 and 7.0. A few, like potatoes, perform best in acid soil, while a very few, like alfalfa, do best in neutral or mildly alkaline soil.

Acid soils are treated with various forms of agricultural lime, mostly ground limestone. Lime replaces hydrogen and aluminum on the cation exchange sites with calcium and converts hydrogen ions to water. The amount of lime needed depends on the amount of pH change required, the buffering capacity of the soil, and the form of lime. Soils too alkaline for the crop being grown may be treated with sulfur.

Salted soils may be saline, sodic, or saline-sodic. Saline soils, which are high in soluble salts but low in sodium, reduce the water available to plants. Saline soils can be treated by flooding to leach out salts. Sodic soils are high in sodium and exhibit poor physical structure. They are treated with gypsum to displace the sodium. Saline-sodic soils contain both soluble salts and sodium. Care must be taken with these soils to avoid leaching the salts while leaving the sodium. Once a salted soil is treated, it must be managed carefully to reduce salt problems.

REVIEW

1. The acid range of the total pH scale is 7-14 (True/False).
2. Neutral pH is 7.0
3. A pH range of 6-7 is satisfactory for most crops.
4. Percolation of mildly acidic water causes an acid soil.
5. High pH soils tend to contain a lot of sodium or calcium.
6. Soluble salts make it easier for plants to absorb water from the soil.
7. Sodic soils generally suffer from poor structure.
8. Ponding may be used to reclaim saline soils.
9. Trickle irrigation makes salinity problems worse.
10. Soil salinity is measured by means of electrical conductivity.
11. _____ _____ ions are acid, _____ are basic.
12. pH measures the amount of _____ in the soil.
13. Impurities in lime _____ (raise/lower) its neutralizing power.
14. Acid loving plants prefer a pH range of _____ ____ to _____ .
15. _____ is probably the best material for acidifying soil.
16. _____ is used to treat sodic soils.
17. Soils which contain large amounts of both sodium and other salts are called _____ .
18. In sodic soils, very high pH is caused by _____ _____ .
19. Of the soil problems described, _____ soils are the ones with the poorest soil structure.
20. _____ ____ soils contain large amounts of salts other than sodium.
21. List three effects of pH on plant growth.
22. List four forms of agricultural lime and briefly mention qualities of each.
23. The SMP test on a sample of mineral soil produces a buffer index of 6.4. How many pounds per acre of an 85% pure hydrated lime should be applied to the soil? (Use figures 10-14 and 10-15).
24. List three effects of soil salinity.
25. In acid treatment of sodic soils, where does the calcium come from?
26. Describe how lime neutralizes acidity.
27. Explain why fine-textured or organic soils require more lime than coarse-textured soils.
28. Explain how gypsum improves the soil structure of sodic soils.
29. What exactly is a soluble salt?
30. In excavation of some sites of ancient civilizations, archaeologists noted evidence of a decline in agriculture. Early in the history of the site, wheat was grown; later, barley became the dominant crop, then agriculture ceases. Can you explain this?

SUGGESTED ACTIVITIES

1. Put a few drops of a strong acid on a piece of limestone, and observe the fizz. Can you explain the bubbles? See figure 10-11.
2. Use pH paper to test a number of household solutions, such as vinegar, lemon juice, tapwater, and ammonia.
3. Check a soil sample for pH by mixing 20 grams of soil with 20 milliliters of distilled water. Mix for 10 minutes, then let stand for 10 minutes. Check the liquid with pH paper.
4. At the beginning of the course, the instructor can mix finally ground lime, sulfur, and gypsum into seperate samples of soil in pots. Keep warm and moist for several weeks. Check for pH in class as in three above. One could also try different forms or fineness of lime, etc.
5. At the beginning of the course, the instructor can mix into plastic pots well-granulated soil samples, being careful to add the soil loosely, not packing it down. Treat one pot with table salt (NaCl), another with ice-melting salt ($CaCl_2$), and leave a pot untreated. Keep warm and moist, but don't allow drainage water to carry off the salts. Later, check to see how readily water will drain through the pots. Is there crust on the soil surface? If so, why?

chapter eleven
Plant Nutrition

Notice the dark semi-circle in this lawn. If this were a color picture, it would be darker green. This is a fairy-ring, caused by a fungus that can inhabit lawns. The ring is the outer edge of a circle created as the fungus grows outward from the center. Why is the ring dark green? Because nitrogen is being released from organic matter and feeding the grass. This chapter will detail the role of nitrogen and the other twelve mineral nutrients.

OBJECTIVES

After completing this chapter, you should be able to:

- discuss nitrogen nutrition and the nitrogen cycle
- discuss phosphorus nutrition
- discuss potassium nutrition
- answer questions about the secondary nutrients
- answer questions about trace elements

TERMS TO KNOW

ammonia volatilization	chelates	enzyme
ammonization	chlorosis	lodging

Many soil factors, such as texture, structure, and water, affect plant growth. Often these conditions are less than ideal but are not easily or cheaply improved. For instance, growers cannot alter soil texture over large areas. Even irrigation is a costly investment.

The supply of soil nutrient elements, on the other hand, can be controlled with more ease. Growers can measure nutrient levels in their soils by bringing soil samples to a testing center. The laboratory, knowing the ideal nutrient levels for each crop, suggests how much fertilizer to apply. Then the grower can fertilize to satisfy crop needs for essential elements.

Let us take a detailed look at the essential elements.

NITROGEN

Nitrogen, more than any other element, promotes rapid growth and a dark green color. Plants need a lot of nitrogen because it is part of many important compounds, including protein and chlorophyll. Plants respond to nitrogen in the following ways:

- Nitrogen speeds growth. Plants receiving adequate nitrogen have vigorous growth, large leaves, and long stem internodes.
- With enough nitrogen, plants can make large amounts of chlorophyll, a dark green pigment. Thus, leaves are dark green on well-fed plants.
- Protein content of plant tissue will be at its best. The higher protein content makes the plant a better source of forage, feed, and human nutrition.
- Plants use water best when they have ample nitrogen.

Plants with too much nitrogen do not grow properly, however. The problems associated with too much nitrogen include:

- Soft, weak, easily injured growth is encouraged. For example, plant stems are weaker and more easily topple, or *lodge*, in the rain. Lodging can turn a good crop into a disaster.
- Soft growth is more prone to diseases.
- Overly rapid growth slows maturity of many crops.
- Too rapid growth also delays the hardening-off process that protects many plants from winter cold. Landscape plants, for instance, are more likely to suffer winter damage when nitrogen is applied too liberally.

In general, nitrogen promotes vegetative growth—the growth of stems and leaves—rather than the reproductive growth of flowers and fruit. Home gardeners see the effect if they overfertilize their tomato plants, promoting lush growth but few fruit. As a rule of thumb, nitrogen is most important for crops grown for their vegetation, such as leafy vegetables, hay, or turfgrasses. It is, however, still important to fruiting plants because plants need good vegetative growth before they can flower.

The Nitrogen Cycle. Of the 16 essential elements, nitrogen undergoes the most movement and change. The series of gains, losses, and changes is termed the *nitrogen cycle*. The central portion of the natural cycle operates by the action of soil microorganisms. To review briefly (see chapter seven), nitrogen comes from nitrogen gas (N_2) in the atmosphere, a form unusable to plants. Symbiotic (figure 11-1) or nonsymbiotic bacteria use that nitrogen to form protein for their own bodies. When these bacteria die, other microbes mineralize the protein *(ammonization)* to ammonium ions (NH_4^+). These ions can be taken up by plants, but most are converted by bacteria (nitrification) to nitrite ions (NO_2^-) and then to nitrate ions (NO_3^-). Nitrates are taken up by plants or microbes (immobilization) or return to the atmosphere as nitrogen gas through the process of denitrification. The solid lines in the simplified cycle pictured in figure 11-2 summarize this portion of the cycle.

The complete nitrogen cycle includes some nonbiological processes as well, shown in figure 11-2 as broken lines. Two other forms of fixation add usable nitrogen to the soil. First, lightning during storms provides energy to combine

Figure 11-1. Symbiotic fixation. These nodules contain *Rhizobium* bacteria that fix nitrogen from the air. (Courtesy of *Crops and Soils Magazine,* American Society of Agronomy)

nitrogen and oxygen gases to form nitrogen dioxide (NO_2). The gas dissolves in water vapor to produce nitric acid. (HNO_3). About 5 to 10 pounds per acre of nitrogen fall to earth yearly in rain and snow from this source. Second, large amounts of nitrogen are fixed from the air in fertilizer factories (see chapter thirteen) and applied to soil by growers.

Two nonbiological losses of nitrogen from the soil may be important as well. The nitrate ion is negatively charged and so is not adsorbed by soil colloids. It is not held in the soil by other means except immobilization. Thus, nitrates easily leach from the soil. Although ammonia does not leach readily, (being adsorbed by soil colloids), it too can be lost by a process called *volatilization*. Ammonium ions react with hydroxyl ions in the following reaction:

$$NH_4^+ + OH^- \rightleftharpoons NH_3 + H_2O$$
$$\text{ammonia (gas)}$$

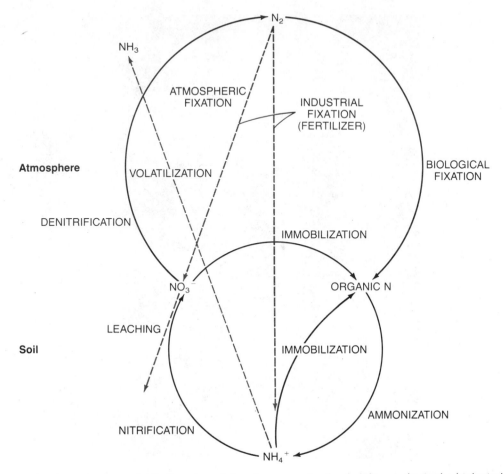

Figure 11-2. In this simplified version of the nitrogen cycle, the dark lines indicate the biological processes described in Chapter 7. The broken lines indicate nonbiological processes described in this chapter.

The smell of an open bottle of household ammonia (ammonia gas dissolved in water) is a result of this reaction. Normally, this is a balanced reaction in the soil, with nitrogen changing back and forth between the two forms. However, the balance can be shifted by soil conditions to cause a loss of ammonia. If the soil dries, for instance, water is lost from the right side of the equation. As a result, the reaction shifts to the right and releases ammonia gas (see appendix one for an explanation). If the soil is made more alkaline (by liming, for instance), the reaction again shifts to the right because of an excess of hydroxyl ions. Thus, ammonia losses may occur in a dry or alkaline soil.

In native habitats, including virgin forests or prairies, gains and losses in the cycle balance over time. However, farming changes the balance greatly in ways that increase nitrogen losses:

- Much of the organic nitrogen contained in plants is removed during crop harvest (figure 11-3), instead of being returned to the soil.
- Cropped soil is more likely to erode, so some nitrogen, along with other nutrients, is carried off the field in running water.
- Irrigation increases percolation of water through the soil profile. Thus, losses of nitrate nitrogen by leaching increase on irrigated land.
- Liming may increase the loss of ammonia by volatilization.

To make up for increased nitrogen losses and to meet the needs of modern high productivity, growers supply more nitrogen by manuring, growing legumes, or by fertilization. Figure 11-4 shows the nitrogen cycle as it operates on modern farms that raise both crops and animals. It is interesting to note that there is a strong trend away from a mixed farming operation toward one that specializes in either cash crops or raising animals. This trend improves efficiency, but it exacts a penalty that is obvious in view of the nitrogen cycle. For the cash crop grower, more money must be spent on fertilizers. For the animal raiser, it means manure becomes a waste disposal problem (see chapter fourteen).

Figure 11-3. As the corn crop is harvested and removed, nitrogen and other nutrients are being removed from the soil. The farmer must replace these nutrients to grow next year's crop.

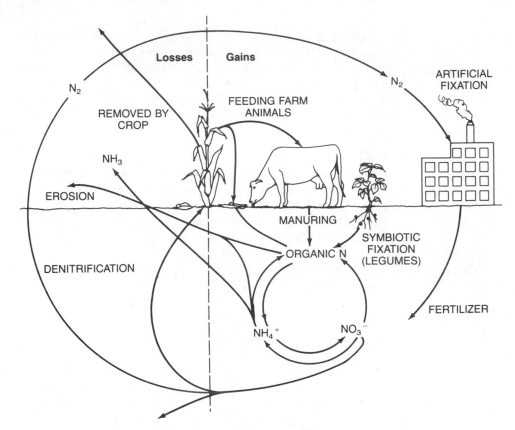

LEACHING AND EROSION

Figure 11-4. Agriculture breaks the natural nitrogen cycle by removing a crop, which removes nitrogen. Tillage also speeds nitrogen losses that are caused by leaching and erosion. Growers adjust nitrogen levels by fertilizing, manuring, and growing legumes.

Forms of Nitrogen in the Soil.   Between 97% and 99% of soil nitrogen is tied-up in organic matter, the soil's storehouse of nitrogen. At any time, only a small percentage of nitrogen has been mineralized to usable forms. On the average, decay makes available about 90 pounds of mineral nitrogen per acre per year (figure 11-5). A 150-bushel corn crop contains about 190 pounds of nitrogen. Obviously, the amount of nitrogen supplied by nature without human help cannot meet the needs of a modern corn crop.

Both mineral forms of nitrogen, ammonium and nitrates, are taken in by plants. In forest and woodland, ammonium is the most common form. Farm crops usually make more use of nitrate, either from nitrate fertilizer or from nitrified ammonium. The two ions behave very differently in the soil (figure 11-6).

Ammonium nitrogen bears a positive charge. Negatively charged soil colloids attract the cation, protecting it from leaching. The nitrate ion, by contrast, moves freely in the soil because of its negative charge. The free movement allows nitrate to diffuse easily through the soil to plant roots.

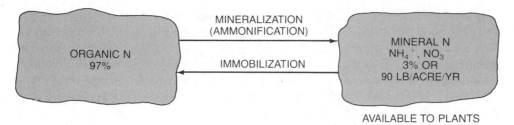

Figure 11-5. Organic matter serves as the storehouse of soil nitrogen (97%). A small percentage of nitrogen exists as the usable nitrate and ammonium ions.

	Organic N	Ammonium N	Nitrate N
Storage	In humus	Adsorbed	Little storage
Losses	Mineralization, erosion	Volatilization, erosion	Leaching, denitrification
Plant Use	Not used	Can be used	Preferred form
Changes	Mineralization	Immobilization, nitrification	Immobilization, denitrification

Figure 11-6. Characteristics of three types of soil nitrogen

However, nitrate losses from soil can be high. Nitrate ions leach out of the soil readily, and some may disappear as nitrogen gas in wet soil.

The amount of ammonium and nitrate nitrogen in the soil depends on the amount and type of nitrogen applied to the soil and the rates of nitrification and denitrification. Some nitrogen fertilizers contain nitrates. Most modern fertilizers, however, mainly provide ammonium nitrogen. Nitrifying bacteria in the soil change this form to nitrates, which are the preferred form of nitrogen for crops. Soil conditions strongly affect the rate of nitrification. Nitrifying bacteria grow best in moist, loose, well-drained soil at a pH of 6.0 to 7.5. Nitrifying bacteria function little below 41°F and reach maximum activity between 85°F and 95°F. Thus, cold, wet, or acid soils can slow the conversion of ammonium nitrogen to nitrate nitrogen.

Low pH and waterlogged soil prevent nitrifying bacteria from thriving. However, denitrifying bacteria grow well in the same conditions, because they do not need oxygen for respiration. These organisms use nitrates instead of oxygen. In the process, nitrates are changed to nitrogen gases that are lost to the air. Most notably, denitrification may cause the loss of nitrogen when growers apply nitrate fertilizers to wet soils.

Because of potential losses of nitrates, it is useful to control the rate at which ammonium fertilizers are changed to nitrate nitrogen. Several chemicals have been tested to inhibit (but not stop) nitrification. In actual practice, however, the results have been extremely variable, sometimes of benefit and sometimes not.

Nitrogen Deficiency. In all plants, slow growth and stunting are the most obvious signs of nitrogen shortage. Because nitrogen is part of chlorophyll, nitrogen-deficient plants lack the dark green color of well-fed plants. This symptom is called chlorosis. Leaves turn light green, then yellow, starting with the lower leaves. In grasses, yellowing starts at the blade tips, progresses down the midvein, and finally the entire leaf yellows. In extreme cases, the leaf dries up, a symptom called *firing*. In broadleaf plants, leaves are small with overall yellowing.

Growers often think of nitrogen as the "growth element." However, it can only work as a partner with other key elements, such as phosphorus.

PHOSPHORUS

Phosphorus also spurs growth but to a lesser extent than nitrogen (figure 11-7). Phosphorus affects plant growth in a number of ways:

- Phosphorus is part of the genetic material (chromosomes and genes) and so is involved in plant reproduction and cell division.
- Phosphorus is part of the chemical that stores and transfers energy in all living things. Without it, all biological reactions come to a halt. In plants, examples of energy reactions include the capture of light energy by photosynthesis and the transport of energy in roots for nutrient absorption.
- Phosphorus spurs early and rapid root growth and helps a young plant develop the roots it needs.
- Phosphorus helps plants use water more efficiently by improving water uptake by roots.
- Phosphorus helps plant resist cold and disease, speeds crop maturity, aids blooming and fruiting, and improves the quality of grains and fruits.

In many ways, phosphorus acts to balance nitrogen. For instance, while nitrogen delays maturity, phosphorus hastens it. Nitrogen aids vegetative growth, and phosphorus aids blooming and fruiting. As a rule of thumb, phosphorus is most important for crops from which we use the floral parts—that is, flowers, fruits, or seeds. Phosphorus is also needed to promote early and rapid root growth. For this reason, it is often the major element in starter fertilizers.

Forms of Phosphorus in the Soil. Soil phosphorus is provided by the weathering of certain minerals, primarily the mineral *apatite*, which is a calcium phosphate mineral. As apatite weathers it gives off a group of anions that can be used by plants. These ions are primary orthophosphate ($H_2PO_4^{-2}$) and secondary orthophosphate (HPO_4^{-2}). For simplicity, the text will refer to them both as phosphates.

Many soils contain large amounts of phosphate, but much of it is not available to plants. Phosphate in insoluble forms that are not free for plant growth is said to be "fixed." The reactions that fix phosphate depend on soil pH. In strongly acid soil (pH 3.5–4.5), insoluble iron phosphates form. Between pH 4.0 and 6.5, much of the phosphorus reacts with aluminum. Calcium phosphates are

Figure 11-7. The wheat plants on the right exhibit a phosphorus deficiency. (Courtesy of Potash and Phosphate Institute)

important between pH 7.0 and 9.0. Maximum availability lies at pH 6.5 in mineral soils, but the range 6.0 to 7.0 is satisfactory for most crops.

Between 25% and 90% of all phosphorus in the soil is part of soil organic matter. Organic matter is an important storehouse of phosphorus. Figure 11-8 summarizes the forms of phosphate in the soil.

A typical acre of soil holds between 800 and 1,600 pounds of phosphate in the plow layer. Of that total, only about four pounds is in solution at any time. As plants remove phosphate from solution, reserve phosphate (fixed or organic) moves from the other forms and becomes soluble. At the height of the growing season, solution phosphate may be replaced from soil stores several times daily.

In many soils, plants cannot tap soil stores fast enough to produce a full crop. For this reason, growers fertilize soil with phosphate to compensate for fixation.

Movement and Uptake in the Soil. Phosphorus moves very little in mineral soil. It moves only by diffusion over a distance as small as 1/4 inch. This limited movement has an important meaning for soil management.

One result is how phosphate is lost from the soil. It cannot flow downward in soil as do nitrates. Instead of leaching, phosphorus is more commonly lost when erosion or blowing soil carries it from the field. Because of the low mobility of phosphorus, it is critical that phosphate fertilizer be placed near where it is to be

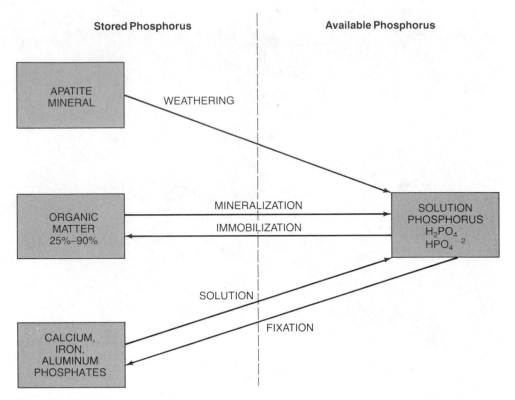

Figure 11-8. Phosphorus comes from the weathering of minerals such as apatite. Most phosphorus is stored in organic and fixed forms. At any one time, only a small amount of phosphate ion is in solution and usable by plants.

used. It is important that phosphates be placed near seed when planted or that it be mixed into soil near plant roots (see chapter thirteen).

The uptake of phosphorus depends on a number of soil conditions:

- Soil pH largely sets the degree of fixation. Phosphorus is most free at a pH of 6.0 to 7.0.
- Dry soil stalls the diffusion of phosphorus to roots. Therefore, plants take up phosphate best in moist soils.
- Oxygen is needed for the breakdown of organic phosphates. Roots also need oxygen to take up nutrients. Thus, a loose, well-drained soil improves the use of phosphorus. Compacted or poorly drained soil reduces access.
- Cold soil slows the activity of microorganisms that place phosphorus in solution, slows the rate of diffusion to roots, and slows root growth. In addition, root respiration also declines, depriving roots of the energy needed to absorb phosphorus. Phosphate shortages are commonly seen on cold, wet soils.

- The total nutrient balance is also important. Nitrogen, for instance, improves phosphorus uptake. Too much zinc seems to lower it.
- Mycorrhizae infection of plant roots helps the plant absorb phosphorus, especially in phosphorus-deficient soils.

A crop uses only 10% to 30% of the phosphate fertilizer applied to it. The rest goes into reserve and will be used by later crops. Many growers, in fact, have built up large reserves of soil phosphorus. Only soil testing can tell growers how much phosphorus crops need.

Deficiency. A shortage of phosphorus can cause stunting, but the plant remains green. Phosphorus-deficient plants often have a purple tint to the leaves and stems, starting on the lower leaves. A shortage of phosphorus may delay the maturity of several crops, including corn, cotton, soybeans, and others. Some crops, like carrots, develop poor root systems. On the other hand, an excess of phosphorus in the soil ties up several plant nutrients, such as iron. Soil testing will indicate the amount of phosphorus a soil needs.

POTASSIUM

Potassium, often called potash, is a key nutrient in the soil. Plants consume more potassium than any other nutrient except nitrogen. No organic compounds in a plant contain potassium, but many life processes need it. Potash activates enzymes needed for the formation of protein, starch, cellulose, and lignin. Thus, it is necessary for the development of strong, rigid plant stems. Potash regulates the opening and closing of leaf stoma (pores in the leaf that pass oxygen, carbon dioxide, and water vapor into and out of the leaf). Therefore, potassium is involved in the gas exchange needed for photosynthesis.

Potassium acts to balance the effects of nitrogen. Nitrogen leads to soft growth, but potassium promotes a tougher growth. The toughness results from thicker cell walls. This increased toughness improves crops in a number of ways:

- Plants well stocked with potash have strong stems that are less prone to lodging (figure 11-9). In corn, reduced lodging also results from the greater number of brace roots (figure 11-10).
- Well-fed plants fight off disease. Potash reduces diseases such as mildew in soybeans, wildfire in tobacco, and leaf and dollar spot in turfgrass.
- Potash makes plants more winter-hardy and less likely to be injured by spring or fall frosts.
- Potash, by regulating the stoma, influences the transpiration rate. A plant well-supplied with potash transpires less and so makes better use of water supplies.

The more potassium in the soil, the more plants take up. However, there is no evidence that supplying potassium beyond its needs will additionally

Figure 11-9. Lodging of corn as a result of a potash deficiency

increase hardiness or toughness. While luxury consumption of potassium does not harm the plant, it is wasteful. In addition, the plant may take up too much potassium but not enough calcium, causing calcium hunger in the plant.

Forms of Potassium in the Soil. Potassium comes from a number of rather common minerals such as feldspars and micas. Weathering lets potassium go directly into the soil solution as the potassium cation. This ion can be easily taken up by plant roots. Little potassium becomes part of soil organic matter, so most is stored in the soil by adsorption and fixation.

Potassium ions bear a positive charge and so adsorb on soil colloids. In most mineral soils, a few pounds of potash are dissolved in the solution of an acre of soil at any one time. In contrast, several hundred pounds per acre occupy the cation exchange sites. This supply is called *exchangeable potassium*. Both exchangeable and solution potassium are used by plants.

Potassium can also be fixed by 2:1 clays (see chapter nine) in a way that removes much of it from the reach of crops. Illite and vermiculite can trap potassium ions between the 2:1 layers, as shown in figure 11-11. This potassium can be released slowly if the potassium level in the soil solution declines. Montmorillinite clay layers are so loose that potassium ions can enter and leave easily, allowing potassium to remain available. Figure 11-12 shows the forms of potassium.

Figure 11-10. Potash promotes brace root development in corn. The potash-deficient plant on the right has the smallest number of brace roots. The plant on the left has received adequate levels of all three nutrients and has the largest number of brace roots. (Courtesy of Potash and Phosphate Institute)

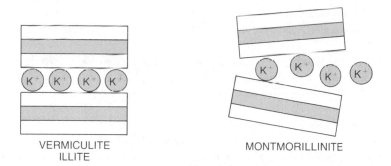

Figure 11-11. Potassium fixation occurs when potassium ions are trapped between the 2:1 layers of illite and vermiculite. The layers in montmorillinite open up enough so that potassium does not become trapped.

Movement in the Soil. Potassium moves more readily in soil than does phosphorus, but it moves less readily than nitrogen. Because potassium is held on clay or other colloids, it is least mobile in fine-textured soil. Potassium can leach out of sandy soils, but it leaches more slowly from clay soil.

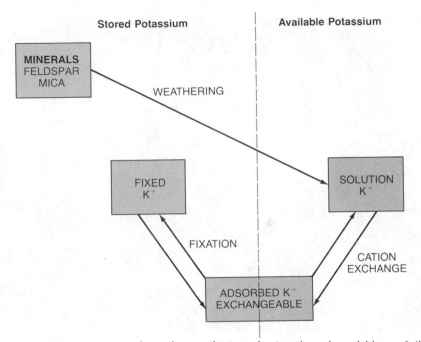

Figure 11-12. Potassium comes from the weathering of minerals such as feldspar. Soil stores potassium by adsorbing some on soil colloids and fixing some in certain clay particles. Both exchangeable and solution potassium are available for plant use.

Most plant uptake of potassium occurs by diffusion. Fortunately, because the element moves more readily than phosphorus, fertilizer placement is less crucial.

Deficiencies.   Growers see potassium hunger signs less often than those of other primary nutrients. Shortages occur primarily in sandy, heavily leached soils, especially if irrigated. Overfertilization with nitrogen can cause plant tissues to lack potassium. Dry, cold, or poorly aerated soil may also slow uptake. Potassium uptake is most rapid near neutral pH.

Plants show a lack of potassium by a "marginal scorch," or burnt look on the edges of the lower leaves. This symptom can be easily mistaken for moisture shortage during hot dry weather or for salt damage. In some cases, the margins merely turn yellow.

SECONDARY NUTRIENTS

Calcium.   Calcium is the third most used nutrient. Plants mainly use calcium to build cell walls. A thin layer of the compound calcium pectate lends strength to plant cell walls (figure 11-13). The crispness of apples comes from calcium pectate

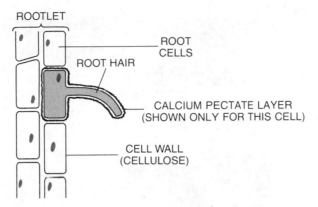

Figure 11-13. A wall of calcium pectate in a root hair gives the cell strength.

in the fruit cells. Because calcium makes strong cell walls, plants need it most where cells are dividing most rapidly—in the root and shoot tips (figure 11-14). Calcium also largely controls soil pH and helps aggregation. Finally, calcium plays a role in the protein formation and carbohydrate movement in plants.

Calcium comes from the weathering of a number of very common minerals and rocks, including feldspars, apatite, limestones, and gypsum. These materials are so common that most soils contain enough calcium to supply plant needs. Soils produced from high-lime parent materials may be calcareous—containing high levels of free lime. A calcareous soil is defined as one that "fizzes" when treated with dilute hydrochloric acid. The bubbles are carbon dioxide given off by the reaction.

Calcium is not fixed in the soil, and it is not held in organic matter. It is the main occupant of the cation exchange complex. Calcium storage depends on the cation exchange capacity of the soil (figure 11-15).

Calcium shortages are rarely seen, but when they occur, they are most likely on leached sandy or acid soils. Dry soil may limit calcium uptake. Excess potassium can create a shortage of calcium in plant cells because plants take up potassium and calcium in a ratio that is the same as the ratio of these elements in the soil. When the potash level is much too high, plants take up too little calcium.

Because calcium is needed to build new cells, the most obvious sign of shortage is problems with shoot tips. The shoot tips may become jelly-like because of weak cell walls. Because fruits need the same substance, they may also show signs of calcium shortages. Many apple growers, for instance, prevent "water core" in apples by spraying trees with calcium. Water core results from the collapse of cell walls in the fruit. A disease of tomatoes called "blossom-end rot" is similar. Blossom-end rot occurs when soil dries out and halts the movement of calcium to tomato roots.

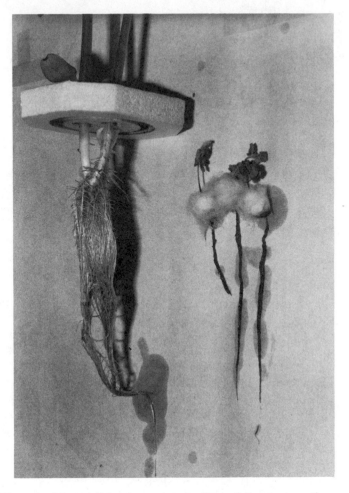

Figure 11-14. Effect of calcium deficiency on roots. Lacking calcium to strengthen cell walls, roots turn mushy and fail to develop, as shown by the plant on the right.

Magnesium. Magnesium resembles calcium chemically and in its action in soil. Its role in the plant differs, however. Magnesium is the essential ingredient in chlorophyll—each molecule has one magnesium atom at its center (figure 11-17). Magnesium also aids the uptake of other elements, especially phosphorus. Like potash, magnesium activates a number of important enzyme systems.

Magnesium weathers from minerals as a cation (figure 11-15). However, clay holds magnesium less strongly than calcium, so it is more easily leached. As a result, low-magnesium soils are more common than low-calcium soils. Highly leached coarse soils are most likely to need fertilization with magnesium, especially if they have been limed with low magnesium lime. High levels of soil potassium may also induce a magnesium shortage in plants.

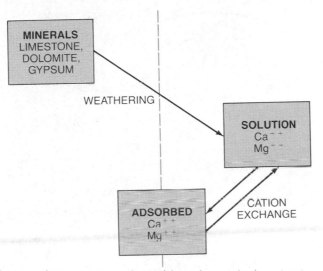

Figure 11-15. Calcium and magnesium are formed from the weathering of rocks such as dolomite. There is no storehouse for either element, except for cation exchange sites and solid minerals.

Hunger signs resulting from low levels of chlorophyll include chlorosis, a yellowing of the leaf.

Sulfur. Crops need less sulfur than the other macronutrients, but it is still a crucial nutrient. Several proteins include sulfur, and it is needed for making chlorophyll. It also aids nodulation of legumes and seed production of all plants. Overall, sulfur improves protein and chlorophyll content, stress tolerance, animal nutrition, and the appearance of plant products. Alfalfa and members of the mustard family (including cabbage, turnip, and radish) need much sulfur. Small grains and corn need less sulfur.

Most soil sulfur comes from the weathering of sulfate minerals such as gypsum. The sulfate anion is the form used by plants. Organic matter contains 70% to 90% of the soil sulfur; it is neither adsorbed nor fixed to any degree. Since it is readily leached, surface layers of soil are often low in sulfur. Figure 11-16 reviews the sulfur cycle.

Interestingly, acid rain supplies some sulfur in many areas. It should be pointed out that this benefit does not cancel out the bad effects of acid rain, however, which include pollution damage to plants, soil acidification, killing of aquatic life in many lakes, and health problems for the population. In many parts of the country, the sulfur from acid precipitation has been cut down slightly by burning low-sulfur coal and by better scrubbing of smokestacks.

Older fertilizer types were another source of sulfur. They contained sulfur as a byproduct of their manufacture. The fertilizers that are most popular now are much purer. Since pollution- and fertilizer-supplied sulfur have both been reduced, shortages are increasingly common. Use of sulfur fertilizer has

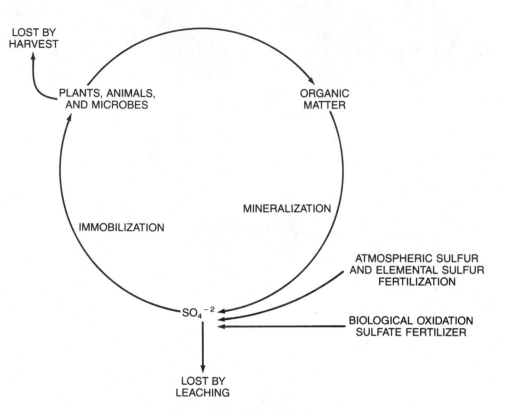

Figure 11-16. The sulfur cycle

increased rapidly, especially in the Southeastern states. Soils that are leached and that have low organic matter are likely candidates for sulfur shortages. Soils high in organic matter or soils located near industrial centers are least likely to be short of sulfur.

Plants that are short of sulfur may be stunted and older leaves will be pale green. These signs of sulfur shortage resemble those of nitrogen deficiency.

TRACE ELEMENTS

Trace elements play many roles in plants. These roles are difficult to understand without knowing plant chemistry. With the exception of boron and chlorine, trace elements are metals. These metals form special organic-metal molecules, called *enzymes*, that control important biological reactions. Enzymes can be thought of as keys that turn biological reactions on and off in living systems. They are not used up in the process. For instance, an iron enzyme controls one step in the formation of chlorophyll, but is not itself part of chlorophyll.

Very little of each enzyme is needed, because each is reused repeatedly. Therefore, very little of the trace elements that are part of enzymes is needed. Without this tiny amount, however, important processes suffer. On the other hand, an excess of a trace element can be toxic to plants or the animals feeding on them. The difference between enough and too much can be quite narrow, sometimes only a few pounds per acre. Growers should apply trace elements with caution, after proper soil and tissue testing.

Trace elements are stored in the soil in a somewhat different manner than the macronutrients. Some trace elements are stored in slightly soluble compounds. Some are involved to a small extent in cation exchange. Many trace elements combine with organic molecules in the soil to form very complex molecules called *chelates*. A chelate is a metal atom surrounded by a large organic molecule (figure 11-17). Chelates are an important form of storage for many trace elements.

Iron. Iron is part of many enzymes necessary in the formation of a number of chemicals, especially chlorophyll. Iron minerals are widespread in soil. They weather to release the iron, in ionic form. Most soils have sufficient iron, but much is in the form of insoluble compounds, such as ferric hydroxide, Fe $(OH)_3$. Organic matter chelates some iron in the soil. Interestingly, some soil microbes living in the rhizosphere emit compounds that chelate iron, probably improving iron uptake by plants.

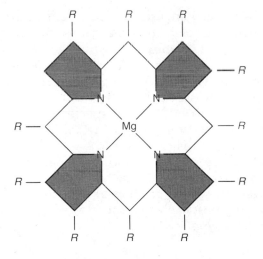

R = ADDITIONAL ORGANIC GROUPS

Figure 11-17. In the chelation of a metal ion, very complex organic molecules, represented here by several connected pentagons, surround the metal ion and form several bonds with it. Chelation protects the ion from other soil processes. Artificial chelates of trace elements are important fertilizers. This illustration represents chlorophyll, which is also a chelate.

The solubility of iron compounds is directly related to pH, declining about 100 times for each rise of one pH point. Acid-loving plants begin to suffer iron shortages when the pH rises above 5.0 or 6.0, depending upon the species. Other plants, like common field crops, suffer from a lack of iron when pH rises above 7.4. Sorghum, soybeans, and flax are field crops sensitive to an iron shortage. Iron hunger is most likely in alkaline or calcareous soils, or with excesses of phosphate, zinc, copper, or manganese. Anything that inhibits nutrient uptake, like cold, wet soils, or drought, may induce iron deficiencies.

Iron chlorosis is the usual symptom of iron hunger. It is easy to see as an interveinal chlorosis on new, growing leaves (figure 9-2). While leaf veins remain green, tissue between the veins becomes light green or yellow. In trees, branches begin to die back. Fruit and ornamental crops most often show these symptoms. Examples include azaleas, pin oaks, and blueberries.

Various treatments are available to overcome a lack of iron: (1) soil pH can be lowered to free the iron; (2) soluble iron compounds such as iron sulphate may be mixed into the soil, sprayed on leaves, or even injected into the trunks of trees; (3) artificially prepared chelates may be used in the same way; and (4) animal manures can be mixed into the soil.

Manganese. Manganese resembles iron in its action. In both nutrients, weathering creates a cation that is tied up as the molecule MnO_2 in nonacid soil. Manganese acts with iron in the formation of chlorophyll. Manganese speeds seed germination and crop maturity. It also helps plants take up several other nutrients.

Manganese deficiencies are usually seen on calcareous soils or on soils that have been overlimed. The solubility of manganese decreases a hundredfold for each rise of one pH point. Soybeans grown on some slightly acid to alkaline soils of the Atlantic Coastal Plain are known to suffer manganese shortages. Dwarfing is a common symptom of manganese deficiency and is often seen in combination with chlorosis. Flecks of dead tissue, along with chlorosis, often appear on the leaf, as shown by some species of maple trees growing on alkaline soil (figure 11-18). When soil pH is below 5.0, so much manganese may be free that it reaches toxic levels.

Deficient soil can be treated by mixing manganous sulfate into the soil. Applying sulfur to lower pH may also be helpful. Leaves may be sprayed with a solution of manganous sulfate or chelate. Oats, soybeans, sugar beets, and several vegetables are most likely to respond to manganese treatments. Liming cures manganese toxicity in acid soil.

Zinc. The zinc cation is weathered out of soil minerals, where it can be adsorbed, form a chelate, or form slightly soluble zinc compounds. Several biological reactions use zinc, including chlorophyll production. Low zinc levels are widespread in many crops, including beans, corn, and rice. Some nutritionists have voiced fears that these shortages could be passed on to human consumers.

Figure 11-18. Manganese deficiency on a maple growing in alkaline soil

The symptoms of a zinc deficiency in plants are quite varied, but dead spots on leaves are a common sign. Chlorosis between leaf veins is also common on field crops.

The availability of zinc is tied to pH in that zinc is most available in acid soil. Zinc shortages are most common on alkaline or recently limed soils. Soils that have lost topsoil by leveling, terracing, or erosion may also be zinc poor. Low levels may also appear on very coarse soils, because the parent materials lacked zinc and the soils tend to be low in organic matter. Cold soils or excess levels of phosphate can inhibit uptake. Like iron, a lack of zinc can be treated by fertilizing soil or spraying foliage with zinc compounds or chelates. Sewage sludge is an excellent source of zinc (see chapter fourteen). Corn, rice, and onions are most likely to respond to zinc treatment.

Copper. Copper is held by cation exchange and combines chemically with organic matter. Some organic-copper complexes are so stable that the copper is unavailable to plants. Copper is part of a number of important enzymes, especially for the formation of chlorophyll. Copper affects how well a plant resists disease and how well it controls moisture. Shortages are not common but when hunger occurs, it shows in a wide variety of signs, including reduced growth. Mild hunger merely reduces yield. Shortages are most likely to be seen in either leached sands or peats and mucks. A few pounds of the blue crystal copper sulfate mixed into the soil usually supplies all the copper that is needed. Carrots grown on organic soils are most likely to need extra copper, but small grains and other vegetables sometimes suffer as well.

Boron. Boron, unlike the trace elements discussed previously, is released by weathering as an anion. The borate anion, BO_3^{-3}, is taken up by plants and gathers in organic matter near the soil surface. Fixation at a high pH and leaching limit the amount of boron plants can use. Shortages sometimes appear if a soil is overlimed. Conditions that limit organic matter decay also limit the amount of free boron.

Boron shortages are seen most often in dry weather. A review of figure 4-14 explains why. When soil is moist, roots feed in the boron-rich surface layers. As the soil dries, roots begin to feed in the more boron-short lower layers. Boron is occasionally present in excess, mostly in the semi-arid western states or where crops are irrigated with high boron water.

Boron participates in many plant roles. It is needed to create a number of proteins and plant hormones. It controls carbohydrate transport. Finally, the development of new cells needs boron. This latter explains a number of shortage symptoms.

An interesting variety of diseases arises from boron shortages. Of the farm crops, alfalfa, clover, corn, cotton, peanuts, tobacco, and sugar beets are most often affected. Many fruits and vegetables also react to low boron levels. Hunger signs include cracked stem in celery, heart rot in sugarbeet, and top rot in tobacco. In many plants, buds on the tips of shoots die. When side shoots grow as a result, their tip buds die in turn. The densely bushy growth that results is called a "rosette." Boron shortages may also show up as incomplete growth of seed or grain, as in hollow-heart of peanuts.

A number of boron fertilizers may be applied to the soil or sprayed on plant leaves. The oldest form, the common laundry product borax, may be applied at the rate of a few pounds per acre. However, even slightly high boron levels hurt plants. Boron should not be used without first testing the soil.

Molybdenum. Molybdenum is needed by nitrogen-fixing bacteria, both those on legumes and free-living bacteria. Thus, molybdenum is important to both legume growth and nitrogen fixation. Plants also need molybdenum to make protein. The molybdate ion, MoO_4^{-2}, gathers in soil organic matter. Unlike other micronutrients, it is most available at a high soil pH. Shortages are most common on acid, leached, and low-organic matter coarse soils.

Several crops, in addition to legumes, respond to treatment. Crops in the mustard family are especially sensitive. Whiptail of cauliflower, for instance, results from a lack of molybdenum. An ounce of a soluble molybdenum material, often mixed with phosphate fertilizer, will usually treat an acre of deficient soil. Frequently, liming releases enough of this trace element to cure shortages.

Chlorine, Nickel, and Others. The function of chlorine, a recently identified essential element, is not well understood. It is known to play a role in photosynthesis. Chlorine is needed in very small amounts and is commonly found in the soil. Chlorine is thought to be never lacking in farm soils. However, it has been shown to increase grain yields in some soils of the Great Plains, most commonly where plant diseases have been a problem.

Nickel was tentatively identified as an essential element in 1984. Soybeans, and possibly other plants, need nickel for the proper use of nitrogen within their systems. It may also help plants resist disease. These are the initial results of research; further research is necessary before nickel is confirmed as an essential element.

A number of other elements contribute to nutrition of certain plants, though they are not currently considered to be universal essential elements. Legumes need cobalt for nitrogen-fixing. Some grasses and horsetail need silicon; it also is needed for best yields in rice and sugarcane. Sodium appears to be required for many plants native to sodium-rich soils. Plants that need sodium also include species that have special types of photosynthesis adapted to hot sunny climates. These include cacti, succulents, and many warm-season grasses.

BALANCED NUTRITION

As important as having enough of each nutrient is the concept of balanced nutrition. This means that each essential element is available to plants, none is deficient, and none is in excess. A few examples will illustrate this point.

Figure 11-10 compared the development of brace roots on corn with different levels of the primary nutrients. Brace-root growth is associated with adequate potassium levels, as the figure shows. However, the figure also shows that the best growth came from a balance of all three primary elements.

Nitrogen is considered the key to rapid vegetative growth, as is phosphorus to a lesser degree. Note in figure 11-19, however, that oat plants well-supplied with phosphorus and nitrogen are still stunted when compared with plants that are well-supplied with all three nutrients.

This chapter also has noted a number of interactions between nutrients that can render one of them unusable. For example, excess potassium can lead to a calcium or magnesium shortage. High levels of phosphates can tie up zinc, iron, and other elements.

Balanced nutrition is possible only with timely soil testing, the subject of the next chapter.

Figure 11-19. The oat plants on the left show the effects of a potash deficiency. Although nitrogen is considered to be the main growth element, balanced nutrition is needed for the best growth. (Courtesy of Potash and Phosphate Institute)

SUMMARY

The 13 mineral nutrients perform many important tasks in the plant. Of the major elements, nitrogen promotes rapid succulent growth. Phosphorus gives early root growth, blooming, and resistance to pest and weather damage. Potassium lends toughness and strength and pest resistance to plants. Plants need a balance of these three nutrients for strong, vigorous, and healthy growth.

Anyone who grows crops should know how nutrients behave in the soil. An important consideration, for instance, is how a nutrient is stored in the soil. Some nutrients, such as nitrogen and boron, are stored predominantly in organic matter. Some nutrients, such as calcium and magnesium, are adsorbed primarily on soil colloids. Many nutrients are part of slightly soluble compounds, including phosporus and iron. Many trace elements, like copper, react with organic matter in the soil to form chelates. Most nutrients are found in several of these forms.

Other important traits of nutrients include their solubility and mobility. The solubility of most nutrients depends on pH. For example, phosphorus compounds are most soluble between pH 6.0 and 7.0. Highly mobile nutrients, like nitrate nitrogen, can leach easily from the soil. Elements that move only a short distance, like phosphates, must be placed where roots or seeds will use them.

Plants grow best when each nutrient is present in the right amount. A lack of any one nutrient causes poor or abnormal growth. In addition, plants need a balance of nutrients. To achieve this balance, soil testing should be completed before fertilization is started.

REVIEW

1. Nitrogen in the soil is stored mostly in organic matter (True/False).
2. Calcium is mostly stored in organic matter.
3. Ammonium nitrogen leaches more readily than nitrate nitrogen.
4. Phosphorus is most available at a pH of 7.0-7.5.
5. Sulfur is an important part of protein.
6. Potassium improves winter hardiness.
7. Calcium is part of cell walls.
8. Pin oaks often exhibit iron chlorosis on acid soils.
9. Lack of copper is most likely on clay soils.
10. Manganese can be toxic.
11. _____ is the nutrient called the growth element.
12. _____ promotes early growth of seedlings by speeding root growth.
13. _____ toughens plants, making the stems stronger and the plant hardier.
14. _____ is the metal contained in the chlorophyll molecule.
15. _____ helps plants make cellulose and lignin, making them stronger and more rigid.
16. _____ is the form of nitrogen lost most quickly from the soil.
17. Iron deficiencies occur mostly at _____ soil pH.
18. _____ soils are most likely to lack several elements.
19. Acid rain supplies some _____, especially near industrial centers.
20. _____ are compounds that hold a metal ion in a large, complex organic molecule.
21. For which primary macronutrient is fertilizer placement most important? Least important?
22. When is denitrification most likely to cause the most loss of nitrogen?
23. Turf fertilizers are mostly what nutrient?
24. What element is involved in the disease of tomatoes called blossom-end rot?
25. Some chemicals are on the market to preserve nitrogen in the soil. What process do they inhibit?
26. Explain why a starter fertilizer is often high in phosphorus.
27. Describe the ways that nitrogen can be lost from the soil.

28. Describe the function of many trace elements.
29. How is potassium stored in the soil?
30. Why is soil testing important?

SUGGESTED ACTIVITIES

Try to grow nutrient-deficient plants to observe the symptoms. This is best done in construction sand. It contains few nutrients so you can control the amount of each. Plant several inch-tall seedlings of tomato, corn, and bean into moist sand in individual pots. Fertilize them with water-soluble fertilizers. Each pot should receive all the elements except one of your selection.

chapter twelve
Soil Sampling and Testing

An important procedure for any grower is soil sampling. Only by testing the soil can wise decisions be made about changing soil pH or about fertilization. There are also employment opportunities in helping growers with such decisions. Read this chapter to discover the basics of sampling and testing soil and plant tissue samples.

(Courtesy of USDA-SCS)

OBJECTIVES

After completing this chapter, you should be able to:

- explain why soils are tested
- sample soils correctly
- describe soil testing
- interpret soil test reports
- explain how plant tissue tests are used

TERMS TO KNOW

composite sample	soil sampling	tissue testing
green tissue test	soil testing	

WHY TEST SOILS?

Fertilizing can increase yields, and increased crop yields add to a grower's income. However, because fertilizers cost money, a grower must add the amount that is most profitable. How does a grower know how much fertilizer to apply for the best return?

Figure 12-1 shows a stylized relationship between plant nutrient levels in plant tissue and productivity. The nutrient level of a plant can be divided into four levels:

- *Level I: Deficient.* The nutrient is clearly deficient; growth and productivity are affected. After the missing mineral is applied, growth response is strong and profitable.
- *Level II: Sufficient.* A critical level is reached which satisfies plant needs. More fertilizer may increase yields slightly, but not enough to pay for fertilizer.

275

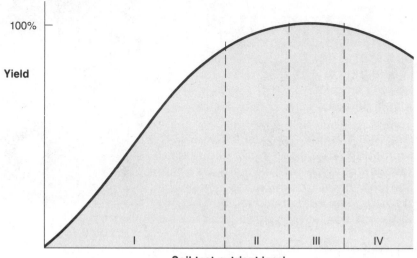

Figure 12-1. The yield of a crop is directly related to the nutrient level of the soil. Fertilization is most profitable for crops in Level I.

- *Level III: High:* Nutrient levels are high, yields are maximum. Additional nutrients would be stored in the plant (luxury consumption). Fertilization could also shift the plant to Level IV, or contribute to water pollution.
- *Level IV: Toxic:* Nutrient levels in plant tissue are so high as to be toxic. Yields decline.

Testing laboratories operate on one of two concepts of fertility levels. Some promote fertilizing the soil to bring it to an optimum level, then adding yearly maintenance amounts to replace those lost by crop harvest. Others suggest frequent soil tests followed by fertilization of the plant to supply its needs. Evidence would seem to indicate that the latter approach leads to more efficient fertilizer use.

Whichever approach is followed by the lab, growers test soil to determine how much of each nutrient is available for plant growth. From the results of the soil tests, they decide how much fertilizer should be applied to reach a sufficient level of each nutrient. Growers can use three methods to find nutrient shortages in plants:

- Visual inspection of crops for deficiency signs may uncover clear shortages. Unfortunately, this method often notes only critical shortages after yield damage has already occurred. In addition, visible symptoms may be quite unreliable. Chlorosis, for example, may result from low nitrogen, nematode feeding, dry or salty soil, diseases, or other problems not related to soil nutrient levels.

- Soil tests measure nutrient levels in soil as well as other soil features. Growers depend on these tests to determine the lime and fertilizer needs for crops. Soil tests have limits, however. Conditions that affect nutrient uptake, such as wet soils, cannot be detected in the laboratory.
- Tissue testing measures nutrient levels in plant tissue itself. This type of testing may uncover problems soil testing misses.

Of the three methods described, soil testing is most important for a majority of crops, especially annual farm crops. A soil test can be performed early in the season to allow the grower to supply the necessary nutrients *before* the crop is planted.

There are three separate activities involved in soil testing. (1) The grower *samples* the soil and sends the sample to a testing center. (2) The soil laboratory *tests* the sample and makes a recommendation to the grower. (3) The grower acts on the recommendations from the testing center. First let's look at soil sampling.

SOIL SAMPLING

Soil laboratories use the most modern testing methods and tools. However, the material to be tested is the sample provided by the grower. This means that the test results will be no better than the sample itself. Some general sampling methods will be described, but local recommendations may vary. Ask the local testing center or Cooperative Extension agent to recommend the best sampling method for your area.

Testing Frequency.   Frequency of soil testing depends on the crop and how it is grown. For most annual farm crops, sampling every two or three years should be adequate. Intensive crops like fruits or vegetables benefit from annual sampling, and greenhouse crops are tested even more often. Soils should be tested before any crop is planted that occupies the soil for longer than one season, such as turf, trees, or perennial forages. This practice allows the grower to mix potash and phosphate into the soil before planting. (Recall that potash and phosphates do not move very far in the soil. Therefore, applying them to the soil surface after a crop is planted is less effective than mixing them into the soil before planting.)

Any change in cropping practices should be preceded by thorough soil testing. For example, if a grower intends to shift from regular to conservation tillage, the soil should be tested before the first year. A grower changing crops, such as a corn grower trying out sunflowers, should also test the soil before planting the new crop.

Growers may sample soil whenever it is not frozen. Fall is probably the best time, because the grower can then use the winter months to respond to the results and order fertilizer.

Selecting the Sampling Area.   Land to be tested should be divided into sampling areas of uniform texture, topography, and cropping history. For example, a coarse-textured field should be sampled separately from a nearby fine-textured field. If half a field received manure last year and half did not, each section should be tested separately. The sampling areas vary in size from a homeowner's garden to a maximum of 20 acres on farm fields. Some labs now recommend using a soil map, if available, to determine sampling areas. Each soil type, in this system, becomes a sampling area. One would need to continue to consider cropping history.

A sampling area will be fertilized as a unit, using the same rate throughout the area. Even in this area, all of the soil is not the same. The differences are averaged by carefully taking many samples from random spots in the sampling area and combining them for a *composite sample*.

Depth of Testing. For field crops grown by conventional tillage, the top six to nine inches of soil should be sampled. For no-till systems, a special pH sample should be taken from the top two inches of the soil. This sample is needed because the no-till system permits an acid layer to form at the soil surface. This layer can harm seed germination. For the ridge system of reduced tillage, samples should be removed from the *sides* of the ridges. Chapter fifteen of this text describes these tillage systems.

Sod or pasture need only be sampled to a 2- to 3- inch depth. Tree crops, on the other hand, may need to be sampled as deep as 18 to 24 inches. Testing laboratories suggest that nitrogen tests be made on samples taken as deep as three feet. The deeper tests tell how much nitrate is in the subsoil.

Sampling Procedure. For each sampling area, the following steps are to be performed:

1. As shown in figure 12-2, gather many topsoil subsamples from random spots in the field. Samples are to be taken no closer than 100 to 300 feet (depending upon the recommendations from the laboratory) from such odd areas as dirt roads, barns, or fencerows. Also to be avoided are dead furrows, fertilizer spills, and other spots with unusual conditions. Large areas need 15 subsamples, smaller areas fewer.
2. Scrape away surface litter at each testing spot and remove a sample of the soil. Augers or soil sampling tubes (figure 12-3) are convenient sampling tools. A spade can also be used. Dig a V-shaped hole, remove a 1/2-inch slice from the side of the hole, and shave away most of the sample on the blade as shown in figure 12-4. Each soil sample should include soil from the entire testing depth. Drop each soil sample in a *clean plastic* bucket as it is collected.
3. Mix all subsamples from one sampling area, and remove about one cup of soil. This *composite sample* represents the average soil in the field. Label the composite sample and let it dry in the air. Do not oven-dry the sample as this will change the testing results by causing normally unavailable nutrients to be measured (by killing microorganisms and possibly charring organic matter).

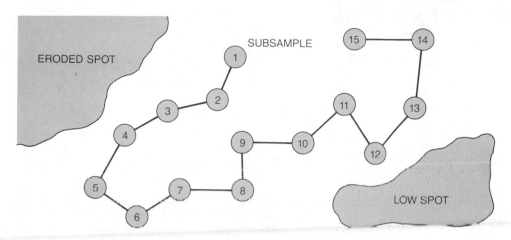

Figure 12-2. When taking soil samples from an area, up to 15 subsamples should be taken at random locations. Samples are not taken from low spots or other unusual areas. These subsamples are then mixed to form a composite sample. (University of Minnesota, Agriculture Extension Service)

Figure 12-3. A soil testing tube is a useful tool for sampling soil.

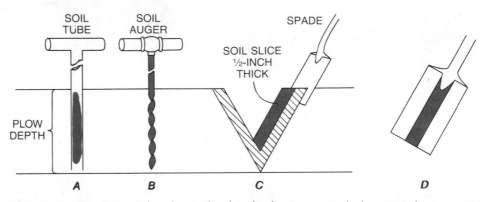

Figure 12-4. Soil samples can be taken to the plow depth using *(A)* a soil tube, *(B)* a soil auger, or *(C)* a common spade. In using a spade, a ½- inch thick slice of soil is taken from the side of a triangular hole. As shown in *(D)*, most of the soil in this sample is removed to leave a one-inch strip on the blade.

4. Fill a mailing container with the dried composite sample. Mark the container according to the instructions provided by the testing center. Complete the sample sheet (figure 12-5), including the intended crop, production goals, cropping history, and other necessary information.
5. Mail the samples to the laboratory, or deliver them to the lab or county extension agent. The sample containers and sheets can be obtained from the soil laboratory or extension agent.

Sampling Greenhouse and Container Plants. The fertility program for greenhouse and container plants is very different from the plan for field plants. Field growers depend mainly on nutrient reserves in the soil, like organic nitrogen or exchangeable potassium. Such growers fertilize to add the extra nutrients needed for best growth. Container growers, on the other hand, use potting mixes (see chapter sixteen) that contain essentially no nutrient reserves. Thus, all of the nutrients the plants need must be supplied by fertilization. Container growers have complete control over the nutrient status of the plant. However, such a program requires a very high level of management, including frequent soil testing. This need is increased by the small soil volume, high watering rates, and soluble salt problems of greenhouse plants.

Greenhouse or container soils should be sampled *before* being planted to the crop. They should then be tested periodically during the growing period. The soil should also be tested at any sign of a growth problem. To sample a crop of potted or bedding plants, first scrape away any material on the soil surface and the top ½ inch of soil. A core of soil is then removed. The core must include soil from the top to the bottom of the pot or flat. Several containers should be tested, perhaps six to a bench, and composite samples prepared. Beds or benches should be tested to eight inches or to the depth of the bench. In some areas, sampling is done through an extension agent specializing in greenhouse or nursery crops.

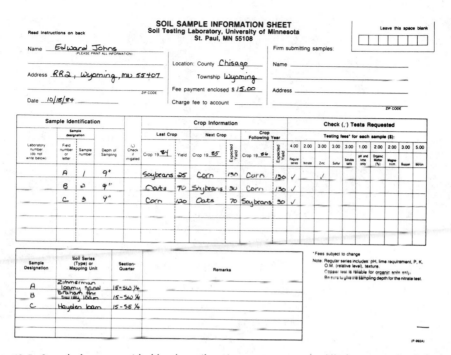

Figure 12-5. Sample forms provided by the soil testing center are to be filled out completely before being sent with the sample that is to be tested. A typical sample information sheet is shown. (University of Minnesota, Soil Testing Laboratory)

The agent may also want to see a sample of foliage to help with the diagnosis.

After a grower has collected and delivered a quality soil sample, he or she has completed the first step of the process. Now let's examine what the testing center does with the samples.

SOIL TESTING

There are two basic ways to test soil samples. The oldest method uses chemical reactions that produce color changes. The exact color depends on the amount of available mineral in the soil. In the case of the pH test, the color depends on soil pH. With the test kits used by home gardeners, a certain amount of soil is placed in a test tube. A measured amount of chemical, or *reagent,* is dripped into the tube (figure 12-6). The color of the resulting mixture is then compared with known standards, and the amount of nutrient is read from the standard.

Simple chemical tests are easy to use, but they are not reliable. One problem is the human element since the results are based on the technique of the tester and what he or she sees. In student labs, for instance, the results of several tests on the same soil sample may be surprisingly different. Further, these tests may not

Figure 12-6. To use a soil test kit, a small amount of the soil sample is added to a test tube. Drops of reagent are then added to the sample. The resulting color is compared with a standard color chart.

provide an accurate measurement of the adsorbed elements that affect plant growth, like potential acidity and exchangeable potassium. However, such tests can be useful in the field, where one cannot drag equipment about or find a source of electricity.

The use of chemicals and color comparison have been largely replaced in laboratory testing by such modern tools as the pH meter (figure 12-7) and the spectrophotometer. These tools rapidly and accurately measure amounts of minerals in soil samples. One type of spectrophotometer, for example, passes a beam of light through a test solution and measures the amount of light absorbed. The more light absorbed, the stronger the solution. The instruments are costly, but they greatly improve lab operations by allowing faster testing and greater reliability.

Laboratory results, however, are only reliable if they have been validated on soils similar to the one being sampled. That is, the tests must be based on research about fertilization and nutrient levels on soils like the sample soil. There are also differences in climates and crops. Usually this means that a grower should use a local laboratory whose recommendations are based on local conditions.

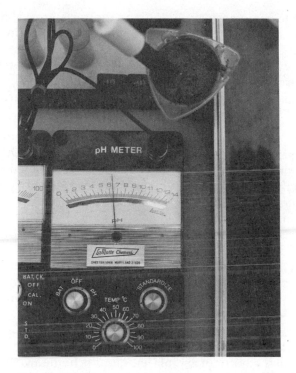

Figure 12-7. Testing laboratories use a pH meter to measure soil pH.

Such labs may be associated with a university or a state agricultural experiment station. A number of private firms also do soil and plant tissue testing.

Few testing labs routinely test for all thirteen mineral nutrients. Rather, a lab offers tests useful *for the area it serves.* Generally, a standard series includes the following tests:

- Texture is determined by simply feeling the soil sample or by mechanical analysis.

- Organic matter is measured by comparing the soil color with known standards. Some labs measure the organic matter with methods we need not discuss here.

- The pH and the buffer pH are measured with a pH meter (figure 12-7). From the pH and buffer pH the technician can suggest lime application rates.

- Phosphorus is washed from the sample with an acid solution. The solution then is put in a spectrophotometer. This method measures soluble phosphorus only; insoluble forms are not measured. The resulting value, called *available phosphorus,* indicates how much phosphorus is free for plant growth. Several different tests may be used, and they do not always get the same results. As long as the tests are validated for your area, all are satisfactory. However, it confuses comparing results from different labs.

● Potassium is washed from the sample with a solution that replaces potash on the cation exchange sites. The resulting value indicates the *exchangeable potassium*, the amount that is readily available for plant growth.

Nitrogen is difficult to test accurately, so there may not be a standard test for it in your area. Testing labs base nitrogen needs on the amount of nitrogen stored in the soil organic matter, the effect of last year's crop, other cropping practices, and the grower's production goals. However, in areas of low rainfall where leaching and denitrification are low, nitrate nitrogen can carry over from the previous year. Here, a nitrate-nitrogen test is valuable. The top two or three feet of soil should be sampled for nitrates.

Other tests may be standard or optional, depending on the laboratory. These include testing for soluble salts, cation exchange capacity, calcium, sulfur, magnesium, and trace elements known to be a problem in the area. Growers of containerized plants, like nurserymen, should ask for a soluble salt test.

After testing the sample, the lab issues a computer-printed report. The report includes test results, interpretation of the results, and fertilizer and lime recommendations. Figure 12-8 shows a sample report for strawberries in the home garden. Note that after the test results were entered, the computer printed a chart interpreting the nutrient and pH level for the homeowner. The computer also compared the present levels with the best levels for stawberries, and advised how much fertilizer to use.

Part of the interpretation is to translate the raw numbers as low, medium, or high. The numbers attached to these categories vary from lab to lab. A "low" category implies that a yield response to fertilization is highly likely, whereas a yield response to a "high" category is unlikely.

Once a grower has obtained a test report, the local county extension agent can help explain the results.

TISSUE TESTING

Plant tissue tests in combination with soil tests give the most complete picture of nutrient status in the plant. In *tissue testing,* nutrients in the plant itself, rather than nutrients in the soil, are tested. Such tests are very useful for "smoking out" trace element problems and may be more reliable than soil tests. Growers can also use them to tell if a symptom arises from a lack of some nutrient or from disease or other problems. Tissue tests may also indicate if some soil condition is hindering nutrient uptake. Some growers employ tissue testing to check the effectiveness of their fertilizer programs.

Plant tissue tests are also very useful for tree and vine crops in nurseries, vineyards, or orchards. The root systems of these plants are much more extensive

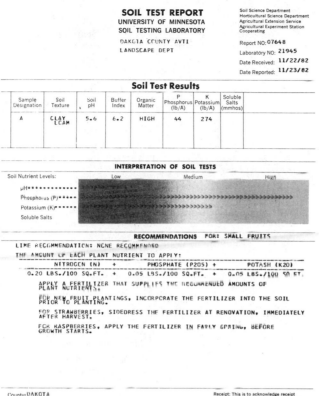

SOIL TEST REPORT
UNIVERSITY OF MINNESOTA
SOIL TESTING LABORATORY

DAKOTA COUNTY AVTI
LANDSCAPE DEPT

Soil Science Department
Horticultural Science Department
Agricultural Extension Service
Agricultural Experiment Station
Cooperating

Report NO: 07648

Laboratory NO: 21945

Date Received: 11/22/82

Date Reported: 11/23/82

Soil Test Results

Sample Designation	Soil Texture	Soil pH	Buffer Index	Organic Matter	P Phosphorus (lb/A)	K Potassium (lb/A)	Soluble Salts (mmhos)	
A	CLAY LOAM	5.6	6.2	HIGH	44	274		

INTERPRETATION OF SOIL TESTS

Soil Nutrient Levels: Low / Medium / High

pH
Phosphorus (P)
Potassium (K)
Soluble Salts

RECOMMENDATIONS FOR: SMALL FRUITS

LIME RECOMMENDATION: NONE RECOMMENDED

THE AMOUNT OF EACH PLANT NUTRIENT TO APPLY:

NITROGEN (N) +	PHOSPHATE (P2O5) +	POTASH (K2O)
0.20 LBS./100 SQ.FT. +	0.05 LBS./100 SQ.FT. +	0.05 LBS./100 SQ.FT.

APPLY A FERTILIZER THAT SUPPLIES THE RECOMMENDED AMOUNTS OF PLANT NUTRIENTS.

FOR NEW FRUIT PLANTINGS, INCORPORATE THE FERTILIZER INTO THE SOIL PRIOR TO PLANTING.

FOR STRAWBERRIES, SIDEDRESS THE FERTILIZER AT RENOVATION, IMMEDIATELY AFTER HARVEST.

FOR RASPBERRIES, APPLY THE FERTILIZER IN EARLY SPRING, BEFORE GROWTH STARTS.

County: DAKOTA
For additional information contact your
County Extension Agent:

Receipt: This is to acknowledge receipt
of payment in the amount of $ 4.00

Customer copy

Figure 12-8. A sample soil test report for a homeowner growing strawberries. Note that the report consists of three parts: results, interpretation, and recommendations. (University of Minnesota, Soil Testing Laboratory)

than are those of annual crops. Thus, it is often difficult to determine exactly where the feeding roots are and at what depth the soil sample should be taken. For these reasons, tissue tests are useful in monitoring the nutritional status of the crop.

Nutrient levels vary sharply in different plant tissues of different ages. Before sending samples to a laboratory, be sure to determine the plant part used and the growth stage required. For example, soybean samples are collected from fully open leaves at first flowering. Apple samples are taken from leaves in the middle of shoots 8 to 12 weeks after full bloom. Figure 12-9 lists the suggested samples and amounts for various crops from one testing laboratory. A grower should follow the instructions of the actual laboratory being used. A general method for sampling plants is as follows:

1. Sample about 10 to 15 plants (or the number requested by the laboratory), using the recommended plant parts (figure 12-9). The parts should be clean

Plants	When to Sample	Plant Parts Sampled	Number of Samples
Field Crops			
Alfalfa	At or before 1/10 bloom stage	Mature leaf blades 1/3 way down plant	45–55
Cereal grains	Seedling stage or	All above-ground part	50–75
	prior to heading	First 4 blades from top of plant	30–40
Corn	Seedling stage or	All above-ground part	25–30
	prior to tassel or	Fully grown leaf below the whorl	15–20
	from tassel to silking	Leaf at or one above or below ear node	15–20
Cotton	Prior to or at first bloom or squares	Youngest fully mature leaves or stem	30–35
Forage grasses	Before seed head or best quality stage	4 uppermost leaf blades	50–60
Soybeans	Seedling stage or	All above-ground part	20–30
	prior to or at first flowering	Fully grown leaves at top of plant	20–30
Tobacco	Before bloom	Top fully grown leaf	25–30
Fruit Crops			
Apple, apricot, cherry, peach, pear, plum	Midseason	Leaves near base, new growth	75–100
Grapes	End of bloom	Petioles from leaves next to fruit clusters	75–100
Orange	Midseason	Spring-cycle leaves, 4–7 months old from nonfruiting terminals	25–30
Strawberry	Midseason	Youngest mature leaves	50–70
Vegetable Crops			
Beans	Seedling stage or	Whole above-ground part	25–30
	before or at first flowering	2–3 mature leaves at top of plant	25–30
All cabbage crops	Before heading	First mature leaves from center to whorl	10–20

Figure 12-9. For tissue testing, specific plant parts and amounts are recommended as samples for selected crops. (Adapted from John E. Bowen, "Plant Parts to Sample," *Crops and Soils Magazine*, 31, No. 3, Dec. 1978, pp. 10–11, by permission of the American Society of Agronomy.)

Plants	When to Sample	Plant Parts Sampled	Number of Samples
Leafy crops	Midgrowth	Youngest mature leaf	30–50
Melons	Before fruits set	Mature leaves at base of main stem	20–30
Peas	Before or at first flowering	Leaves from 3rd node down from top of plant	30–50
Potato	Before or during early bloom	3rd to 6th leaf from growing tip	20–30
Root crops	Before root or bulb enlargement	Center mature leaves	25–35
Sweet corn	Before tasseling or	Entire mature leaf below whorl	20–25
	at tasseling	Entire leaf at ear node	20–25
Tomato (field)	Before or during early bloom	3rd or 4th leaf from growing tip	20–25
Ornamentals			
Chrysanthe-mums	Before or during early bloom	Top leaves on flowering stem	20–30
Trees and shrubs	Current years growth	Fully mature leaves	30–75
Turf	During growth season	Leaf blades	2 cups

Figure 12-9. *(continued)*

of soil or dust. Sample only the intended species. Do not, for example, mix both clover and grass samples from a pasture. Do not include dead materials and avoid damaged parts unless they are the intended sample.
2. If necessary, use water to clean dust or soil from the leaves.
3. Air-dry the samples before shipment.
4. Fill out the information sheet completely. Include any recent soil test results, if suggested by the laboratory.
5. Ship the samples to the laboratory in a heavy *paper* sample bag. Leaves may mold if shipped in a plastic bag unless they are quite dry.

Green Tissue Tests. A simpler form of tissue test is the *green tissue test* or *plant sap test*. In this test, plant sap in the leaf petioles or young stems is tested for nutrient levels. Portable test kits are available that can be used in the field. Some kits use test papers treated with testing reagents. To use these, squeeze plant sap on the test spots. Other test kits use glass vials and spot plates (figure 12-10). These tests require bits of leaf petioles to be mixed with a liquid reagent in a vial and the color noted on a spot plate.

As in tissue testing, several plants should be tested. Best results are obtained by comparing test results from both deficient plants and nearby healthy plants.

(A)

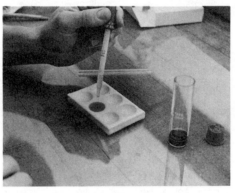

(B)

Figure 12-10. A portable plant sap test kit can be used to measure the concentrations of nutrients in the sap of crop plants. As an example, geraniums are being tested by cutting up the petioles *(A)* and treating them with a reagent *(B)*.

SUMMARY

Growers need to make effective use of fertilizers. Soil testing is the best tool growers have to decide how much fertilizer is needed to avoid both under- and overfertilization.

The first step is to sample the soil. Each area to be sampled should be uniform. Many subsamples are collected and mixed to form a composite sample from which a small amount is removed and sent to a private testing laboratory or university soil testing center. A sampling information sheet accompanies the sample to provide data the laboratory needs to make useful recommendations.

Soil testing laboratories use modern equipment to measure nutrient levels quickly and precisely. A computer then prints the test results, interprets the data, and makes fertilizer and lime recommendations.

Plant tissue tests are used less often, but when added to soil tests, a more complete picture of plant nutrient levels is obtained. Growers who wish to try tissue tests should consult with the testing laboratory for instructions.

Commercial laboratory testing provides more reliable results, but portable test kits are available to growers for both soil and plant sap tests.

REVIEW

1. Nitrate samples should be taken 12" deep (True/False).
2. Soils should only be sampled in the spring.
3. When sampling, include soil from the surface to the depth of the sample.
4. People that grow plants in containers should ask for a soluble salt test.
5. Tissue samples may be taken at any time.
6. Soils for tree crops should be sampled deeper than for field crops.
7. Before shipping a soil sample, dry it in an oven.
8. Typically, both exchangeable and non-exchangeable potassium are tested for.
9. Most labs do not routinely check for all mineral nutrients.
10. Greenhouse soils are tested more often than others.
11. For most field crops, soils are tested every _____ years.
12. Topsoil is sampled to a depth of _____ inches for field crops.
13. In each sampling area, many subsamples are collected and mixed to make a _____ sample.
14. The phosphorus content measured in a soil test is called _____ phosphorus.
15. Lime recommendations are based on the _____ and the _____ test.
16. _____ is the primary nutrient not routinely included in soil tests.
17. _____ are the most reliable for micronutrients.
18. Nitrate-nitrogen tests are most valid in _____ climates.
19. A tool for measuring pH is a _____ .
20. _____ is often the most convenient time to sample soil.

21. Note the number of sample areas included in each of these fields:

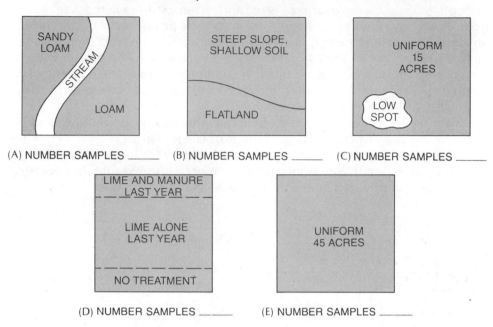

(A) NUMBER SAMPLES _____   (B) NUMBER SAMPLES _____   (C) NUMBER SAMPLES _____

(D) NUMBER SAMPLES _____   (E) NUMBER SAMPLES _____

22. Using figure 12-8:
    a. What is the soil texture? Is that coarse, medium, or fine?
    b. What is the soil pH? What words would you use to label that pH?
    c. How many pounds of phosphorus are there per acre? How high a level is this?
    d. Should the berry grower lime?
23. What happens if you add nutrients to a soil already high in that element?
24. List four tests commonly part of a standard series.
25. What are two useful portable tests mentioned in the text?
26. Why do container plants need more frequent testing?
27. When should field crop growers test soil?
28. Explain why a tissue test can add to a soil test.
29. When is fertilization most cost effective?
30. Why should a grower not rely on deficiency symptoms as a way to detect nutrient shortages?

SUGGESTED ACTIVITIES

1. Sample a test area and send the sample to a local testing lab.
2. Use a soil test kit to test the same sample in your school. Compare the results of your test with the lab results.
3. Tour a soil testing facility.

chapter thirteen
Fertilizers

As this display of fertilizers suggest, there is quite a variety of fertilizer products. What exactly is in all these bags? A grower, for whom fertilizer is a "tool of the trade," should know. This chapter will explain fertilizer contents and more.

OBJECTIVES

After completing this chapter, you should be able to:

- distinguish forms of fertilizers
- describe fertilizer sources for each nutrient
- perform important fertilizer calculations
- describe how to use fertilizers
- list two effects fertilizers have on the soil

TERMS TO KNOW

banding	fertilizer filler	pop-up fertilizers
broadcast	fertilizer grade	pressurized liquid
chelates	fertilizer ratio	prills
complete fertilizers	fluid fertilizers	pulverized fertilizers
fertigation	foliar feeding	sidedressing
fertilizer	fritted trace elements	split application
fertilizer analysis	granules	starter fertilizers
fertilizer burn	mixed fertilizers	topdressing
fertilizer carriers		

Before plant nutrients were identified, growers knew that some materials helped plants prosper. Early farmers probably noticed the lush growth of grass around animal droppings and began using them to raise better crops. Lime, ashes, dead fish, and ground bones have all been used. Even human waste ("night soil") has been an important fertilizer in some cultures. Most American growers now use chemical fertilizers to supply the elements their crops need.

Fertilizer is defined as any material applied to soil or plants to supply essential elements. Some states define by law minimum requirements for a material to be sold as a fertilizer. Fertilizers can be grouped into three categories— mineral, organic, or inorganic (figure 13-1):

- *Mineral fertilizers* are simply ground rocks containing nutrients. Dolomitic lime, for example, is a fine source of calcium and magnesium. Most minerals dissolve very slowly, and so their usefulness as fertilizers is limited.
- *Organic fertilizers* are organic materials, such as animal manure, that contain nutrients. Many can be considered "slow-release" fertilizers because they do not act quickly. Nutrients are released slowly over the growing season as the organic matter decays. Many organic materials contain low amounts of nutrients, and may therefore be expensive nutrient sources. In some states, "dilute" organic materials may not be legally labeled as fertilizers because their nutrient content is too low.

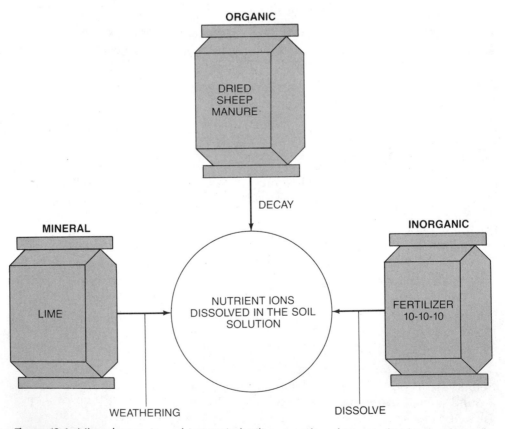

Figure 13-1. Mineral, organic, and inorganic fertilizers weather, decay, or dissolve to release the nutrient ions that are taken up by plants.

- *Inorganic fertilizers* or *chemical fertilizers* are made by industry. Some of these fertilizers are mined but others are entirely manufactured. Most dissolve quickly in the soil for fast growth response. Some fertilizers, however, are made to be slow-release. Inorganic fertilizers usually have a higher proportion of usable nutrients than mineral or organic materials. Because of their high nutrient content, inorganic fertilizers tend to be the least expensive nutrient sources. Most modern growers mainly use chemical fertilizers.

The materials used as fertilizers are provided in a number of forms, giving growers several choices of application methods. The forms can be divided into four main groups: pressurized liquids, fluids, dry fertilizers, and slow-release fertilizers.

Pressurized Liquids. Anhydrous ammonia is the primary *pressurized liquid*. Ammonia is a gas at normal temperatures and pressure, but it changes to a liquid when cooled to -28°F. It can then be stored in large high-pressure or refrigerated tanks. Smaller-wheeled tanks are filled from the storage tanks and are driven to the field to be fertilized.

The liquid is applied by injecting it into the soil (figure 13-2). Pressure in the tank forces the liquid through tubes to special chisels that are pulled underground. When it reaches the soil, the liquid evaporates rapidly.

Fluid Fertilizers. *Fluid fertilizers* are also liquids, but they are not under pressure. In the case of nitrogen, they are often called *nonpressure solutions*. The

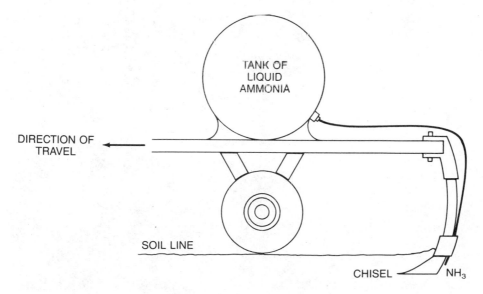

Figure 13-2. Anhydrous ammonia, an example of a pressurized liquid, is injected into the soil through chisels. The liquid ammonia evaporates in the soil where it clings to clay and humus particles until it is used by plants or converted to nitrates.

most common fluid fertilizers are *solutions* (figure 13-3). In a true solution, fertilizer dissolves in water to form a clear liquid. The fertilizer will not settle out of the solutions. Solutions can only be made of chemicals that are soluble in water.

Some chemicals that do not dissolve in water can be made into fluids called *suspensions*. A finely ground fertilizer is coated with a special clay and mixed with water. The treatment helps the fertilizer grains hang in the water to form a cloudy liquid. If the liquid is not kept stirred, however, the grains will settle out.

To compare the two liquids, mix salt in a jar of water and cornstarch in a second jar of water. Salt forms a solution in water, cornstarch forms a suspension. See what happens when both jars are allowed to remain undisturbed for an hour.

Fluid fertilizers are popular because they can be applied in many ways: sprayed on a field, injected into the soil, mixed into irrigation water, or even sprayed on crop foliage. They can also be mixed with fluid lime, herbicides, and other crop chemicals to save application time. However, care must be used when mixing chemicals to ensure that they can be mixed without problems.

Figure 13-3. Greenhouse growers have long applied fertilizers in liquid form. Fluid fertilizers are now becoming important for field crops as well.

Dry Fertilizers. The familiar dry fertilizers are still widely used, although the use of fluid fertilizers is growing more rapidly. Dry fertilizers are applied to soil, where they dissolve quickly in soil water to release nutrients. Dry fertilizers are available in three types:

- *Pulverized fertilizers* are made by crushing fertilizers into a powder. They are dusty, making them unpleasant to handle or spread evenly. Some pulverized fertilizers absorb moisture from the air causing them to cake during storage.
- *Granules* are much easier to use. The manufacturer treats the material so it has large, more evenly sized grains. Granules spread evenly and easily, with much less dust. However, some dust, or "fines," still causes problems. Granules are coated to reduce moisture absorption.
- *Prills* are smooth, round, and dust-free. They are made by a different process than granules. Prills have superior flowing and spreading qualities and are free of fines. Growers find them easy to use. Prills are also coated to prevent caking during storage.

Slow-Release Fertilizers. Slow-release fertilizers are also dry. The nutrients they contain dissolve into the soil solution slowly. Most of these fertilizers are made to dissolve in a time period ranging from several weeks to a few months. Slow-release fertilizers are too costly for common use, but they are widely used in horticulture. For example, they are popular turf fertilizers. Slow-release fertilizers benefit growers by releasing nutrients only as fast as crops can use them, so little is lost to leaching.

FERTILIZER MATERIALS

Few fertilizers are pure elements. In fact, in a pure form, many would be dangerous. (Pure potassium, a soft metal, reacts explosively with water.) Rather, most consist of compounds that release nutrients in forms useful to plants. These compounds are called nutrient *carriers.*

Nitrogen Carriers. Organic nitrogen exists in several forms (figure 13-4). Decay changes the organic nitrogen to ammonia, which, in turn, changes to nitrates. Many organic nitrogen sources are too expensive for farmers and are used primarily by home gardeners. However, some sources, such as manure, continue to be used by many farmers.

The first commercial nitrogen fertilizer was actually organic—guano, or bat and bird droppings. Guano was simply gathered and shipped to growers. Later, deposits of saltpeter (sodium nitrate) were mined in Chili. Today fertilizer companies manufacture most nitrogen carriers by the Haber process.

The *Haber process* uses nitrogen from the air. Natural gas is used as a source of hydrogen, which is combined with gaseous nitrogen. This reaction, in the presence of heat, pressure, and an iron catalyst, produces ammonia:

$$3H_2 + N_2 \longrightarrow 2NH_3$$

Ammonia can be used as a fertilizer or changed to other forms, as shown in figure 13-5.

Organic Material	Percentage, Dry Weight Basis		
	N	P_2O_5	K_2O
Bat guano	10.0	4.0	2.0
Blood meal	12.0	2.0	1.0
Fish meal	10.0	6.0	—
Cotton seed meal	6.0	3.0	1.5
Soybean meal	7.0	1.2	1.5
Bone meal, raw	3.0	22.0	—
Bone meal, steamed	1.0	15.0	—
Wood ashes	—	1.0	4.0

Figure 13-4. Organic fertilizers do not give a rapid response but release nutrients through the decay of organic matter over a longer time period.

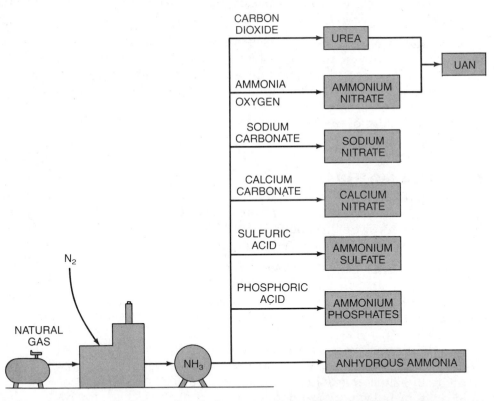

Figure 13-5. The Haber process is the source of most chemical nitrogen fertilizers.

- *Anhydrous ammonia,* which is 82% nitrogen, results directly from the Haber process. The word "anhydrous" means "without water." It is the main pressurized liquid and is applied as shown in figure 13-2. Anhydrous ammonia is the cheapest, strongest form of nitrogen. For that reason, its use is rising. If it is not injected deep enough, especially in sandy soil, it may evaporate and be lost from the soil. Ammonia destroys lung tissue if inhaled, so care should be taken during transport and application to ensure that none escapes.

- *Aqua ammonia*, consisting of 20% nitrogen, comes from dissolving ammonia in water to form a low-pressure solution. The use of aqua ammonia has been declining in the last 20 years as growers switch to other fluid fertilizers.
- *Ammonium nitrate*, containing 33% nitrogen, is half ammonium nitrogen and half nitrate. This is a good all-round dry material that is easy to handle and apply. It absorbs moisture from the air, causing it to cake. To prevent ammonium nitrate from hardening, it should not be left in open bags or piles. The use of ammonium nitrate began to decline in the early 1970s because of the increasing use of urea.
- *Ammonium sulfate*, which is 21% nitrogen, contains sulfur as well as nitrogen. It is a dry fertilizer. Ammonium sulfate is very acid-forming and is ideal for acid-loving plants. It is not suitable as a starter fertilizer for plants that grow best at a neutral pH Any ammonium fertilizer, including ammonium nitrate or sulfate, can lose nitrogen by volatilization when spread on recently limed or calcareous soils. The nitrogen changes to ammonia gas, which escapes into the air. To prevent the loss of nitrogen, the material is best mixed into the soil.
- *Nitrate of soda* (sodium nitrate, 16% nitrogen) is commonly used on tobacco. Unlike most other nitrogen sources, it raises soil pH because of the sodium (chapter ten, reaction c). *Calcium nitrate* is a similar but less salty material. Both of these materials are dry fertilizers.
- *Urea*, containing 46% nitrogen, is a manufactured organic material. In the soil urea rapidly breaks down to ammonia. If urea is left on top of the soil, the ammonia escapes into the air. The urea must be mixed into the soil. Urea is rapidly becoming the most popular dry fertilizer, because it can be produced more cheaply than ammonium nitrate. It is also used to make fluid fertilizer.
- *Urea-ammonium nitrate* (UAN), is a *nitrogen solution*. UAN is made by mixing liquid urea and ammonium nitrate to make either a 28% N solution or a 32% N solution. The use of UAN is increasing rapidly.
- *Urea-formaldehyde (UF), IBDU,* and *SCU (sulfur-coated urea)* are slow-release materials. These are used primarily to fertilize turfgrasses and potted plants. They are too costly for general use.

Figure 13-6 summarizes the characteristics of these and other nitrogen forms.

Phosphorus Carriers. Phosphorus fertilizers are obtained from the mining of *rock phosphate* in Florida and other states (figure 13-7). Phosphate rock contains the mineral apatite, or calcium phosphate. The ground rock can be applied directly to the soil, but it is usually treated with acid to break down the apatite into simpler compounds (figure 13-8, 13-9). Figure 13-4 lists some organic sources of phosphorus and Figure 13-10 summarizes these carriers:

Nitrogen Carrier	% Nitrogen	% Ammonium	% Nitrate	Effect on pH
Anhydrous ammonia	82.0	82.0	0	Very Acid
Ammonium nitrate	33.0	16.7	16.6	Acid
Ammonium sulfate	21.0	21.0	0	Very Acid
Aqua ammonia	24.0	24.0	0	Acid
Nitrate of soda	16.0	0	16.0	Basic
Calcium nitrate	15.5	0	15.5	Basic
Urea	46.0	46.0*	0	Acid
UAN (nitrogen solution)	32.0	22	10.0	Acid
Urea formaldehyde†	37.0	37.0	0	Acid
Sulfur-coated urea†	39.0	39.0*	0	Acid
IBDU†	30.0	—	—	—

* Urea becomes ammonia in the soil.
† Slow-release fertilizers

Figure 13-6. Common carriers of nitrogen. Other sources are listed under potash and phosphate sources.

Figure 13-7. During the mining of phosphate rock in Florida, water is added to make a slurry, which flows through pipes to the processing plant shown in figure 13-8. (Courtesy of Potash and Phosphate Institute)

Figure 13-0. When the slurry arrives at a phosphate rock processing plant, it is treated to produce phosphate fertilizers, as shown in figure 13-9. (Courtesy of Potash and Phosphate Institute)

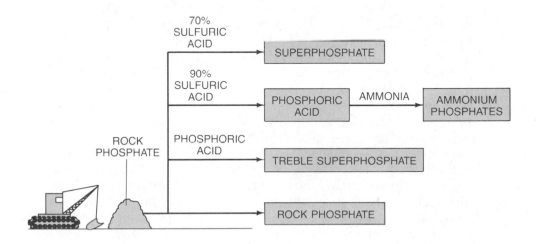

Figure 13-9. Phosphate rock is the beginning point in the production of phosphate fertilizers. The acid treatment breaks down the apatite rock into simpler compounds such as superphosphate and phosphoric acid.

Phosphate Carrier	Percentage Available		
	P$_2$O$_5$	P	N
Rock Phosphate	25–35	11–15	—
Superphosphate	20	8.7	—
Treble superphosphate	46	20	—
Monoammonium phosphate	48–55	21–24	11–13
Diammonium phosphate	46–53	20–23	18–21
Ammonium polyphosphate	36–62	16–27	10–15
Phosphoric acid	53	23	—

Figure 13-10. Common phosphate carriers

- *Rock phosphate*, containing between 25% and 35% phosphate, may be spread on the soil, but it will dissolve slowly. It is an example of a mineral fertilizer. Rock phosphate works best if it is finely ground and used on acid soils (pH below 6.0) or on soils high in organic matter. Acid soils break down the apatite.
- *Superphosphate*, containing 20% phosphate, results from the reaction of rock phosphate with sulfuric acid (figure 13-9). This material is half gypsum (calcium sulfate) and half calcium phosphate. Because it is lower in phosphate than other carriers, it is no longer used by most growers. Growers find it most useful when sulfur or calcium are lacking in the soil.
- *Treble superphosphate*, with 46% phosphate, is also treated rock phosphate. It is higher in phosphorus than regular superphosphate and contains no sulfur or calcium. Treble superphosphate is a popular fertilizer.
- *Phosphoric acid*, containing 53% phosphate, is a highly corrosive liquid in its pure form. Phosphoric acid is often used to prepare fluid fertilizers.
- *Ammonium phosphates* are made by mixing phosphoric acid with ammonia. Two similar compounds result from the reaction, monoammonium phosphate (48% phosphorus), and diammonium phosphate (53% phosphorus). These are often referred to as MAP and DAP. Both compounds are water soluble. They are used either as dry fertilizers or as fluids. *Ammonium polyphosphates* (35% to 62% phosphorus) are similar materials and are used to make fluid fertilizers. The phosphorus in all three is converted to orthophosphates in the soil, so they are basically the same in their effects. Interestingly, it has been found that the presence of ammonium ions improves the uptake of phosphate by plants, making these chemicals very effective phosphorus sources. Note that ammonium phosphates also supply nitrogen. Ammonium phosphates are rapidly becoming the most popular phosphate fertilizers and are widely used in bulk blends.
- *Bone meal* and *manure* are both organic sources of phosphate. Bone meal is made by grinding bones that are a byproduct of the meat-packing industry. Homeowners use bone meal as a phosphate and calcium fertilizer.

Potassium Carriers.   Potash mines in the states of New Mexico, Utah, California, and in Canada produce most of the potash used by American growers (figure 13-11). The deposits are mixtures of several potassium, sodium, and magnesium salts. Producers treat the salts to separate and purify the potash (figure 13-12). Figure 13-4 lists some organic potash sources, and figure 13-13 summarizes the primary carriers.

- *Muriate of potash* (potassium chloride) is 60% potash. It accounts for 97% of all potassium fertilization. It costs less than other carriers and dissolves readily in water. Potassium chloride is sold primarily as a dry fertilizer. It can also be used to make a weak solution (8% potash) and a stronger suspension (25% potash). Because it contains chlorine, other sources may be better for crops that are sensitive to chlorine.
- *Sulfate of potash* (potassium sulfate) contains 49% potash and is used to a minor extent in dry fertilizers. It can be used to make a suspension. It is preferred for tobacco because of the sensitivity of this crop to chlorine. It also adds sulfur to the soil.
- *Nitrate of potash* (potassium nitrate) contains 13% nitrogen and 44% potash. It is a common fertilizer for container plants. Although it is mostly used in a dry form, it can also make a weak solution.
- *Sulfate of potash-magnesia* is also useful for chlorine-sensitive crops. It consists of 22% potash, 11% magnesium, and 22% sulfur. It may be used for soils that lack sufficient amounts of magnesium and sulfur.

Figure 13-11. This continuous mining machine digs potash deposits from the ground. The potash is then sent to a processing plant. (Courtesy of Potash and Phosphate Institute)

(A)   (B)

Figure 13-12. *(A)* Potash ore is processed in a plant such as this. *(B)* In the process, KCl is separated by flotation from other salts. (Courtesy of Potash and Phosphate Institute)

| | Percentage Available | |
Potash Carrier	K_2O	Nitrogen
Muriate of potash	60	—
Sulfate of potash	49	—
Potassium nitrate	44	13
Sulfate potash magnesia	22	—

Figure 13-13. Common potash carriers

- *Wood ashes* and manure are also good potash sources.
- *Granite meal* is a mineral fertilizer used by some growers who prefer not to use chemical fertilizers. This material is a finely ground, gritty waste product of the monument and building stone industry. Granite dust is too insoluble to be of any immediate use to plants but may be thought of as adding to the "bank" of soil potash. Figure 2-4 lists nutrients in a typical granite.

Secondary Elements. Mineral fertilizers supply most of the calcium, magnesium, and sulfur required by plants. The most important fertilizers include lime, gypsum, and sulfur. The finer the grind, the more quickly these materials act. Many of the minerals affect soil pH. Figure 13-14 lists several sources of secondary elements.

Trace Elements. Each trace element is available in a number of chemical forms. Most trace elements, however, are commonly used in the following forms, which are summarized in figure 13-15:

- *Sulfate salts* are inexpensive and dissolve easily in water. They can be used as dry or fluid fertilizers. Trace elements are also available in other types of salts and oxides.

Material	Percentage Available Ca	Mg	S	Effect on pH
Calcitic lime	31.7	—	—	Basic
Dolomitic lime	21.5	11.4	—	Basic
Gypsum	22.5	—	12.0	Neutral
Hydrated lime	46.1	—	—	Basic
Burned lime	60.3	—	—	Basic
Magnesia	—	55.0	—	Basic
Magnesium sulfate	—	11.0	14.5	Neutral
Potassium magnesium sulfate	—	11.0	22.0	Neutral
Flowers of sulfur	—	—	30–100	Acidic

Figure 13-14. Common sources of secondary elements

Trace Element	FTE*	Sulfate Salts	Chelates	Others	Treatment
Boron	X			Borax	BC* borax
Copper	X	X	X	Oxide	BC or B* sulfate
Iron	X	X	X		F* chelate, acidify soil
Manganese	X	X	X	Oxide	BC or B sulfate
Zinc	X	X	X		B chelate
Molybenum	X			Sodium molybate, molybdic acid	Mix with NPK, liming soil

*Key: FTE = Fritted trace elements   BC = Broadcast   B = Banding   F = Foliar feeding

Figure 13-15. Sources of trace elements. The most commonly used forms of each nutrient are marked "X." The treatment is the most common effective type.

- *Fritted trace elements* (FTE) are a safe way to apply trace elements. FTE are made by adding salts to molten glass, which is poured into cold running water. The glass cools and shatters; the pieces are then ground into fine powder. Frits are dry fertilizers that dissolve slowly in the soil.
- *Chelates* (pronounced key-lates) are a special form of trace element fertilizer that guards elements from being fixed in the soil. An atom of the nutrient is surrounded by a complex organic molecule that keeps it from reacting in the soil until it is used by plants (figure 11-17). Chelates are water soluble.

MIXED FERTILIZERS

Growers can apply a fertilizer that contains a single nutrient. To do so, however, would mean a fertilization operation for each nutrient. It is often more convenient to use fertilizers that contain several nutrients. Such fertilizers are made by mixing several of the carriers just described. How does a grower determine how much of each nutrient is contained in a mixed fertilizer?

Fertilizer Analysis and Grade. The contents of a bag of fertilizer may be listed in two ways. Some bags list the *fertilizer analysis,* which simply lists the fertilizer elements in the bag and their percent content (figure 13-16). Such a list could include any of the 13 mineral elements. All bags of fertilizer should show the *fertilizer grade,* which indicates the primary nutrient content of the fertilizer. Grade lists the content as a sequence of three numbers that tell, in order, the percentage of nitrogen (N), phosphate (P_2O_5), which is also called phosphoric acid, and potash (K_2O). The grade is often referred to as "N-P-K," which stands for nitrogen, phosphorus, and potassium, in that order.

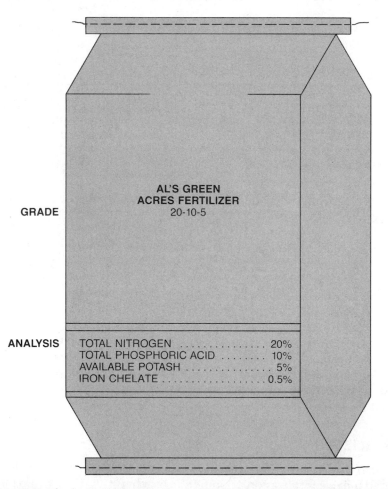

Figure 13-16. The fertilizer *grade* is expressed by three numbers. In order, the numbers indicate the percentages of nitrogen, phosphoric acid, and potash. Since all three primary nutrients are available in this fertilizer, it is a complete fertilizer. Also shown is the *analysis,* a listing of all the nutrients and the amount of each contained in the fertilizer.

For example, a fertilizer with the grade 0-0-60 is 60% potash with no nitrogen or phosphate. To decide how much potash is in a bag or ton of the fertilizer, simply multiply the weight times the percentage. Thus, one ton of muriate of potash contains the following amount of potash:

$$\text{Potash} = \frac{2{,}000 \text{ pounds} \times 60\%}{100} = 1{,}200 \text{ pounds}$$

A fertilizer containing only one element is called a *single-grade* fertilizer. Many fertilizers contain two or three nutrients and are called *mixed fertilizers*. *Complete fertilizers* have all three of the primary elements. Note that "complete fertilizer" does not mean that all 13 mineral nutrients are included.

To determine the amount of each nutrient in a complete fertilizer, the percentage of the nutrient is multiplied by the weight of the fertilizer. For example, in a 50-pound bag of 20-10-10:

$$\text{Nitrogen} = \frac{50 \text{ pounds} \times 20\%}{100} = 10 \text{ pounds}$$

$$\text{Phosphate} = \frac{50 \text{ pounds} \times 10\%}{100} = 5 \text{ pounds}$$

$$\text{Potash} = \frac{50 \text{ pounds} \times 10\%}{100} = 5 \text{ pounds}$$

Additional information may also be found in the analysis, like the percent nitrogen which is ammoniacal and the percent which is nitrate. Some fertilizers, especially those blended for turf, may contain nitrogen sources that dissolve slowly. These will be identified as water-insoluble nitrogen (WIN) or slow-release nitrogen (SRN).

Increasingly, grade may also identify a secondary nutrient as a fourth number in the traditional NPK. For example, calcium nitrate may carry the grade 15-0-0-39Ca, meaning the material is 39% calcium. Similarly, one may find sulfur (S) or magnesium (Mg) as a fourth number.

Contents of Fertilizers.   Fertilizer grades never total 100%. For example, a 10-10-10 fertilizer is 30% nutrient and 70% other ingredients. What are those other ingredients? Primarily, the remainder of the fertilizer is the weight of the other elements that are part of the carrier, such as hydrogen and oxygen.

A small percentage of fertilizer is *filler* and *conditioner*. Fillers may be sand, clay granules, ground limestone, or ground corn cobs. They are used to bring a load of bulk fertilizer to a weight of one ton. Conditioners improve the quality of the fertilizer and make it easier to use.

Fertilizer Ratio. *Fertilizer ratio* is another useful term. It states the relative amounts of nitrogen, phosphate, and potash in fertilizers. Ratios are useful when comparing two fertilizers, as shown in the following examples:

	Grade	Ratio
(a)	10-10-10	1-1-1
(b)	20-20-20	1-1-1
(c)	6-12-12	1-2-2
(d)	5-15-30	1-3-6

Note that *(a)* and *(b)* have the same ratio. This means that one fertilizer can be used in place of the other. Applying one ton of 10-10-10 is the same as applying 1/2 ton of 20-20-20.

Being able to obtain fertilizers of different ratios is very useful. The grower simply selects a fertilizer with the ratio suggested by soil test reports. For instance, if the test report suggested 100 pounds of nitrogen, 50 pounds of phosphate, and 75 pounds of potash per acre, a single fertilizer with the ratio of 4-2-3 would be ideal.

Elements and Oxides. The way fertilizer grade is listed leads to some confusion. Most people think of fertilizer grade as "NPK." This is read as nitrogen, phosphorus, and potassium. Actually, nitrogen is listed as the element, but the other two nutrients are listed in their oxide forms. The true grade should be listed as $N-P_2O_5-K_2O$, which is read as nitrogen, phosphoric acid, and potash.

As an example of the confusion, consider the fertilizer 20-10-10. The numbers lead one to expect 200 pounds of phosphorus in a ton of this fertilizer. Actually, one ton contains only 88 pounds of phosphorus. The amounts of nutrients in a ton of 20-10-10 can be listed in the elemental and oxide forms:

	Oxide	Element
N	400	400
P	200	88
K	200	166

When reading soil test reports or other recommendations, always check to see which form is being used. To convert between the amounts of phosphorus/phosphoric acid and potassium/potash, the following formulas are used:

(a) $P \times 2.29 = P_2O_5$
(b) $P_2O_5 \times 0.44 = P$
(c) $K \times 1.2 = K_2O$
(d) $K_2O \times 0.83 = K$

As an example of the use of one of the formulas, determine how much actual potassium is contained in one ton of 0-0-60:

$$K = \frac{2,000 \times 60\%}{100} \times 0.83 = 996 \text{ lb}$$

In recent years, there has been an effort to change the way in which fertilizer grade is written. If the nutrients were listed in the elemental form (N-P-K), much confusion would be avoided.

MIXING FERTILIZERS

Calculations for Blending.   Growers may buy a premixed fertilizer, but a limited number of ratios are available. Fertilizer can be custom blended to mix carriers to obtain the analysis and ratio that best suits the needs of the grower. To blend fertilizers, it is necessary to determine how much of each carrier is needed to produce the final mixed fertilizer. The following formula can be applied to each of the carriers:

$$Z = \frac{A \times B}{C}$$

where

Z = pounds of carrier for each element
A = pounds of mixed fertilizer needed
B = percentage of the element needed
C = percentage of the element in the carrier

As an example, let's determine how a ton of 10-10-20 can be blended from the following carriers:

Ammonium nitrate	33-0-0
Treble superphosphate	0-46-0
Muriate of potash	0-0-60

Calculations: $Z = \dfrac{A \times B}{C}$

$$\text{Ammonium nitrate} = \frac{2,000 \times 10}{33} = 606 \text{ lb}$$

$$\text{Treble superphosphate} = \frac{2{,}000 \times 10}{46} = 434 \text{ lb}$$

$$\text{Muriate of potash} = \frac{2{,}000 \times 20}{60} = 666 \text{ lb}$$

$$\text{Total carriers} = 1{,}706 \text{ lb}$$

A total of 1,706 pounds of carrier will be blended. To bring the total to 2,000 pounds, 294 pounds of filler will be added to the mix.

SELECTING FERTILIZER

Growers can choose from an array of fertilizers. Factors influencing the selection include the crop to be fed, the time of year, the application method, and the cost.

For most crops, the form of the fertilizer is not critical. One choice is between nitrate and ammonium nitrogen. Plants absorb both ions, but the preference is for the nitrate form. However, under warm, moist conditions, ammonium ions will nitrify to nitrate nitrogen in four to six weeks. For that reason, ammonium and nitrate usually have the same effect on crop growth. On the other hand, nitrates are lost more easily from the soil.

In a few cases, either ammonia or nitrates work best. There are a few simple rules to guide the grower in selecting the best form:

- Nitrates are preferred for early spring planting of cool-season crops.
- A better tobacco leaf results when nitrates are used.
- Ammonia is better for fall fertilization, because less nitrogen will leach out of the soil before spring.
- Ammonia is better for paddy rice. In a flooded paddy, nitrate nitrogen quickly converts to nitrogen gas and is lost.
- Fertilizers for container plants should, while containing both, favor nitrates over ammonium nitrogen. Roots growing in a pot are easily damaged by excess ammonia.

Some crops, including tobacco, are sensitive to chlorine. For these crops, low chlorine fertilizers should be chosen. In some applications, growers need to be concerned about a fertilizer's affect on soil pH or salinity—to be discussed later in this chapter.

Growers also base their fertilizer selections on the means to be used to apply the fertilizer. For example, fertilizers applied through the irrigation system must be water soluble. Several other recommendations will be noted later in the chapter.

The selection of fertilizers commonly depends upon the price—the least costly fertilizer per pound of plant food is the one commonly selected. The cost can be calculated as follows, using nitrogen as an example:

$$\text{Price/lb N} = \frac{\text{Price per ton}}{2{,}000 \times \%N} \times 100$$

For instance, the price of nitrogen in a ton of ammonium nitrate (33-0-0) that costs $200 would be:

$$\text{Price/lb N} = \frac{200}{2{,}000 \times 33} \times 100 = \$.33 \text{ per lb nitrogen}$$

The same calculation can be made for a single bag of fertilizer. Simply substitute the weight of the bag and the price per bag. Similarly, the cost of potash and phosphate may be computed by substituting their values for nitrogen in the formula. These figures allow a grower to compare the cost of different fertilizer elements.

APPLYING FERTILIZER

Fertilizers can be applied before a crop is planted, while it is being planted, after it is growing, or in some combination of the three methods. The time of application has different effects on the crop.

Preplant feeding brings all of the soil in the field to a good nutrient level before a crop is planted. On a heavy soil with little leaching and a high cation exchange capacity, this one feeding may supply all the nutrients needed for the season. While this is the main fertilization for most crops, there are three drawbacks:

- Phosphate is not concentrated near the young seedling. Phosphates do not move much in the soil, and young seedlings are limited in their ability to forage for nutrients. It is often hard, therefore, for young roots and phosphate supplies to come in contact. The same is true, to a lesser extent, of potash.
- In coarse, low cation exchange capacity soils, much of the nitrogen applied in a preplant feeding will leach away before plants can use it.
- Applying the fertilizer before planting does not match the needs of the crop. Generally, the crop will need most fertilizer when growing rapidly later in the season.

Fertilizer applied while planting, called *starter fertilizer,* allows phosphate to be placed near the seed. Fertilization *after* planting solves the other two problems. This is usually done by dividing the year's fertilizer into two or more parts; one part is applied before planting and the rest is used later in the season in one or more applications. For example, corn may be fertilized before planting, then again 30 days after planting. This *split-application* can reduce the loss of nitrogen by leaching and allows a grower to apply fertilizer when the crop has the greatest need for it. Fertilization with irrigation allows the grower to make several applications to closely match growth needs. Figure 13-17 presents a model

Crop Stage	Fertilization
(1) Preplant	1/6
(2) 8 leaves	1/6
(3) 12–15 leaves	1/2
(4) Early tassel	1/6

Figure 13-17. A model nitrogen fertilization schedule for irrigated corn on sandy soils. A soil test can suggest how much total nitrogen should be applied. The total amount is divided into four applications at the fractions indicated. This method reduces leaching losses and supplies nitrogen according to crop needs.

schedule for *fertigating* (fertilizing with irrigation) corn. (Confer with extension agents for local recommendations.)

For perennial crops like hay or fruit, later fertilizations must follow a preplant application. A single preplant feeding will not meet the nutrient needs of the crop in later years. Thus, fertilizer is added yearly.

Now let's look at the methods used to apply fertilizers at these different times.

Fertilizing Before Planting

Broadcasting. The simplest way to fertilize before planting is *broadcasting.* Machinery, and sometimes aircraft, is used to spread dry fertilizers evenly on the soil surface (figure 13-18). Fluid fertilizers also can be sprayed on the soil. For phosphate and potash, the material should then be "plowed down," or mixed into the soil, before the crop is planted. This step is important because these nutrients do not leach very far into the soil, and, if left on the surface, will not reach the root zone. Nitrogen will leach into the soil, especially if irrigation is used to help it move. Broadcasting is quite popular because bulk blends can be applied rapidly.

Soil Injection. This is also known as *knifing* or *chiseling* and can also be used before crops are planted. Most commonly, anhydrous ammonia is chiseled into the whole field. This can be done with special ammonia applicators. Some tillage equipment can be adapted to apply ammonia. Fluid fertilizers can also be chiseled into the soil. Some growers use special equipment that injects both anhydrous ammonia and a solution like phosphoric acid at the same time. This method not only saves time, but the ammonia helps make phosphate more available to plants.

Fertilizing at Planting

Banding. The most common method of applying starters is called *banding.* A seed planter places a band of dry fertilizer two inches below and one inch to the side of the seeds (figures 13-19 and 20). Banding is the most efficient way to apply phosphate, and sometimes potash. The placement is not so close as to hurt the seed, but is close enough that young roots can quickly find

Figure 13-18. The dry fertilizer applicator broadcasts or spreads granules or prills over a field and then the fertilizer is plowed into the soil. (Courtesy of Potash and Phosphate Institute)

the band of fertilizer. Since the phosphate fertilizer is packed in a narrow band, there is less soil/fertilizer contact, thus reducing phosphate fixation. Fertilizers used for banding usually have an N:P ratio of 1:2 or 1:4—the nitrogen aids in the uptake of the phosphorus. Banding is an excellent way to apply phosphate, but it slows down the planting operation. Since planting often must be done quickly, many growers prefer not to band.

Banding is most called for when certain soil conditions restrict phosphate nutrient uptake, as follows:

- cold soil at planting time, common in northern states.
- wet or compacted soil, causing low-oxygen conditions.
- acid or calcareous soils that tend to fix phosphate
- low soil test value for phosphate.

Pop-up Fertilizers. Fertilizer in this form is placed in the row with the seeds, rather than beside the seed as in banding. These fertilizers are quite effective in cold soils. Generally, only small amounts are applied to prevent damage to the seed. Fertilizers for pop-up use should

- be complete fertilizers high in phosphate.
- be water soluble.

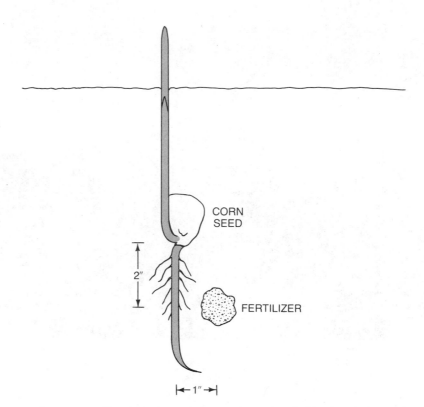

Figure 13-19. In the process of banding, the seed planter places a band of fertilizer below and to the side of the seeds. The fertilizer is in the path of young roots. This method lessens the soil/fertilizer contact that fixes phosphorus. Placement of the fertilizer immediately below the seeds may injure the emerging roots.

- have a low salt index (salt can injure the seedling).
- not produce any free ammonia, because ammonia will also injure the seedling; this excludes urea, UAN, and diammonium phosphate.
- be either fluid (figure 13-21) or dry fertilizers.

Fertilizing after Planting

Topdressing. Topdressing is the same as broadcasting, except that the fertilizer is spread over a growing crop and is not mixed into the soil. Either dry or fluid fertilizers can be used. Farmers often topdress perennial crops like hay to replace the lost nutrients. This method is also used to feed grains, range, and lawns (figure 13-22).

Sidedressing. Sidedressing is a way of making a second application of fertilizer part way through the growing season. Sidedressing is done by fertilizing along the crop row. Commonly, this is done by knifing ammonia into the soil. Sidedressing is the most popular way to make split applications.

Figure 13-20. Banding fertilizer (Courtesy of *Crops and Soils Magazine,* American Society of Agronomy)

Fertigation. A third way to fertilize a growing crop is to inject fertilizer into irrigation water (figure 13-23). Special pumps force very strong fertilizer solutions into the irrigation system. Obviously, this method will only be as efficient and uniform as the irrigation system. Fertigation works best in sprinkler or trickle irrigation, but it can be used with surface irrigation. The best fertilizers for fertigation are clear solutions, most commonly the nitrogen source UAN. Other nutrients are applied less often by irrigation. Ammonium phosphates and potassium nitrate can be used to supply phosphate and potash. Chelated trace elements also can be used. Ammonia and aqua ammonia are difficult to use because ammonia tends to evaporate.

Foliar Feeding. Growers sometimes fertilize by spraying solutions directly on the leaves of the crop. The nutrients are absorbed through the stomata (openings in leaves that allow gases to move in and out). The quickest response of any method of fertilization is obtained with foliar feeding and so this method may be used as a quick cure for a deficiency. The results are usually short-lived; thus, it may be necessary to repeat feedings several times. Usually it is used along with a complete program of soil fertility.

Figure 13-21. Starter fertilizer solutions are applied next to seeds. (Courtesy of National Fertilizer Solutions Association)

Plants use a lot of the major elements, so it is hard to supply enough by foliar feeding. Sprays strong enough to supply much nutrient value can both burn the leaves and damage application equipment.

The most practical use of foliar sprays is to solve trace element shortages. Plants need trace elements in small quantities, but problems often arise because trace elements are tied up in the soil. Spraying the leaves bypasses soil problems. The most common example is spraying iron chelates to relieve iron shortages.

Figure 13-24 compares the types of fertilizing methods with the types of fertilizers best suited to each method.

Effects of Fertilizer on Soil pH. Fertilizers may change pH. For instance, some acid-forming fertilizers shorten the time between lime applications. One may even pick fertilizers to help change pH in a desired way. For instance, acid-loving plants may be fertilized with acid fertilizers. Acidification by fertilizers in other crops may be counteracted by liming. The effects of fertilizers on pH are summarized as follows:

● Potassium fertilizers do not cause a lasting pH change.

Figure 13-22. A fluid fertilizer applicator is topdressing hay to replace nutrients removed during the harvest. (Courtesy of Potash and Phosphate Institute)

Figure 13-23. During fertigation, a fertilizer solution is pumped out of the tank and into the wheel-move irrigation system. (Courtesy of National Fertilizer Solutions Association)

Types of Fertilizers

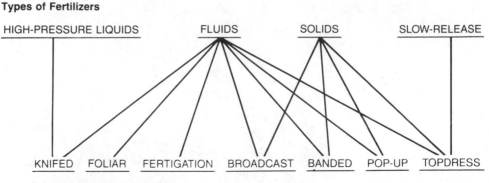

Methods of Applying Fertilizers

Figure 13-24. The types of fertilizers are matched to the application methods to which they are best suited. Fluid fertilizers are the most versatile form because they can be applied by any method.

- Superphosphates also do not cause a lasting effect on pH. Phosphoric acid is highly acidic and has been used to counter a rising pH. Ammonium phosphates are also acidic.
- Nitrogen fertilizers containing ammonia or urea cause soil acidity. Ammonium sulfate, which contains both sulfur and ammonia, is highly acidic.
- Calcium and sodium nitrates both have a basic reaction in the soil.

Effect of Fertilizers on Soluble Salts. Most fertilizers are salts. A high concentration of fertilizer salts prevents seed sprouting, slows water uptake by plants, and injures plants. This is why the banding method places fertilizers no closer than two inches from the seeds.

Misapplying or spilling fertilizer may injure a plant (figure 13-25). This causes death or browning of some leaves, a condition known as *fertilizer burn*. Homeowners who fill their fertilizer spreaders on the lawn risk patches of dead grass. Some water-soluble fertilizers are known as nonburning because they contain materials with very low salinity values.

Starter fertilizers place salts very close to sprouting seeds, which can be very sensitive to damage from fertilizers. These fertilizers, such as the pop-up type, should have low salinity.

Fertilizer salinity is of greatest concern to container plant growers. The constant high rate of feeding easily results in a buildup of fertilizer salts. For this reason, feeding with a product low in salts is best for greenhouse growers. Farmers with saline soils may also choose such fertilizers. Figure 13-26 gives the salt index and relative salinity of several common fertilizers.

Figure 13-25. Fertilizer burn resulted when this tree seedling was watered with too concentrated a fertilizer solution. Affected plants often turn brown.

Fertilizer	Salt Index	Relative Salinity	Analysis	Calcium Carbonate Equivalent/20 lb N
Sodium nitrate	100	100	16.5-0-0	+36
Ammonium nitrate	105	49	35-0-0	−36
Ammonium sulfate	69	54	21-0-0	−107
Anhydrous ammonia	47	9	82-0-0	−36
Urea	75	27	46-0-0	−36
Monoammonium phosphate	30	13	12-55-0	−107
Superphosphate	8	17	0-20-0	0
Treble superphosphate	10	9	0-46-0	0
Potassium nitrate	74	24	13-0-46	+36
Potassium chloride	116	39	0-0-60	0

Figure 13-26. Effects of fertilizers on soil salinity. The salt index compares different fertilizers with the effect of sodium nitrate. The relative salinity compares the effect per unit of nutrient. For example, sodium nitrate and ammonium nitrate have nearly the same index, but a grower needs to apply only half as much ammonium nitrate (35% N) as sodium nitrate (16% N). Thus, ammonium nitrate has half the saline effect. The table also shows acidity and alkalinity. The minus figures in the last column state how many pounds of calcium carbonate are needed to counteract the acidity of 20 pounds of nitrogen. The plus figures indicate an alkaline material equivalent to the number of pounds of calcium carbonate shown. (Adapted from A. C. Bunt, *Modern Potting Composts*, Pennsylvania State University Press, 1976 and Miller, *Fundamental Facts about Soils, Lime, and Fertilizer*, Maryland Fact Sheet #186, 1971)

SUMMARY

A fertilizer is a substance used to supply essential elements. Fertilizers may be finely ground minerals, organic materials, or chemicals made by industry.

The Haber process fixes nitrogen from the air to make ammonia. Ammonia, in turn, is the base for most other nitrogen carriers. Phosphate and potash result from mining. Factories treat rock phosphate to produce superphosphate and purify potash deposits to make potassium fertilizers.

Ground-up materials supply most of the secondary elements. They may also be obtained from other fertilizers. For instance, superphosphate also contains calcium and sulfur. A wide variety of compounds deliver trace elements, including sulfates, oxides, fritted trace elements, and chelates.

Fertilizers come in a number of physical forms that allow many methods of use. Dry blends come in large grains that can be scattered on the soil, as in broadcasting or topdressing. They can also be banded next to seeds or used as pop-up fertilizers.

Fluid fertilizers, applied as liquids, are becoming more important. They can be sprayed on the ground for broadcasting or topdressing, injected into the soil, added to irrigation water, or sprayed on plant leaves. The high-pressure liquid, anhydrous ammonia, is used to prepare a field for planting or to sidedress a row crop.

Slow-release fertilizers, unlike other forms, release nutrients slowly. They find their greatest use in turfgrasses and in growing potted plants.

Mixed fertilizers contain two or three primary elements. The fertilizer grade lists the percentage of each primary element in the form of nitrogen, phosphate, and potash. These numbers can be used to determine how much of each nutrient is contained in a fertilizer, how much they cost, and fertilizer ratios. Growers can also determine how much of each carrier must be blended to make a mixed fertilizer.

Growers have many fertilizers from which to choose. Obviously, the fertilizer should fit the needs of the crop and the method of use. Cost is often the most important factor.

REVIEW

1. Blood meal is an example of an organic source of nitrogen (True/False).
2. Fluid fertilizers are versatile.
3. Most fertilizers contain materials other than the pure nutrient elements.
4. Anhydrous ammonia is ammonia mixed into water.
5. Treble superphosphate is less widely used than superphosphate.
6. Complete fertilizers contain all thirteen mineral nutrients.
7. Banding places fertilizers near the planted seed.
8. Sidedressing is done before planting.
9. Foliar feeding is most effective for micronutrients.

10. Ammonium fertilizers produce an acid soil reaction.
11. Dry fertilizers with the best spreading and handling properties are called _____ .
12. A fertilizer with the analysis 15-15-20 is an example of a _____ fertilizer.
13. Blueberries grow best on acid soils, so _____ is a good nitrogen fertilizer.
14. There are _____ pounds of elemental phosphorus in a 50 pound bag of treble superphosphate.
15. There are _____ pounds of nitrogen in a ton of 31-0-0 IBDU.
16. The fertilizer ratio of an 18-6-12 is _____ .
17. If you wanted to increase calcium in the soil without changing pH, you could use _____ .
18. Fertilizer spread around a field before planting has been _____
19. According to figure 13-26, _____ is the most saline nitrogen fertilizer.
20. Damage to plants by excess fertilizer, causing salt damage, is called _____ .

21. Figure the cost per pound of nitrogen for each of the following:
    a. ammonium nitrate at $150 per ton.
    b. anhydrous ammonia at $210 per ton.
    c. urea at $5.00 per fifty pound bag.
22. How much of each of the following must be mixed to make one ton of a 5-10-20: urea, superphosphate, muriate of potash, and filler?
23. Name an organic, mineral, and inorganic source of phosphorus.
24. Identify whether each of the following has a acid (A) or basic (B) effect on the soil:
    a. sodium nitrate
    b. ammonium nitrate
    c. phosphoric acid
    d. dolomitic lime
25. Suggest three fertilizers you could mix to make a complete fertilizer for fertigation.
26. What situations call for banding phosphorus?
27. Explain differences between "organic" and "inorganic" fertilizers.
28. What is in a bag of fertilizer?
29. What are several ways to solve micronutrient problems? (Consider also chapter ten.)
30. What value is there in split applications?

SUGGESTED ACTIVITIES

1. Make test plots on your school grounds. Use them to test various fertilizers.
2. Use a soil test of your school grounds and local recommendations to develop a fertilizer program for an important crop in your area.

chapter fourteen

Organic Amendments

This pasture scene on a Wisconsin farm demonstrates a destination of manure the author, an avid trout fisherman of streams such as this one, hates to see. This chapter will discuss proper ways to use manure, sewage sludge, and compost. It will also discuss environmental problems associated with these materials and with other fertilizers.

(Courtesy of USDA-SCS)

OBJECTIVES

After completing this chapter, you should be able to:

- explain the benefits of organic amendments
- describe how to use animal manure
- describe how to use sewage sludge
- explain large-scale composting
- list environmental side-effects of fertilizers and amendments

TERMS TO KNOW

Biological oxygen demand	eutrophication organic amendment	sewage sludge

An *organic amendment,* as defined here, contains both plant nutrients and large amounts of organic matter. Organic amendments are used to both fertilize and amend the soil. In this chapter, we stress their use as fertilizers.

Organic amendments benefit the grower in a number of ways that inorganic fertilizers may not.

- Organic amendments contain a combination of nutrients, including secondary and trace elements. Part of these nutrients are readily used by plants and part will be used by plants as the nutrients are slowly released by decay.
- The organic matter acts as the main soil storehouse of many nutrients, including nitrates, phosphates, sulfates, and others.

- Organic amendments contain large amounts of organic matter to improve the physical condition of the soil and increase its cation exchange capacity.
- Organic amendments support the growth of living organisms in the soil. The organic matter feeds earthworms, mycorrhizae, nitrogen-fixing bacteria, and other useful soil organisms.
- Organic amendments often produce a greater yield than a complete fertilizer applied at the same NPK rate, especially on sandy soils. Further, improved yields may continue years after organic amendments are applied, unlike the shorter-term benefits of inorganic fertilizers.

Society also benefits from a grower's use of manure and sewage sludge as fertilizers. These materials are potentially harmful waste products that can be most safely disposed of by spreading them on the land. A good example is sewage sludge, which can be dumped in a river or ocean, buried in a landfill, burned with the ashes buried, or spread on land. Of these options, land-spreading is safest. Soil dilutes the sludge and filters out harmful chemicals. Soil microbes break down harmful compounds, and human disease organisms die in the soil. In fact, much of the current interest in organic amendments stems from clean-water laws passed in the early 1970s.

Of the many organic materials available, three account for the greatest use: animal manures, sewage sludge, and compost.

ANIMAL MANURE

It is ironic that for many farms today manure has become a waste disposal problem. This is an abrupt change for throughout history people have long relied on animals as a source of soil nutrients. However, many farms do still use this resource. For these farms, manure has several benefits.

- Manure is a fertilizer with good amounts of nitrogen and potash. Phosphorus and calcium are present as are lesser amounts of sulfur and magnesium. Most manures also have traces of several micronutrients. Figure 14-1 provides examples of the nutrient content of several manures.
- Manure adds organic matter to the soil. Organic solids make up 20% to 40% of manure. This matter decays readily because of its high nitrogen content. Nitrogen tie-up occurs only if the manure includes a lot of straw or wood shavings used as animal bedding.
- Manure itself is low in phosphorus, but it helps prevent phosphorus from being tied up in the soil, thus making it more available for plants.
- Manure has longer-lasting effects than an equivalent amount of chemical fertilizer. Improved yields may continue years after manure stops being added to the soil.

Content of Manure. Manure includes both solids and liquids, which, for the most part, are the feces and urine of the animal. The solid part may also include bedding. As figure 14-2 shows, the solid part of the manure contains most of the phosphate. Most of the potash is in the liquid part.

Urine holds about half the nitrogen in manure, primarily in the form of urea and similar compounds. The rest of the nitrogen is contained in the animal feces. Half of this part has already turned to humus in the stomach of the animal. The humus portion will release its nitrogen slowly during further decay.

Several factors determine the amount of nutrient in manure, including the type of animal. In general, sheep and poultry manure has a high nitrogen content; the manure of cattle, pigs, and horses has a lower nitrogen content. The amount and type of bedding also influences nutrient content since it thins out the manure. If manure contains a large amount of high C:N ratio bedding, nitrogen tie-up can even occur in the soil for a time. The amount and type of rations and the age and health of the animal are also important factors.

Figure 14-1 lists average values of nutrient content for several animal manures. To change these values to the standard percentages used in commercial fertilizers, divide by 20. This operation changes pounds per ton to percent. For example, poultry manure contains 25, 11, and 10 pounds per ton of nitrogen,

Animal	Pounds/Ton					
	N	P_2O_5	K_2O	S	Ca	Mg
Dairy cattle	10	4	8	1	6	2
Beef cattle	11	8	10	1	3	2
Poultry	23	11	10	3	36	6
Swine	10	3	8	3	11	2
Sheep	28	4	20	2	11	4
Horse	13	5	13	—	—	—

Figure 14-1. Nutrient content of several animal manures, in pounds of nutrients per ton of manure.

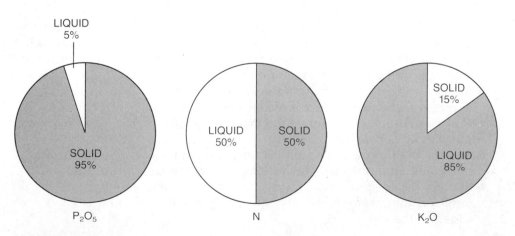

Figure 14-2. Most of the potash in manure is contained in the urine, and the phosphate is contained primarily in the feces. Nitrogen is distributed equally between the two parts.

phosphate, and potash respectively. Dividing these values by 20, its NPK becomes 1.2-0.5-0.5. This is much weaker than commercial fertilizer, mainly because manure is largely water and organic carbon. Manure must be applied in quantities of tons per acre rather than pounds per acre.

Part of the secret to using manure is to keep its nutrient value intact and prevent large losses.

Nutrient Losses from Manure.   Urine contains about 50% of the nutrient value of manure. If this part of the manure is lost, most of the potassium and much of the nitrogen will also be lost. Urine is lost when it seeps into the ground through barn floors or in feedlots. A great deal of urine simply drains away from manure heaps.

Sharp nitrogen losses occur if the manure begins to decay before it is spread. As much as 90% of the nitrogen can be lost within three weeks if manure is poorly handled. The losses occur when urea changes to ammonia gas during decay (see the discussion of volatilization in chapter eleven). The loss is most rapid when it is warm and the concentration of urea is highest. Water in the manure dilutes the urea and slows the change. The following storage conditions promote nitrogen loss:

- High air temperatures speed decay, with resulting nitrogen loss.
- Heat in a manure pile during decay also speeds up losses.
- As manure dries out, urea becomes more concentrated. Therefore, as manure dries out during storage, ammonia enters the air rapidly.
- Freezing also speeds up losses. As the water in the pile begins to freeze, urea is concentrated in the remaining unfrozen water. The higher concentration speeds up losses.

Nutrients can be lost even after manure is spread in the field. Ammonia continues to escape unless the manure is mixed quickly into the soil. Runoff and leaching increase the loss. Spreading manure on frozen, sloping land increases the chances of manure being lost to runoff.

Decay organisms respiring in a manure pile change organic carbon to carbon dioxide gas. Organic matter decay explains why manure piles shrink over time. Since the organic matter would be a desirable addition to the soil, it can be considered a loss as well.

Figure 14-3 summarizes the ways in which nutrients are lost from manure.

Handling Manure.   The best way to handle manure is to spread it immediately on unfrozen ground and then plow it into the soil. In this way the grower can prevent the loss of ammonia gas that occurs during storage and in the field. However, in some regions, this technique is not practical in every season.

If manure cannot be mixed into the soil immediately, it should be stored properly and then applied when it can be plowed into the soil. The actual loss of nitrogen varies with handling and storage systems. As noted in figure 14-4, piles in an open lot, exposed to sun, rain, and air movement, will lose

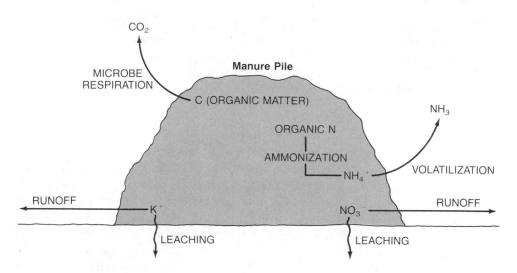

Figure 14-3. Much of the nitrogen and potash can be easily lost during manure storage.

Method	Nitrogen Loss
Solid Systems	percent
Daily scrape & haul	15-35
Manure pack	20-40
Open lot	40-60
Deep pit (poultry)	15-35
Liquid Systems	
Anaerobic deep pit	15-30
Aboveground storage	15-30
Earthen storage pit	20-40
Lagoon	70-80

Figure 14-4. Nitrogen losses from animal manures as affected by method of handling and storage. (From *Utilization of Animal Manure as Fertilizer;* Sutton, Nelson, and Jones; Purdue University)

about half their nitrogen. Long-term storage in lagoons is even worse. Better is short-term storage of solid or liquid manures in proper storage structures. Good storage facilities have concrete floors and walls and a roof to stop drainage losses and slow down the drying of the manure.

Liquid manure handling systems are the best way of saving nutrients. In these systems, growers store liquid animal wastes in concrete pits (figure 14-5) or in ponds. The bottoms of the pits are sealed to prevent percolation into the soil. The manure is about 90% or more liquid and can be handled by pumps. The liquid manure can be spread on fields by machinery (figure 14-6) or even by gun irrigation (figure 14-7).

Figure 14-5. A livestock waste-holding tank under construction. (USDA, Soil Conservation Service)

Freshly spread liquid manure should be plowed into the soil immediately. In warm weather, 20% of the nitrogen volatilizes within six hours. The best system uses a tank that knifes the liquid into the ground. Knifing stops the loss of gaseous ammonia and reduces odors (figure 14-6). However, salts and ammonia released by injected manure can damage germinating seeds planted near injected zones. Injection just before a tillage operation that will mix the soil avoids the problem.

Figure 14-8 summarizes nitrogen losses from various methods of application.

Application Rates. Manure should be spread at rates that follow local recommendations. In general, manure should be spread in a thin layer over as much land as possible, rather than piling it thickly over a small area. In this way, its benefits go farther while the problems of overapplication are avoided.

Unlike commercial fertilizer, the nutrients in manure are not immediately available to plants. About half of the nitrogen usually becomes available during the first season. In early spring, when soils are still cool, mineralization is especially slow, so a starter fertilizer may be a useful supplement to manure. Like any ammonia fertilizer, manures depress pH. Approximately 1 ton of lime will counteract the acidity in 10 tons of manure.

Figure 14-6. Injecting liquid manure into a Nebraska cornfield. Injection stops the loss of ammonia gas, reduces odors, and makes pollution from runoff less likely. (Courtesy of Ag-Chem Equipment Co., Inc.)

Manure is fairly salty, especially if common salt (sodium chloride) is fed to animals to increase appetite and reduce kidney stones. Therefore, the addition of too much manure on a field can hinder plant growth. Excess manure can also cause nutrient pollution and high nitrate forages. Both of these problems are discussed later in the chapter.

Animal manure is most familiar to growers, but "human manure," or sewage sludge, is also used by many.

SEWAGE SLUDGE

The spreading of human waste on soil has a long history. For centuries oriental societies added "night soil" to fields. Between the 1500s and 1800s, sewage was spread on fields in part of Germany. In the United States, sewage began to be put on land in the early 1970s in response to federal clean air and water laws. The latest figures suggest that about 40% of the sewage sludge produced in the United States is now spread on land.

Spreading wastes on soil attracts interest for two reasons: (1) it is probably the safest way to dispose of sewage wastes, and (2) it is a useful fertilizer for farmers. In a sense, this technique closes a nutrient loop—farm fertilizers are absorbed by

Figure 14-7. Gun irrigation is used to spray liquid manure or sludge on a field. (USDA, Soil Conservation Service)

Method	Types of Manure	Nitrogen Loss
Broadcast without incorporation	solid	15-30%
	liquid	10-25%
Broadcast with incorporation	solid	1-5%
	liquid	1-5%
Injection	liquid	0-2%
Irrigation	liquid	30-40%

Figure 14-8. Nitrogen losses from manure to the air as affected by method of application. (From *Utilization of Animal Manure as Fertilizer*, Sutton, Nelson, and Jones; Purdue University)

crops, the crops are eaten by people in towns and cities, and some of the nutrients are flushed into the sewage system. If the sewage is then spread on farm fields, the nutrients return to the land and the nutrient loop is closed. There are 18,000 sewage treatment plants in the United States producing more than 5.5

million tons of sludge a year. There are probably tens of thousands of farms that are close enough to treatment plants to use some of the sludge.

Sewage Treatment. Sewage is treated to separate chemicals and solids from the water in sewage. The purified water can be discharged into rivers or used to irrigate land. The remaining solids or sludge must also be disposed of in some safe manner. Sludge can be very watery (as much as 90% water) or very solid (with as little as 10% water).

A number of treatment methods are used to process sewage. All of these methods involve several steps. In the first steps, solids and chemicals are removed from the liquids. The resulting raw sludge harbors many human disease organisms (pathogens) and cannot be used safely. The raw sludge is then treated by chemicals such as lime or digested to reduce odors and the number of pathogens. This fully treated sludge can be used on farmland.

Nutrient Content. Sludge contents vary greatly from place to place and even from day to day in the same treatment plant. The variation results from the different treatment methods used and the kinds of industries and homes that feed into a sewage system. Figure 14-9 shows some *average* figures from samples taken in several states.

Note that sludge contains a lot of carbon, which becomes humus in the soil. Sludge also has good amounts of nitrogen, phosphorus, calcium, sulfur, iron, and zinc. Sludge is short of potash. In most cases, if sludge is applied at the rate that supplies all the nitrogen needs of a crop, more than enough phosphate will be added but not nearly enough potash.

Sludge Problems. Sludge is a valuable amendment but it has a number of problems that are avoided by following local, state, and federal Environmental Protection Agency (EPA) rules. The problems encountered with sludge are heavy metals, human pathogens, organic poisons, nutrient pollution, soluble salts, and odor.

Heavy Metals. Many elements known as heavy metals can be toxic to plants or animals. The best known example is lead poisoning in humans. Sludge may contain fairly large amounts of the metals, starred in figure 14-9, if any industries feed into the sewage system. The two critical elements are cadmium and zinc, with cadmium being the most critical. These metals may injure plant growth or enter the food chain, ending up in the human diet. When absorbed by plants, they tend to gather in *leafy* plant parts, rather than in fruits or grains.

Human Pathogens. Even fully treated sludge contains some bacteria, viruses, and stomach parasites. Most of these die shortly in the soil. The presence of these pathogens gives rise to four possible dangers:

Components	Percentage	Pounds/Ton
Organic carbon	30.4	608
Inorganic carbon	1.4	28
Total nitrogen	3.3	66
Ammonium N	1.0	20
Nitrate N	—	—
Organic N	2.3	46
Total phosphorus	2.3	46
Inorganic P	1.6	32
Organic P	0.7	14
Total sulfur	1.1	22
Calcium	3.9	78
Iron	1.1	22
Magnesium	0.4	8
Potassium	0.3	6
Zinc*	0.17	3.5
Copper*		1.7
Lead*		1.0
Manganese		0.52
Nickel*		0.16
Boron		0.07
Molybdenum		0.06
Cadmium		0.03

*Heavy metals which may be a health concern

Figure 14-9. An average sludge contains the elements listed. It should be recognized that the values for a specific sludge may vary widely from these averages. For example, of the 200 samples that make up this average, the calcium content ranged from 0.10% to 25.0%. (Adapted from Sommers, L. E. "Survey of Sludge Composition," #NC-118. Cooperative State Research Service, 1976, USDA, Cincinnati, Ohio)

- The organisms may contaminate food grown on sludge-treated fields. In turn, the food may infect people who ingest it.
- The organisms may infect animals grazing on treated land or eating hay from treated land.
- The organisms may be carried into surface waters by runoff, or they may leach into groundwater. Either situation could endanger water supplies.

Despite these potential dangers, there is no recorded case to date of a disease outbreak in humans resulting from land-spread sludge.

Organic Poisons. Some sludges that contain industrial wastes may include dangerous organic compounds like PCB (polychlorinated biphenyls). It may not be possible to use such sludges on land.

Nutrient Pollution. Nitrogen and phosphorus may wash into surface water or leach into groundwater. (This is also a problem with manure and fertilizer.)

Soluble Salts.   Sewage sludge is fairly saline. In areas of adequate rainfall, any addition to soil salts is short-lived. In areas where soil salinity is already a problem, sludge may add to it.

Odor.   Fully treated sludge has a much less disagreeable odor when compared with raw sewage. There is some disagreeable odor, however, and this must be considered when using sludge.

Application Guidelines.   States control sludge use to protect the public health. If state and federal guidelines are followed, there should be no danger in using sludge as a soil amendment. It is the responsibility of treatment plant personnel primarily to ensure that these guidelines are met. Local laws usually consider the sludge itself, the disposal site, the crops to be grown, and application rates.

Sludge suitable for spreading on land should not have too high a level of heavy metals, especially cadmium. It should have a low level of organic poisons like PCBs. The federal Environmental Protection Agency and state agencies set the limits for metals and PCBs. Only fully treated sludge may be used—partially treated sludge has too many pathogens. The sludge must be analyzed for its nutrient and metal content to ensure that the right amount is spread on land.

Most farms are able to accept sludge, but there are restrictions. To prevent runoff, slopes should be less than 6% (9% in some states). The soil should be well-drained and preferably of medium texture. The soil should have a root zone of at least 20 inches. The water table should not come closer than 3 feet from the surface of the soil. The sludge should not be applied near sensitive areas, like homes, lakes, or streams. Most of all, the pH should be between 6.5 and 8.2, or the soil should be limed to that reading. The metals are mostly tied up at this pH level and so are less likely to be taken up by plants.

The best land for sludge spreading—from a health point of view—is not used to grow edible crops. Examples are nurseries or sod farms. Sludge is also very effective for reclaiming degraded land like old mines. However, since metals tend to gather in leaves, crops in which the seeds and fruits are eaten are also safe. Corn, in particular, excludes metals from the grain, so corn (excepting silage) is suitable for receiving sludge. Some crops should not be grown on treated land, including leafy vegetables, root crops, foods eaten raw, and tobacco. Sludge can be applied on pasture, but animals should not graze the pasture until all the sludge has washed off the foliage. Hay should not be harvested if there are still traces of sludge on it.

Some sludges are also processed as turf fertilizers. Milwaukee, for instance, treats sludge, bags it, and sells it as Milorganite. Sludge has also been applied to forest lands. Composting sludge with wood chips produces a product useful in greenhouses and nurseries.

Application.   Most states limit how much sludge can be put on land. The rates are designed to stop nutrient pollution and excess levels of metals in the soil. Sludge is usually applied *at the rate that satisfies the nitrogen need of the crop.*

Figure 14-10. The truck is loaded with liquid sewage sludge at the treatment plant and is driven to the farm where the sludge is injected about 10 inches into the soil. (Courtesy of Ag-Chem Equipment Co., Inc.)

However, the rate cannot exceed the maximum allowed amount of cadmium, usually about 1.8 pounds per acre. Sludge can be applied yearly until the maximum amount of any of the heavy metals has been reached. Then sludge use should be stopped. It is the treatment plant's job to keep track of these amounts.

Sludge can be handled much like manure. Watery sludge can be injected into the soil (figure 14-10), or irrigated (figure 14-7), or spread on the surface. The injection method reduces odors and the chance of runoff and keeps ammonia from being lost. Surface-applied sludge should be tilled in as soon as possible. If it is not tilled in to the soil, liquid sludge tends to seal the soil surface and keep water from seeping in. Solid sludge can be spread using a manure spreader or a truck spreader designed for the purpose (figure 14-11).

Sludge is usually delivered and spread by people and equipment from the treatment plant. It is also usually free or at least inexpensive. Thus, for little or no money and effort, a grower can use sewage sludge to improve the soil.

COMPOST

Composting is defined in chapter eight as a method of causing the decay of organic matter in a pile above the ground. That chapter described a typical homeowner compost pile. Composts, however, have commercial uses as well. For instance, growers of container nursery stock compost bark chips for use in potting mixes. Some sewage sludge is also composted with bark chips before it is

Figure 14-11. After solid sewage sludge is spread on the soil, it is tilled into the soil as soon as possible to reduce odors, ammonia loss, and the chance of pollution. (Courtesy of Ag-Chem Equipment Co., Inc.)

spread on farms. Lawn care companies generate leaves and grass clippings, and some cities pick up leaves. Composting these wastes, then using the humus as amendments, functions well as a disposal method.

Composting has two main benefits over simply spreading uncomposted organic matter like manure or sludge on the soil. First, composting improves very high C:N ratio materials like sawdust or woodchips. If wood chips are mixed into the soil without composting, their decay ties-up nitrogen. Second, a compost pile gets hot enough (up to 150°F) to kill most disease organisms, weed seeds, and insects. Of course, the pile must be mixed a few times to make sure matter on the outer edges of the pile is pulled into the hot interior of the pile.

Commercial composting, compared to the home compost, is an exacting process. It works best if a nitrogenous material (for instance, turkey manure or sludge) is mixed with a carbonaceous material (wood chips or leaves). If possible, shredding will reduce particle size and speed decay. The materials are mixed, moistened, and piled into long piles called windrows. Decay is monitored by soil probe thermometers poked into the pile. Front loader operators occasionally turn the pile to mix the components and stir oxygen into the pile. The operator can decide when the pile is done by simple examination and following internal temperature, which will fall as decay slows.

Composted soil amendments are of value to growers if there is a close source of compostable materials. The author has visited one large-scale composting operation that can serve as an example. This firm sells compost to local farmers, with extra hauling charges for anyone farther than 25 miles away. They compost locally obtained turkey manure and wood shavings which is analyzed and priced for sale. This firm prices its compost at 10% below the current market

Component	Pounds/Ton
Total nitrogen	44.00
Phosphate (P_2O_5)	68.00
Potash (K_2O)	38.00
Calcium carbonate	160.00
Magnesium	8.00
Sulfur	12.50
Sodium	5.56
Iron	6.80
Aluminum	4.24
Manganese	0.73
Copper	0.59
Zinc	0.50
Organic matter	1000.00

Figure 14-12. Average analysis of wood shaving/turkey manure compost (Courtesy of Agri-Brand Compost, Holden Farms Inc.)

value of the NPK it contains. The compost is hauled and spread by lime trucks (figure 10-17).

Figure 14-12 shows a typical analysis for the compost. The firm has performed studies over several years to compare crop yield and production costs on plots treated with their compost and nearby plots fed the same amount of NPK in commercial fertilizer. They claim that the compost plots have been more profitable. After four years, the compost plots have a higher cation exchange capacity, more organic matter, and more nitrogen, phosphate, and calcium. However, the fertilized plots had a higher potash content.

Much of the current interest in composting arises from the high rate of manure and sludge generated with no nearby means of disposal. The operation just described composts turkey waste from turkey raisers who have no use for it, and sells the compost to cash crop growers who produced no manure of their own. Composting reduces the volume of organic matter to 65% to 75% of its original volume, reducing transportation costs and making this transfer more practical. Thus, such an operation completes the nitrogen and carbon cycles that are broken when growers specialize.

FERTILIZER AND THE ENVIRONMENT

Fertilizers aid the productivity of American farms, but they have not been without their problems. Fertilizers, manures, and sludges all can cause pollution and human health problems. The farmer who is environmentally aware acts to keep these problems to a minimum.

Animal and Human Health. The main health problem is the effect of nitrates on animal and human infants, and ruminant animals like cattle. Small amounts of nitrates in drinking water cause an infant anemia called "blue-baby disease," in which the ability of the blood to carry oxygen is reduced. In some rural parts of the United States, water is no longer fully safe for human or animal infants.

Groundwater may be polluted by water percolating through fertilized soil. This has occurred most often on irrigated outwash sands with water tables near the surface. These soils, naturally droughty and infertile, tend to be heavily irrigated and fertilized. Natural rainfall or irrigation can then leach the nitrates into the shallow groundwater. A prime example is an area of outwash sands in central Wisconsin, where leaching of fertilizer nitrogen has been identified as one source of nitrates in well water (McWilliams, Lindsey; "A Bumper Crop Yields Growing Problems," *Environment* 26(4): 25-34, 1984). Nitrate levels have been measured there at nearly twice the levels allowed by the Environmental Protection Agency.

Avoiding the nitrate pollution problem from farm fields means improving fertilizer and irrigation efficiency. Careful split-applications of nitrogen based on soil tests are essential. Crop rotation with legumes can also lower nitrogen fertilization requirements. In addition, irrigation can be carefully scheduled to avoid excess percolation (see chapter six).

Nitrate problems result not from fertilizers but from cattle feedlots. Because there are a lot of cattle in a small space and no plants in the feedlot to remove nitrates, high soil nitrate levels result (figure 14-13). The nitrates can seep into the groundwater or wash into streams or lakes.

Interestingly, some plants tend to store nitrates in their leaves and may store large amounts of nitrates when fertilized with nitrates. These "nitrate accumulators" include spinach, corn, and sorghum. This may result in high-nitrate forages that can harm livestock. It is uncertain if any human health problems have also occurred.

Phosphate Pollution.   Anyone who has gone swimming in a scummy lake during hot August days has felt the result of phosphate pollution. Usually low phosphate levels in water limit the growth of aquatic plants and algae. If phosphorus, and to a lesser degree, nitrogen, wash into a lake, algae and plant growth explodes, taking oxygen out of the water. The result is green water, a scummy surface, dead fish, and a lake unsuitable for recreation. This process is called *eutrophication,* the acceleration of the natural aging of a lake.

Phosphates come from a number of sources, including sewage. Phosphate fertilizers contribute to pollution when erosion washes soil into lakes and streams. Erosion control is the farmer's key to keep from adding to the problem. Banding or incorporation of phosphorus also reduces phosphate runoff.

Homes and cabins on lake shores often add more phosphate to lakes than local farms. Homeowners feed their lawns with phosphate-containing fertilizers. Most of it sits on top of the soil, ready to wash into the lake when it rains. Lakeside residents should use phosphate-containing fertilizers only if they are really

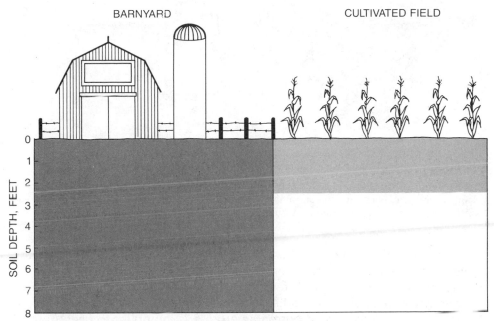

BARNYARD   CULTIVATED FIELD

SOIL DEPTH, FEET

0
1
2
3
4
5
6
7
8

(The shading indicates the relative amount of nitrate-nitrogen in the soil.)

Figure 14-13. Much nitrate groundwater contamination results from livestock feeding operations, rather than crop fertilization. (Source: University of Minnesota Extension Bulletin #168)

needed. Figure 14-14 (A) shows where plant nutrients are likely to reduce water quality in the United States.

Biological Oxygen Demand. Biological oxygen demand (BOD) occurs when organic materials, like sewage, manures, or food processing wastes wash into bodies of water. Microorganisms feeding on the wastes deplete oxygen in water, causing fish kills and losses of other aquatic life. Farmers avoid contributing to BOD in nearby water by handling manures properly to keep it in their fields and out of lakes and streams (remember the picture that began this chapter?). Figure 14-14 (B) shows where organic wastes have been a problem.

Energy Costs. Fertilizers have a high energy cost. The mining and processing of phosphate and potash fertilizers consume some energy. But the highest energy cost is involved in making ammonia. Each ton of chemically fixed nitrogen uses 1.5 tons of natural gas. Several percent of the nation's yearly fuel bill goes to making fertilizer.

The high energy bill for farming raises the cost of farming. In the long run, farming would be threatened if fuel supplies begin to run out or are cut off. Organic farmers cite this factor as one reason for avoiding chemical fertilizers. It

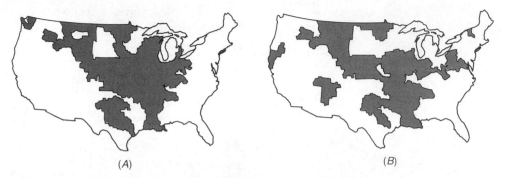

Figure 14-14. *(A)* Areas of the United States where watersheds are troubled by nutrient pollution, mainly in the form of nitrates and phosphates. *(B)* Areas where watersheds are troubled by organic wastes, including sewage and animal wastes. (Source: USDA *1980 Soil and Water Appraisal*)

has also spurred interesting research. For instance, some researchers are using genetic engineering to develop field crops that can fix their own nitrogen.

Farmers can reduce their fertilizer energy bill by using other nitrogen sources where possible. These other sources include animal manure, sludge, and legumes grown in rotation. Controlling erosion is important to keep nutrients from washing off fields. Also, plants can make better use of fertilizer if the soil is in good condition and the pH is in the right range.

SUMMARY

Manure, sludge, and compost provide a double benefit to growers—they contain nutrients to promote crop growth and organic matter to improve the soil. Applying these materials to farm fields is a good alternative to other disposal methods.

Manure is highest in carbon, nitrogen, and potash. It also contains phosphates and secondary and trace elements. Proper handling of manure reduces nutrient losses and lowers the chance of polluting surface or groundwater. Manure is best spread on unfrozen land as soon as possible after collection and then tilled into the soil. If this is not practical, the grower should store the manure in sealed, covered pits. Liquid systems also work well to preserve nutrients.

Sewage sludge is processed "human manure" and can be handled much like animal manure. However, possible health problems from heavy metals and human pathogens mean that it must be used according to state and federal EPA guidelines. Sludge is mostly applied according to the nitrogen rate to seed, grain, and fruit crops. It is important to keep the soil pH above 6.5 to fix heavy metals.

Composting is a way to reduce the C:N ratio of organic materials and to kill harmful organisms. For example, composting sludge with wood chips stops nitrogen tie-up from the chips while killing pathogens in the sludge.

Any nutrient source—fertilizer, sludge, or manure—can harm the environment if used improperly. Nutrients and pathogens can wash into surface waters or leach into groundwater, causing pollution and human and animal health problems. Using inorganic and organic fertilizers in the suggested ways and rates and avoiding erosion are important ways to reduce these problems.

Farm fertilizers increase national energy use, and force agriculture to rely heavily on fossil fuels. For most growers, making the most efficient use of fertilizer, sludge, manure, and legumes is the best answer.

REVIEW

1. Organic amendments include nutrients not found in a typical complete fertilizer. (True/False.)
2. Phosphorus is the nutrient present in manure in the highest amounts.
3. Nitrogen loss by volatilization is high in an open manure pile.
4. Manure can never harm crops.
5. One concern about sludge is heavy metal content.
6. Sludge is applied at a rate that satisfies nitrogen needs for a crop.
7. A compost should be made of phosphorus-rich and nitrogen-rich materials.
8. Composting is one treatment for sludge.
9. Nitrates are the key contributors to eutrophication.
10. Manures can contribute to BOD.
11. Most manure is highest in the nutrients _____ and _____ .
12. The element in sludge of greatest human health concern is _____ .
13. According to figure 14-1, the fertilizer ratio of dairy cattle is _____ .
14. According to figure 14-9, the fertilizer ratio of an average sludge is _____ , showing that the nutrient _____ is very low.
15. Nitrogen in manure can be lost in the form of _____ gas.
16. Eutrophication is mostly caused by the element _____ .
17. Organic materials in water create _____ .
18. "Blue-baby" disease is caused by _____ in drinking water.
19. The amendment that should not, for health reasons, be applied to carrots, is _____ .
20. An important form of nitrogen in animal urine is _____ .
21. List five benefits of organic amendments.
22. Why should a soil with a pH of 5.5 be limed before adding sludge?
23. List five crops to which sludge should not be applied.
24. What conditions promote volatilization of nitrogen in manure?
25. What are the two broad types of materials included in a compost pile?
26. How do you lower nutrient losses in manure?

27. Explain why composting sludge and wood chips together makes a better product than either separately.
28. Describe the kind of land that could accept sludge.
29. Some cities burn sludge instead of spreading it. Think about benefits and problems of burning, then decide which you think is best. Defend your choice.
30. What can effect the nutrient content of manures?

SUGGESTED ACTIVITIES

1. Find out what your state's guidelines are on sewage sludge use.
2. Visit the closest sewage treatment plant. How do they treat sewage and dispose of sludge?
3. Visit a nearby farm with a manure lagoon or other manure storage facility.
4. Build a compost pile.

chapter fifteen
Tillage and Cropping Systems

A common sight on American farms, the moldboard plow has long been the primary tillage tool in many parts of the world. Some form of tillage has been in use for thousands of years; tillage tools are one of the most important inventions of humanity. Today tillage methods are undergoing rapid change, with the introduction of various forms of conservation tillage. This chapter will discuss modern tillage and the cropping systems that go along with it.

(Courtesy of *Crops and Soils Magazine*, American Society of Agronomy)

OBJECTIVES

After completing this chapter, you should be able to:

- explain the reasons for and effects of tillage
- describe conventional and conservation tillage
- list several cropping systems
- briefly describe organic and sustainable agriculture

TERMS TO KNOW

allelopathy	fallow	rangeland
conservation tillage	finishing harrow	reduced tillage
conventional tillage	lister plows	row crops
crop rotation	moldboard plow	saline seeps
disc plow	mulch till	secondary tillage
double cropping	pitting	small grains
dryland farming	primary tillage	tillage

To produce crops, a grower places seeds in contact with the soil, provides nutrients, controls pests, and manages soil water. These activities usually involve some form of tillage. There are many ways to work the soil and different situations require different methods. Each method has an effect on the crops and the soil. This chapter will look at some standard tillage and cropping systems.

USES OF TILLAGE

Tillage is working the soil to provide a favorable environment for seed placement and germination and crop growth. In the United States, mechanization and research has led to a variety of tillage tools and tillage systems. Regardless of the method of tillage used, a grower has three basic goals: (1) weed control, (2) alteration of physical soil conditions, and (3) management of crop residues.

Weed Control. Tillage used for weed control can be divided into two time periods: before crop planting and after crop planting. Before planting, tillage prepares a weed-free seedbed that greatly simplifies weed control during the growing season. Tillage destroys young seedlings that are easily buried or ripped out of the soil. Repeated tillage operations may also weaken perennial weeds. After planting, cultivation continues to destroy or bury emerging seedlings. Cultivation may also break hard surface crusts on some soils. However, deep cultivation or cultivation late in the season may sever surface roots and reduce crop yields.

The importance of tillage for weed control has declined with the increase in both herbicide use and tillage systems designed around herbicide use. Some herbicides require shallow incorporation into the soil, which is usually accomplished by shallow tillage.

Physical Soil Conditions. Tillage alters physical soil properties, such as structure, moisture, and temperature. Tillage during seedbed preparation stirs and loosens the soil, improves aeration, and creates a suitable medium for growth. Deep tillage, or subsoiling, may temporarily break up subsoil compaction.

However, tillage causes a long-term decline in physical structure. Partially, the decline is due to losses of soil organic matter that result from tillage. Also, repeated tillage operations crush some soil aggregates. Wheel traffic compacts the soil, especially wet soils, and tillage pans may form. Soil aggregates on the surface of a bare soil shatter from raindrop impact, causing crusts that hinder seed germination and shed water. The bare soil resulting from many forms of tillage erodes easily. Recent changes in tillage aim to reduce these adverse side effects.

Tillage also affects the moisture level and temperature of soil. Tilled soil usually warms up earlier in the spring, allowing earlier seeding and better germination. In areas where soil tends to be wet or cold in the spring, crops may be planted on ridges created by tillage. The ridges warm and dry faster than the rest of the soil.

Shallow cultivation of crust-forming soils may improve crop yield even where herbicides are used to control weeds. By breaking up crusts, cultivation improves water infiltration and reduces runoff. Such cultivation should be just deep enough to break the crust.

Crop Residue Management. After most crops are harvested, residues like stalks or leaves remain in the field. The amount of residue depends upon the type of crop, how well it grew, and how it is harvested. For example, a good corn crop leaves about 8,500 pounds of residue per acre for a 150-bushel corn crop, and about 5,600 pounds of residue for a 100-bushel crop. If the corn is harvested for silage rather than grain, little residue is left in the field. Figure 15-1 lists residues for several crops.

There are several ways in which growers can manage crop residues, depending on objectives. Moldboard plowing buries crop residues, resulting in a clean field that is easy to plant and cultivate. Present trends are to maintain some residue on the soil surface to save moisture and prevent erosion. In semiarid grain growing areas, special tillage tools, including rodweeders and sweeps, till under the surface to kill weeds but leave residues on the surface to protect against wind erosion. Conservation tillage in more humid climates leaves residues on the surface to protect against water erosion. Figure 15-2 lists the amount of residue left on the soil surface from various tillage tools.

In addition to crop residues, tillage may also incorporate phosphorus, potash, and lime into the root zone. Tillage may also be used to incorporate sewage sludge, manures, and nitrogen sources like urea that may volatilize if left on the soil surface.

Crop	Approximate Residue per Bushel Grain (lb/acre)		Sample Yield (bu/acre)		Sample Residue (lb/acre)
Barley	80	×	50	—	4,000
Corn	56		125		7,000
Flax	80		15		1,200
Oats	60		32		4,300
Rye	100		30		3,000
Sorghum	60		50		3,000
Soybeans	50		40		2,000
Wheat	100		40		4,000

Figure 15-1. Crop residues in pounds per acre for several crops are given in the last column. To obtain these values, the number of pounds of residue per acre for each bushel of grain produced is multiplied by the sample yield in bushels per acre for each crop. The sample yields may not represent the yields in your area. (Courtesy USDA, Soil Conservation Service Technical Guides)

Implement	Estimated Percentage of Residue Remaining after Each Operation
Inverting tools	
Moldboard plow	5
Lister plow	20
Mixing tools	
Field cultivator	80
Chisel plow, spear point	80
Chisel plow, twisted point	50
Rototill to 6 inches	25
Rototill to 3 inches	50
Tandem disc to 6 inches	25
Tandem disc to 3 inches	50
Spring-tooth harrow	60
Spike-tooth harrow	70
Subsurface tools	
Blades or sweeps	90
Rodweeders	90

Figure 15-2. The percentage of residue remaining after one pass over the field for various cultivating implements is given. If two or more tillage operations are practiced, each operation after the first one uncovers as well as covers some residue—about half as much. (Courtesy USDA, Soil Conservation Service Technical Guides)

Seedbed Preparation. The three reasons for tillage come together in preparing a seedbed. The objective of preparing a seedbed is to ensure that the soil meets the needs of the germinating seed. The seed needs a moist soil at the right temperature with sufficient air for seed respiration. The soil should be loose enough for good aeration, but compact enough around the seed for good soil/seed contact. It should be free of clods that prevent proper seed/soil contact and seedling emergence (figure 15-3). The soil in a seedbed should also be warm and moist but not too wet.

The smoothness of the seedbed and the amount of allowable crop residues depends upon the seed size and the type of planter. Large seeds, like corn and soybeans, can germinate in a fairly rough, cloddy seedbed. Very small seeds, like alfalfa seed, germinate best in a very fine, firm seedbed. A seedbed free of crop residue is easiest to plant in, but conservation tillage demands that crop residues be left on the surface to control erosion. Most older seed planters can only be operated on a fairly smooth, clean seedbed. Many modern planters can plant through some crop residues and clods, preparing correct soil conditions near the seed.

CONVENTIONAL TILLAGE

Conventional tillage, the primary form of tillage since the invention of the moldboard plow, involves two stages. First, *primary tillage* breaks up the soil and

Figure 15-3. A conventional seedbed is smooth and free of crop residues. It is excellent for seed germination but is prone to water and wind erosion. (Courtesy John Deere Company)

buries crop residues. Primary tillage is often accomplished with an *inverting implement*, like the plow or lister plow, that inverts or tips over the top few inches of soil. *Secondary tillage* produces a fine seedbed by a series of operations that break up the soil into smaller and smaller chunks. Secondary tillage involves *mixing implements* like harrows. The following discussion describes these operations in more detail.

Plowing.   The traditional primary plowing tool is the *moldboard plow* (figure 15-4). The moldboard shears off a section of soil, tips it upside down, and fractures it along several planes. In the process, any organic material on the soil surface is buried. The moldboard plow leaves the surface very rough with a series of ridges and furrows.

Moldboard plows work best in moist soil, where the moisture level is near field capacity (see chapter four). If the soil is too dry or too wet, the operation takes more power and the results are poor. For wet or dry soils, a *disc plow* works better. A series of 3 to 10 large (2 to 2 1/2 feet) discs are mounted on a frame at an angle to the direction of travel. The discs cut into the soil as they rotate and roll the soil over.

Subsoilers like the one shown in figure 15-5 are used to shatter tillage pans or natural soil pans. Subsoiling should be done when the soil is dry, because if pans

Figure 15-4. The 10-bottom moldboard plow flips over the top seven or eight inches of soil to mix the soil and bury crop residues. (Courtesy John Deere Company)

are moist they do not shatter. Deep plowing can temporarily help water infiltration and root penetration into the subsoil. Usually, however, compacted layers reform as the soil is exposed to further wheel traffic and tillage.

Harrowing. Harrowing is usually a two-step process. In the first stage, ridges left from plowing are smoothed out and large clods broken up. Then smaller lumps are pulverized and a fine seedbed is produced.

Growers commonly begin the operation with a disc harrow (figure 15-6). The typical *tandem disc* has four gangs of discs set like the four arms of an "X." The front two gangs turn the soil inward, and the back two turn it back out. A disc tends to compound compaction problems because it shatters soil aggregates and does not dig deep enough to loosen compaction. Spring-tooth harrows and field cultivators (figure 15-7) may be used rather than the disc. A long, springy C-shaped tooth with a spear point or broad shovel digs into the soil, dragging clods to the surface and breaking them up.

A finishing harrow, or drag, completes the job of pulverizing the soil. Figure 15-8 shows a drag being pulled behind a spring-tooth harrow.

Figure 15-5. A subsoiling chisel plow is used to break up a tillage pan or other compacted layer. (Courtesy Year-A-Round Cab Company)

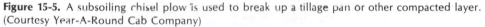

The steps just described are often modified. If the soil has good tilth, deep tillage by plowing may not always be needed. In such cases, the tandem disc shown in figure 15-6 is heavy enough to be used alone. Growers often combine operations, hitching several tillage tools behind the tractor. Any time a grower can eliminate a pass through the field, compaction is reduced and time and fuel are saved.

Lister Plowing. *Lister plows* are equipped with two moldboards mounted back to back, pointed in opposite directions. Lister plowing results in a pattern of ten-inch high ridges and furrows across the field. In humid regions, crops may be planted on the warmer, drier ridges. In arid areas, they are planted in the moist soil of the lister furrow. Listing can also help protect the soil from wind erosion. Listing on the contour captures water to improve water use and reduce water erosion.

Preparation of Furrow-Irrigated Fields. Additional steps are needed to prepare furrow-irrigated fields. After the standard primary and secondary tillage, the grower carefully levels the field with a blade to ensure the proper grade for flow

Figure 15-6. This tandem disc with 24-inch blades is heavy enough for both primary and secondary tillage. (Courtesy John Deere Company)

Figure 15-7. A field cultivator can be used for weed control on fallow ground, for secondary tillage, and for primary tillage where low levels of surface residue are present. (Courtesy John Deere Company)

Figure 15-8. The last step in preparing a smooth, fine seedbed is the use of a finishing harrow. This step is especially important for most small-seeded crops. Note that the harrow is attached to a field cultivator. This means that one trip through the field does two jobs. Combining operations saves time and reduces compaction. (Courtesy John Deere Company)

of surface-applied water. Then the field is listed with a special tool to create ridges and furrows.

Timing and Depth of Plowing.   Farmers in the eastern United States (see figure 5-3) can plow either in the fall or in the spring. Fall plowing gives the farmer a head start on spring planting by warming and drying the soil. Freezing and thawing on fine-textured soil breaks up large lumps, making it easier to develop a good seedbed. The benefits of fall plowing are especially important with fine-textured soils having somewhat poor drainage.

Spring plowing leaves crop stubble in the field over winter to capture snow and reduce erosion. To conserve soil, plow in the spring unless there are overriding reasons for fall plowing.

In the western United States, where moisture preservation is critical, plowing immediately after harvest gives more time for the soil to store moisture, as long as weeds are controlled. However, leaving the soil bare increases the risk of erosion.

The standard plow depth of seven to eight inches gives the best results. Shallow plowing results in a poor seedbed, and deeper plowing takes more power without noticeably improving yields.

CONSERVATION TILLAGE

Conservation tillage is a program of crop residue management aimed at reducing erosion (figure 15-9). Rather than plowing under crop residues, some or all of the residue is left on the soil surface. A field at planting time under conservation tillage has at least 30% (corresponds to about 1,000 pounds of stover or 500 pounds of straw per acre) of the soil surface covered by crop residues. This practice reduces erosion by 40% to 50%. Figure 15-2 lists estimates of the amount of residue after one trip across a field for several tillage tools. Sometimes the term *reduced tillage* is used because fewer tillage operations are needed. Interestingly, many farmers fail at conservation or reduced tillage because they overuse the tillage equipment and bury too much of the residue.

Figure 15-9. Runoff from conventional and conservation tilled field. The runoff sample from a conservation tilled field (bottom sample) contains far less soil than the sample off a field prepared by conventional tillage (top sample). (Courtesy of USDA-ARS)

	Tillage System			
Operation	*Conventional*	*Chisel Plow*	*Tillplant*	*No-Till*
		gal/acre		
Shred stalks	0.6	0.6	0.6	0.6
Moldboard plow	1.8	—	—	—
Chisel plow	—	1.1	—	—
Disc (1st time)	0.6	0.6	—	—
Disc (2d time)	0.5	—	—	—
Plant	0.3	0.3	0.3	0.3
Spray	0.1	0.1	0.1	0.1
Cultivate	0.3	—	0.4	—
Total	4.2	2.7	1.4	1.0

Figure 15-10. Average amounts of diesel fuel consumed in producing a crop for representative tillage systems. Herbicides alone are used to control weeds in the no-till and chisel systems. Cultivation in the till-plant (ridge-till) system is intended primarily to rebuild the ridges. (Adapted from Bulletin #E-780, Cooperative Extension Service, University of Michigan, East Lansing, MI)

Other benefits of conservation tillage are obtained from fewer trips across the field. These benefits included less time in field work and lower fuel costs (figure 15-10). At times, compaction is reduced because of less wheel traffic. Conservation tillage may also require fewer implements, thus reducing equipment costs per acre. Conservation tillage provides a better habitat for pheasants and other wildlife. For instance, no-till fields have been shown to provide improved habitat for nesting ducks in North Dakota and nesting bobwhite quail in Tennessee.

Because of the soil conservation and economic benefits of conservation tillage, its use has spread rapidly in the past two decades. The Conservation Tillage Information Center reported that in 1983 about one-third of United States cropland was under conservation tillage. As technology improves, conservation-tillage use is expected to grow.

Conservation tillage is an umbrella term covering several different tillage methods.

Mulch-Till or Chisel-Plow. A chisel plow (figure 15-11), which loosens the soil but does not invert it, is used for primary tillage. Chisel plowing to eight inches leaves the soil rough with about 50% residue cover (figure 15-12). A secondary tillage operation of light discing reduces residues to 30% to 50%. Seeds are then planted through the remaining residues. After planting, cultivation and herbicides control weeds (figure 15-13).

Strip-Till. With no primary tillage, a specialized implement plants into an untilled field. The planter tills a band of soil and plants seeds into the band. Another type of implement sweeps residues off a strip into the middle of the rows. The planting operation bares about one-third of the soil surface, leaving about 50% of crop residues.

Figure 15-11. The chisel plow is the primary tillage tool in the mulch-till system. (USDA, Soil Conservation Service)

Figure 15-12. Residues and ridges left by using a chisel plow on soybean grounds (USDA, Soil Conservation Service)

(A)   (B)

Figure 15-13. Surface residues remaining in a soybean field planted in last year's corn residue after planting in mulch-till. *(A)* Before the first cultivation, there is a heavy cover of corn residues. *(B)* After one cultivation, the residue is mixed in the top few inches of soil with moderate to low levels of surface cover. (USDA)

Figure 15-14. In the ridge tillage system, planting is done on the top of the ridge where the soil warms up and dries out more quickly in the spring than in other conservation tillage methods.

Ridge-Till. The ridge-till method also plants in a cleaned strip. The seed is planted on a six-to-eight-inch high ridge created by cultivation during the previous year (figure 15-14). These ridges warm up more quickly than soil in other conservation tillage systems. About two thirds of the previous year's crop residues remain immediately after planting.

No-Till. In this method, the soil is barely disturbed. Specialized planters cut a slot through the residues, insert the seed and fertilizer, and close up the slot. About 90% of the soil surface is untouched. Contact, systemic, and preemergent herbicides are used to control weeds with no cultivation.

DIFFERENCES BETWEEN CONVENTIONAL AND CONSERVATION TILLAGE

Conventional and conservation tillage differ in more than the obvious ways. Most growers ask first about yields. Current research shows equivalent yields from conventional and conservation tillage if known technology is

| Tillage System | Yield (Bushels/Acre) | | Silt Loam |
| | Clay Loam | | |
	1982	8-yr avg	1979-1980 avg
Moldboard plow	175	154	163
Mulch-till (chisel)	164	144	156
Ridge-till	166	149	159
Strip-till	160	145	—
No-till	139	129	154

Figure 15-15. Corn yields under continuous corn is influenced by tillage systems on two northern soils. The clay loam is tiled and is located on the University of Minnesota, Southern Experiment Station at Waseca, MN, and the silt loam is on the University of Wisconsin Lancaster Experiment Station at Lancaster, WI. On the moderately fine-textured, somewhat poorly drained soil, conventional tillage and ridge till gave the highest yields under continuous corn. No-till has performed less well on the soil. In the well-drained, medium-textured soil, conventional tillage has little yield advantage and no-till is more feasible. (Adapted from "A Report on Field Research in Soils," Agricultural Experiment Station, University of Minnesota, Misc. Publ. 2 - 1983 and "4-State Field Day," Lancaster Experiment Station, College of Agricultural and Life Sciences, University of Wisconsin-Madison, 1982.

Tillage System	Corn Yield (bu/acre)
Conventional tillage	142
Chisel plow	138
No-till, after harvest for silage	142
No-till, after harvest for grain	144

Figure 15-16. Average corn yield after tillage treatments for continuous corn in Ohio. The soil is a silt loam. In this soil, no-till in the high residues of corn harvested for grain gave the best yields. (Ohio Agricultural Research and Development Center)

applied to each tillage system and the systems are matched to crops and soil types. Figures 15-15 and 15-16 provide some comparisons of the systems.

Conservation tillage requires different management practices compared with conventional tillage.

Equipment. Conservation tillage places some requirements on the equipment. For example, residues should be spread evenly behind the harvest equipment. Planters must penetrate the residues, place the seed, cover the seed, and ensure seed/soil contact in a rough seedbed.

Fertility. Because there is less soil mixing in conservation tillage (especially in no-till), the form and placement of fertilizers are affected. Lime, phosphates, and potash tend to stay near the soil surface. However, since the residues provide a mulch, soil near the surface tends to remain moist, promoting the growth of roots near the surface and improving uptake from that layer. Conservation tillage, especially no-till, can reduce nitrogen availability by increasing nitrogen tie-up in surface layers, increasing leaching, and by reducing average soil temperatures. The injection of nitrogen deeper into the soil and nitrification inhibitors will reduce these problems.

The pH of the top two inches of soil tends to drop rapidly, especially in no-till, affecting seed germination, crop growth, and herbicide activity. Careful testing for pH is required for this layer, followed by a topdressing of lime as needed.

Matching Tillage to Soil Type. Soils tend to be cooler and wetter in conservation tillage, especially with the no-till method. On fine-textured soils in northern states, cooler soil may delay planting and hamper seed germination. No-till is a poor choice on cold, poorly drained fine-textured soils; for these conditions, the ridge-till method is a better choice (figure 15-15). On excessively drained coarse soils, no-till can improve yields by preserving moisture. Local extension agents can provide advice on the best system for each grower.

Weed Control. With less tillage, there is greater reliance on herbicides for weed control. Tillage will kill any weed seedling, but herbicides are more selective. This makes weed identification and herbicide selection more critical. Also, surface-applied chemicals are more suitable for conservation tillage than those needing to be incorporated into the soil.

Pest Control. Conservation tillage, especially the no-till method, alters the environment presented to disease and pest organisms. Diseases that overwinter on crop residues, like small grain leaf diseases, can be a greater problem in conservation tillage compared with conventional tillage where plowing buries these infected residues. This factor increases the need to select resistant crop varieties and to rotate crops.

Some researchers recommend that some insects will be a greater problem in conservation tillage, because overwintering insects are not buried by plowing. Others state that research has not proved this claim. Students should follow the advice of local Cooperative Extension agents.

Drawbacks to Conservation Tillage. While conservation tillage has been successful and has been widely adapted by growers, it does have some problems. These include

- A higher level of management and skill are needed than for conventional tillage.
- Conservation tillage is not adapted to all soils, climates, or crops. Further development may expand the range to include almost all situations.
- Perennial weeds sometimes have been a problem, especially in some southern states.
- No-till may lead to soil compaction, making occasional subsoiling useful.
- The acidic soil surface interferes with herbicide activity, calling for careful monitoring of soil pH and more frequent liming.
- Conservation tillage greatly increases dependence on herbicides. The consequences of this to the environment are not yet totally known. At the

time this text was being written, a major herbicide widely used in conservation tillage was coming into question because of possible health and environmental problems. Also, if the cost of herbicides were to rise, so would the cost of conservation tillage.

CROPPING SYSTEMS

Farmers have three ways to decide what crops to grow each year. First, they may "plant to the market." That is, they look at crop prices and decide what crop will be most profitable. Second, they may simply grow crops that are best suited to their operation. This may mean crops suited to the local soil and climate, or crops suited to the skills of the grower, to available equipment and livestock, or to market demand. The third approach is based on a crop rotation pattern.

Continuous Cropping. In continuous cropping, a farmer grows the same crop each year. Continuous cropping is favored by many farmers because they can grow the most profitable crop. This method also allows the grower to specialize in the crop best suited to local soil or climate conditions. In general, however, yields often decline under continuous cropping. At the same time, expenses for fertilizer, herbicides, and pesticides tend to rise compared with expenses for a crop rotation system.

Crop Rotation. Crop rotation means that a series of different crops is planted on the same piece of ground in a repeating cycle (figure 15-17). Many farmers do not rotate crops because it means planting some less profitable ones. Often a farmer has no use for certain crops in common rotations. For instance, a farmer who feeds no animals has little use for hay unless a buyer can be found for it.

However, crop rotation has important benefits for those who practice it. Crop rotation:

- Aids the control of diseases and insects that rely on one plant host. A grower's pesticide bill can be reduced as a result. One rotation, double cropping soybeans and wheat, has reduced soybean cyst nematode damage by 90%. Some researchers have suggested that yield declines that often accompany continuous cropping result from a buildup of soil pathogens.
- Helps control weeds. Many weed species grow best in certain crop types, so alternating crops suppresses the weeds. For example, small grains and hay tend to smother weeds that grow well in a row crop. This effect can lower a farmer's herbicide bill. Some rotations suppress weeds by *allelopathy*. That is, one plant emits chemicals from the roots that suppress growth of other plants. For instance, soybeans planted into wheat residues suffer fewer weed problems because of allelopathic effects of the wheat.
- Supplies nitrogen if certain legumes like alfalfa are in the rotation. This can lower a farmer's fertilizer bill.

Figure 15-17, In this typical Corn Belt rotation, there is grain stubble in the foreground, then a strip of hay, and then a strip of corn. (USDA, Soil Conservation Service)

- Improves organic matter and tilth of the soil. Some deep rooted crops like alfalfa also improve subsoil conditions.
- Reduces erosion compared with continuous row crops, as long as the rotation includes small grains or hay. This topic is covered in more detail in chapter eighteen.

Generally, crop rotations involve some combination of three kinds of crops: row crops, small grains, and forages. The specific crops and crop sequence varies from place to place.

Row Crops. *Row crops,* where adapted, are usually the most profitable crops. Row crops are planted in wide rows and cultivated for weed control, with the help of herbicides. The crops are fertilized by broadcasting, banding, and sometimes sidedressing. Fertigation is often used in irrigated fields. Row crops usually leave the soil bare, making it erosion-prone. As a result, row crops are best suited to fairly level ground. Conservation tillage, crop rotation, and other conservation practices greatly reduce erosion from row crops (see chapter eighteen). Some common row crops, and their interaction with soil, are described as follows.

Corn. Corn grows best on well-drained, slightly acid loams. Corn is a heavy nitrogen feeder and is often fertilized with anhydrous ammonia. Corn leaves a heavy residue, high in lignin, that does not decay rapidly.

Sorghum.   A relative of corn, sorghum is grown much like corn. It tolerates heat, dry soil, and salinity better than corn and is widely grown in the hot dry areas of Texas and Oklahoma.

Soybeans.   Soybeans and other dry beans grow best in well-drained soil with a pH between 6.0 and 7.5. Since these crops are legumes, they need little nitrogen. They have a high potassium requirement, but commonly they are not fertilized and rely instead on fertilizer left from the previous crop. Soybeans leave a light, fragile residue that makes the soil more friable. The crop does little to control erosion. Some soybeans may be planted on a narrow spacing to reduce erosion a little.

Cotton.   Cotton is rather soil-tolerant, growing between pH 5.2 and 8.0 and accepting fairly saline soil. Cotton is adapted to warm climates. Cotton, too, leaves little residue so is erosion-prone.

Tobacco.   Tobacco needs high levels of potassium, magnesium, and calcium for good quality leaves and proper burning. Tobacco is fertilized with ammonium nitrogen and low chloride fertilizers for good burning quality.

Small Grains.   *Small grains,* like oats or wheat, are planted in closely spaced rows seven or eight inches apart. As a result, they quickly cover the soil surface. Land planted to small grains loses less soil to erosion. Small grains also leave a large amount of residue. If this residue is left in the field, it helps control erosion in conservation tillage systems. The dense growth of small grains also competes with weeds that infest row crops, and several small grains suppress weeds by allelopathy. A number of small grains are important crops, allowing a wide choice for farmers with different growing conditions.

Soil nitrogen and potash must be properly balanced for good grain yields. A good supply of nitrogen promotes growth and improves the protein content of the grain. Excess nitrogen or low potassium, especially in moist soils, causes lodging. Fertilization is usually carried out by preplant broadcasting and may be followed by topdressing of the growing grain. Banding has become popular, especially in conservation tillage systems.

Wheat, the major small grain, is best adapted to fairly dry areas on medium- to fine-textured soils. *Oats* prefer a moister, medium-textured soil. *Rye* is the best small grain for sandy soils. *Barley* is more tolerant of saline soils than most crops and is commonly grown where soluble salts are a problem.

Perennial Forage.   *Forages* are harvested for their green matter and fed to animals. They may be harvested as hay or used for grazing in pasture or range. Forages improve soil tilth, add organic matter, and control erosion. Taprooted plants, like alfalfa, help break up soil pans. Legume forages also fix nitrogen that can be used by later crops.

Forage Legumes.   Alfalfa, the most important legume, grows best on moist, near-neutral, loamy soils. It requires very high amounts of all primary elements, calcium, sulfur, and boron, but it fixes its own nitrogen. It may be necessary to treat the soil with lime for good growth. Before planting, fertilizer and lime are broadcast and tilled into the soil. Established alfalfa is topdressed with phosphate and potassium to replace nutrients removed during harvest. *Sweet clover* is better adapted to dry or saline soils. A number of other legumes, like vetch or trefoil, may be used.

Forage Grasses. Approximately 40 grass species are important forages in the United States. Thus, there is at least one grass species to fit almost any situation. The best production and protein content result from high nitrogen fertilization.

Double Cropping. *Double cropping* is the practice of harvesting two or more crops from the same piece of ground in one year. A common example is planting soybeans into winter wheat stubble. Sometimes a cover crop is planted in the off-season, then killed with herbicides. The crop is then planted into this mulch. Double cropping is easiest in warm climates with long growing seasons. The use of double cropping has grown with the use of conservation tillage. The second crop can be planted right behind the harvest of the first crop, omitting the time-consuming seedbed preparation. This advantage has made double-cropping more practical for northern states.

Multiple cropping keeps the soil covered with vegetation for a larger part of the year. Better erosion control results. Two crops grow more green matter than one, so the practice can also help maintain organic matter in the soil. Where one of the crops is a legume, the nitrogen addition is welcome. Two crops also draw more heavily on soil nutrients and water, so plant foods and water must be more carefully managed.

DRYLAND FARMING

The term *dryland farming* is applied to farming in low-rainfall areas without irrigation. In the United States, dryland farming is practiced in states west of a line from western Minnesota to eastern Texas, following the 96th meridian. We will discuss two characteristic dryland farming systems: small grain-summer fallow rotation and rangeland grazing.

Summer Fallow. Many dry areas lack enough water to produce good crops each year. As a result, crop rotation of small grain-summer fallow is used. In the crop year of the rotation, small grains are grown because they are relatively tolerant of low moisture conditions. After the grain is harvested, the soil is left fallow for the next year. More complex systems include a three-year rotation of winter wheat-fallow-sorghum, common on the southern Great Plains.

Summer fallow is the practice of leaving the soil crop- and weed-free to store moisture. During the fallow period, weeds are controlled by cultivation or herbicides. By controlling weeds, no moisture is lost from the soil because of transpiration. Some water is lost by evaporation, but not all. After a rain, water seeps into the soil. As the soil dries, some moisture moves to the surface by capillary rise, where it evaporates. However, once the surface dries, it seals the rest of the water in the soil. After the next rain, more water is sealed in the soil. Generally about 25% of the rainfall on a fallow field will be stored for the following crop.

The effectiveness of summer fallow can be improved by reducing water runoff. On slopes, contour tillage helps reduce runoff. Using tillage tools that leave crop residues on the surface also helps prevent runoff. Chemical fallow, which leaves grain stubble standing and crop residues undisturbed, saves an additional one-half to two inches of moisture.

Three problems arise from the practice of summer fallow. During the fallow year, wind erosion can be quite serious. This problem will be discussed in detail in chapter eighteen. Second, crop-fallow rotations lead to a long-term decline in soil organic matter. The other problem is the development of *saline seeps.*

Saline seeps (figure 15-18) appear in almost two million acres of the Great Plains of the United States and Canada. Saline seeps appear where glacial till overlays an impermeable rock layer. During fallow, increased percolation picks up salts and carries them deeper into the soil. When the salty water reaches the tight layer, it spreads out sideways, usually flowing downslope. Finally, the salty water seeps to the surface on a lower field. The water evaporates, leaving the salt on the soil surface.

Figure 15-18. Monitoring the water table below a saline seep. Nearby plantings of alfalfa are lowering the water table. (USDA, *Agricultural Research*)

Reclaiming a saline seep begins by finding the origin of the salty water—the recharge area. Shallow ditches, land leveling, or contouring can be used to divert excess water from the recharge area. More commonly, growers seed the recharge area and the soil around the seep to alfalfa. With its high water demand and deep root system, alfalfa actually lowers the water table several feet. During the reclamation period, salt-tolerant barley may be grown if needed.

Researchers have studied ways to reduce fallow problems by devising annual cropping systems. For instance, in the northern Great Plains about a quarter of the annual precipitation falls as snow—if all this were captured rather than allowed to blow off, it could equal the moisture saved during a fallow year. Studies using tall wheatgrass strips—as shown in figure 5-14—to capture snow have indeed shown improved profitability for annual wheat cropping with reduced problems like wind erosion (B.G. McConkey, et. al. "Perennial grass windbreaks for continuous wheat production on the Canadian prairies", J. Soil Water Cons. 45 (4): 482-485. 1990).

Rangeland. Range is an uncultivated area used for grazing. Grazing is the best use of most rangeland because rangeland is too dry, or too rocky, or too infertile for other uses. Most rangeland, if cultivated, would erode badly. Generally, little is done with range because water shortages make improvement unprofitable.

Care is needed, however, to keep range productive. The Soil Conservation Service rates the condition of rangeland in four classes: poor, fair, good, and excellent. A Government Accounting Office report of 1988 estimated that of the land managed by the Bureau of Land Management, 34% was in good condition, 41% in fair, and 25% in poor condition. Rangeland in poor condition produces only one-quarter of the forage it could produce. Range in poor condition also loses three to five times as much soil to erosion compared with range in excellent condition.

Controlled grazing is the key to stopping erosion. The number of grazing animals should be only as large as the land can safely carry. Animals should not occupy a single range for a long period and they should not feed on specific rangeland during periods of slow plant growth.

Range can be fertilized to increase growth. Experiments at the Northern Great Plains Research Center indicate that 30 pounds of nitrogen applied to an acre of crested wheatgrass boosted yields by a factor of 2.5. Researchers at the center suggest fertilizing range grasses at the rate of 40 pounds of nitrogen per acre per year to improve grass stands and beef production (USDA, *Agricultural Research*, May 1984).

Rangeland plants are usually native grasses, broadleaf plants, or shrubs. A rancher can improve forage yield and quality by seeding improved grasses. No tillage is involved since specially constructed seeders drill the seed through existing vegetation (figure 15-19). Some of these grasses are less drought-resistant than native grasses; many of them suffered severe drought damage during dry years in the late 1970s and early 1980s.

Figure 15-19. Renovation of rangeland by seeding improved grasses. (Courtesy John Deere Company)

Limited water management can help overcome the water shortage that is the primary inhibitor of forage growth. Digging closely spaced shallow ditches across slopes has helped reduce runoff. Disc plows may be modified to make small pits in the ground to capture water: this practice is called *pitting*.

ORGANIC FARMING AND SUSTAINABLE AGRICULTURE

Organic farming is usually defined as farming in which no inorganic fertilizers (as defined in chapter thirteen) or synthetic pesticides are used. The term "organic farming" itself has little meaning. Many organizations of such farmers now use other terms. A major theme shared by organic farms is working to keep a healthy soil by controlling erosion and keeping organic matter levels high.

In 1980 the United States Department of Agriculture (USDA) published a study of organic farms. One of its conclusions was that more information is needed. However, the study draws some interesting, if guarded, conclusions. The researchers found that there are many varieties of organic farms. Typically, organic farmers outside the dryland wheat-growing region adhere to strict rotations of row crops, small grains, and legume forages. Many have a mixed operation of crops and animals. Nitrogen is provided by the legumes, manure, and sometimes by amendments from the outside. Phosphorus and potassium are provided by manure and sometimes mineral fertilizers or other outside sources. Many organic farms that had been farmed by conventional methods are using up potash and phosphorus that remain from earlier fertilization.

Organic farms depend on tillage. Most use the chisel plow rather than the moldboard. Without herbicides, they do more cultivating than other farmers. The rotary hoe is a favorite tool. Also, because they use no herbicides, most of the conservation tillage methods cannot be used, except for mulch-till. Crop rotations also help in weed control. The USDA study also reached the following conclusions:

- Conventional farms had better yields of corn and wheat, but organic farms had slightly better yields of soybeans and oats. Organic farms tend to do better than conventional farms in dry years and worse in wet years.
- Conventional corn-soybean farms tend to be more profitable, mostly because of a more profitable crop mix. Aside from the differing crop mix, profits were similar. Organic farms yielded slightly less but production costs were lower.
- Organic farms had good nitrogen levels and good weed control.
- Organic farming is practiced on farms of all sizes.
- Most organic farm products enter regular marketing channels, rather than "organic food" channels.
- Many farms are not purely organic, but use fertilizers and pesticides sparingly in problem areas.
- Erosion is reduced on organic farms.
- Organic farms use only one-third the energy of conventional farms.

This last conclusion is of some interest. Conventional farms use much more energy than is harvested in the form of food. This energy use (except on irrigated land) is primarily in the form of fertilizers and pesticides. Conventional farms depend on the continual use of fossil fuels and would suffer more than organic farms if fuel prices rise sharply or the supply of fossil fuels is threatened.

LISA, or *Low Input Sustainable Agriculture,* a hybrid between organic and conventional farming, has received much attention in the 1980s. Some colleges now even have degree programs in the subject. Two main goals drive sustainable agriculture: First, to minimize off-farm inputs like fertilizers and pesticides, and to maximize on-farm resources, such as legumes for nitrogen and erosion control. Second, to strive not for top yields, but for the most profitable yields by reducing input costs.

Those who practice sustainable agriculture are driven by several concerns. Among these are the rapidly rising costs of production that squeeze the grower, worries over the long-term effects of using up energy and resources, and pollution. The strategies are not unique; many are covered in this text. Examples include crop rotation, legumes for nitrogen fixation, conservation tillage, and others. The difference is the broad ways these are all brought together to achieve the goals mentioned above.

SUMMARY

Tillage has three goals: (1) weed control, (2) alteration of physical soil conditions, and (3) management of crop residues. Tillage also has a number of side effects, however, especially an increase in erosion and compaction.

Conventional tillage buries crop residues to produce a smooth, residue-free seedbed. Conservation tillage leaves residues on the soil surface to prevent erosion and preserve soil water.

Three cropping systems are used by growers: continuous cropping, crop rotation, and multiple cropping. Continuous cropping (or a simple corn-soybean rotation) allows a farmer to grow the most profitable crops yearly. Crop rotation, on the other hand, is better for the soil and helps control erosion.

In low-rainfall areas, small grains are grown in rotation with summer fallow. During fallow, weeds are controlled by cultivation or weed killers to store moisture for the following crop. Problems with summer fallow include erosion and saline seeps.

Animal grazing is the most practical use of dry, steep, or rocky land in the West. Controlled grazing, seeding of improved grasses, and sometimes fertilization and water management keep range in good condition.

Organic farming replaces chemical fertilizers and pesticides with crop rotation, manuring, cultivation, and mineral fertilizers. Organic farmers focus on having a "healthy" soil. LISA aims to reduce some of the problems of standard agriculture by using techniques that lower off-farm inputs and increase use of resources found on the farm.

REVIEW

1. There is no value to cultivation except weed control (True/False).
2. Subsoiling should be done on moist soils.
3. From the soil conservation perspective, spring plowing is better than fall plowing.
4. Special planting equipment is needed for conservation tillage.
5. Crop rotation can reduce weed problems.
6. High nitrogen fertilization is needed for legumes.
7. Saline seeps can be a problem where summer fallow is practiced.
8. Most rangeland could be better used growing crops.
9. The soil tends to be warmer under no-till than with conventional tillage.
10. Sustainable agriculture tries to reduce off-farm inputs.
11. Conservation tillage requires at least _____ percent residue cover.

12. The moldboard plow is an example of a _____ tillage tool.
13. The system that depends most on herbicides for weed control is _____ .
14. Two salt-tolerant crops mentioned in this chapter are _____ and _____ .
15. The best conservation tillage system for soils that tend to stay cool and damp is _____ .
16. In conservation tillage, fuel and labor costs are likely to _____ .
17. In grain-growing on the semi-arid Great Plains, a weed control tool that does not bury residues is the _____ .
18. Conventional plowing that creates ridges for planting is called _____ .
19. The largest amount of residues are left on the surface in the _____ system.
20. The conservation system that uses primary tillage uses the _____ plow.
21. List three basic goals of tillage.
22. List three benefits of crop rotation.
23. List three benefits of conservation tillage.
24. What is allelopathy?
25. What purpose does a harrow serve?
26. Describe changes in the top few inches of soil in a no-till field.
27. Explain why all the water is not lost from the soil by evapotranspiration during summer fallow.
28. Explain why organic farms use less energy than conventional farms.
29. What is the case for fall plowing? Against?
30. This chapter briefly mentioned reasons why some folks advocate sustainable agriculture. Write a short essay expanding on their reasoning, using information presented in earlier chapters.

SUGGESTED ACTIVITIES

1. If you are not already familiar with the tillage tools described in this chapter, visit an equipment dealer to look at them.

2. If you have not already used various tillage tools, arrange to learn how to use them.

3. Survey and visit local farms to find out what tillage and cropping systems they use.

4. Study local fertilizer recommendations for the crops grown in your area.

chapter sixteen

Horticultural Uses of Soil

Here's a building contractor who cares little for the trees by the home he is building. Equipment has compacted soil around the tree's roots, construction debris is piled around the tree, cutting off aeration. This tree will probably die. An understanding of soil and plant interactions are as important here as they are on the farm. This chapter will describe soil management as practiced by landscapers and others that practice horticulture.

OBJECTIVES

After completing this chapter, you should be able to:

- state how to select soils for horticultural crops
- describe fertilization practices for horticultural crops
- describe how growers manage their soils
- solve the special problems of container soils
- describe how soil influences landscaping

TERMS TO KNOW

coarse aggregate	perlite	vermiculite
perched water table	soil-based potting mix	xeriscaping
perforation	soil-less potting mix	

Webster's New Collegiate Dictionary (8th edition, 1979) defines horticulture as the "science and art of growing fruits, vegetables, flowers, or ornamental plants." The information presented so far in this text about using soils applies to horticultural crops as well as to other crops. Let's see how soil is used by growers of different crops, starting with vegetable growers.

VEGETABLE CULTURE

Vegetables are the most important horticultural crop in terms of total value. They make an important contribution to the human diet, supplying some minerals and vitamins that are missing in grains and meats. Vegetables are grown throughout the United States, but southern regions grow most of our produce because of their long growing season.

Soil Selection.   Vegetable growers often select a specific soil type that suits the needs of the crop to be grown and his or her marketing needs. Many growers in northern areas choose coarse soils because they warm up rapidly in the spring, allowing early planting and early harvest when prices are best. In general, vegetable growers favor coarser soils than do other farmers. The soil types and their uses are briefly reviewed:

- *Coarse-textured soils* are best for very early crop growth, especially for cool-season crops like lettuce or carrots. Several crops, like melons, grow best on sandy loams. For best yields, these soils are usually irrigated.
- *Medium-textured soils* are good for all crops. Where yields are more important than an early harvest, medium soils are better than coarse ones.
- *Fine-textured soils* are less desirable for vegetables. They tend to stay wet too long, hampering field operations. Heavy soils are very poor for root crops.
- *Organic soils*, especially mucks, are favorites for cool-season and root crops (figure 16-1). They are ideal for carrots, onions, and celery. Warm season crops like tomatoes seldom thrive in organic soils because they tend to be cold, and are often in "frost pockets."

Soils selected for growing vegetables should be loose, friable, and high in humus. For most crops, a slightly acid soil is best (except for potatoes, which grow best between pH 4.8 and 5.4). The most essential factor for success is good drainage. Poorly drained soil warms up slowly and cannot be planted early. Vegetable crops keep their quality for a very short time once they are ready for harvest. Thus, they must be picked regardless of soil conditions. This is especially true for crops grown on contract to a cannery—the day the crop is ready, harvest machines should be in the field. In this case, growers can ill-afford a muddy soil.

Soil Management.   Vegetable growers prepare their fields much as do other growers. If needed, drainage is installed. Most growers use conventional tillage; that is, the field is plowed and harrowed before planting, or the field is prepared for furrow irrigation. There is some trend by vegetable growers toward conservation tillage. However, many vegetables have very fine seeds that are not well suited to the rough seedbed and high residue levels of conservation tillage. Also, live plants are planted for tomatoes, cabbages, and other crops, and transplanters do not handle the thick residues well. Herbicides and/or frequent tillage control weeds.

Figure 16-1. Radishes growing on an organic soil. (USDA, Soil Conservation Service)

It can be difficult for a vegetable grower to keep organic matter levels high, because many vegetables simply do not grow the bulk of green matter that field crops do. A large part of the plant may even be harvested, as in lettuce or cabbage. Growers may harvest both roots and tops, such as in green-top carrots or beets. To make up for the lack of crop residues, many growers use animal or green manures as often as possible.

Plant nutrition presents problems as well. Vegetables have a wide variety of nutritional problems. Many vegetable soils, being coarse, are naturally infertile. Irrigation makes the problem worse by leaching out nutrients. Muck soils are usually low in nitrogen, and especially in phosphates, potash, and lime (which are nutrients deriving from minerals.) Finally, many vegetables are sensitive to low levels of trace elements.

In general, growers feed vegetable crops by the same methods as those described in chapter thirteen. One difference is the use of starter solutions on vegetable transplants. Crops like tomatoes and peppers are transplanted into the field rather than seeded. The transplanting equipment has a tank to hold a weak fertilizer solution. The solution soaks the planting hole of each plant as it is transplanted. Starter solutions are high in phosphate (to overcome phosphate immobility and stimulate rapid rooting). The fertilizers used typically have ratios of 1-2-1.

FRUIT CULTURE

Soil Selection. Tree fruits require well-drained soil at least three or four feet deep. Shallow soils cannot support a tree crop during dry periods, unless irrigation is provided. Soils with a high water table also restrict the deep rooting of fruit trees. Some fruits are grown in soils that tend to stay wet after a rain. A serious health decline is common in these trees during rainy weather as soil-borne fungi attack weakened roots.

Fruit plants tolerate a wide pH range. The recommended pH range is 6.0 to 6.5, because it suits both the fruit plant and any cover crops that may be grown with the fruit. Blueberries, however, need a pH of 4.3 to 4.8.

Many fruits are clean cultivated to remove competition from weeds or sod. The plants must be grown on fairly level ground that is unlikely to erode, unless a slope is terraced. Apples, when grown in sod, can be grown on rather steep land. In fact, orchardists favor hillsides for growing apples because cold air drains off hills on frosty nights. (Cold air is heavier than warm air and thus sinks.)

Fruits also tolerate a wide range of soil textures. Apples and pears do best on moderately fine-textured soils, while stone fruits, such as plums and peaches, prefer a coarser texture. Grapes grow on any soil, but the sweetest grapes are grown on sandy or gravelly soil (figure 16-2). Berry plants do best on moderately coarse soil.

Soil Management. All fruit crops are perennials. Many occupy the same ground for 10 to 20 years. Therefore, the soil must be properly prepared before

Figure 16-2. Grape vineyard operators often favor a coarse, gravelly soil because it produces a high quality, sweet grape.

planting—there is no way to try again without a major financial loss. The site should be carefully selected. Any major soil changes like terracing, drainage, or leveling should be completed. A soil test should be taken. Based upon the results of the test, the needed amounts of potash, phosphate, and lime are broadcast and plowed into the soil.

Nitrogen is the most important nutrient for the established fruit crop. However, excess nitrogen or shortages of other elements causes poor fruiting, poor fruit quality (figure 16-3), and late ripening. In addition, the tree is more likely to be injured by diseases or cold. Growers usually fertilize by topdressing in the late fall or early spring with a high nitrogen fertilizer. Plant tissue testing can be useful in determining if elements other than nitrogen are needed on an established crop.

Many fruits are very sensitive to low levels of trace elements. In fact, shortages occur in several fruit growing regions of the United States. The best method of determining crop needs is to perform a soil test followed by a tissue test. Trace element problems can be solved by mixing trace elements into the base NPK fertilizer or by foliar feeding with chelated elements. Chapters eleven and thirteen suggest some trace element fertilizers. In some cases, liming or acidifying the soil solves micronutrient problems.

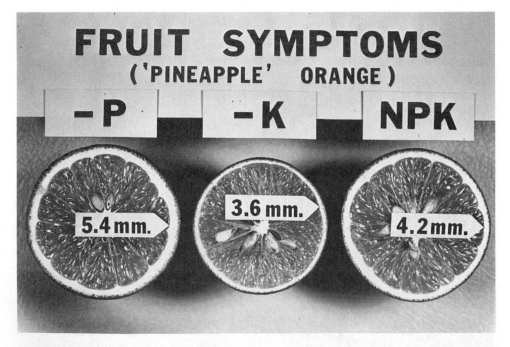

Figure 16-3. Nitrogen is the most important nutrient in tree fruits; however, shortages of other elements can harm fruit quality. Here a shortage of potash resulted in a thin orange rind. (Courtesy Potash and Phosphate Institute)

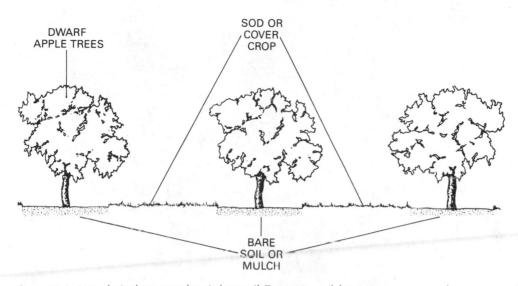

Figure 16-4. Many fruit plants grow best in bare soil. To prevent soil damage, growers may plant cover crops or sod between the rows of fruit trees.

Fruits are hard on the soil if they are clean-cultivated, as many are. Clean cultivation results in a loss of organic matter, poor tilth, and too much erosion. To overcome the problem, growers use manures, cover crops, mulches, and sod. Figure 16-4 shows an example of how these treatments are used. The figure shows part of a dwarf apple tree planting. The grower wants to make sure the apples do not compete with weeds or sod for nutrients and water. One way to do this is to keep the whole soil surface bare, or clean cultivated. Another way to control weeds, with less soil damage, is to keep a strip of ground in the tree row free of weeds or grass by mulching, cultivation, or herbicides. Between the rows, the grower can plant a cover crop like oats. These areas can be mowed during the summer and tilled in the following spring. For apples, the best method is to plant a permanent sod between the tree rows. The vegetative cover between the rows protects the soil and is a source of organic matter. Mulching in the row also provides benefits like saving moisture, adding organic matter, and protecting the soil and fruit.

Mulching can also be very useful for small fruits (figure 16-5) to suppress weeds and save moisture. The mulch means that pickers can work in the field after rain or irrigation; mulch also keeps fruit off the ground.

NURSERY FIELD CULTURE

Soil management is a constant challenge for the nursery grower. The very process of growing nursery stock is very hard on the soil. The soil is clean-cultivated, and the only crop residues are a few leaves (figure 16-6). The soil itself

Figure 16-5. These strawberries are mulched to protect both the soil and the fruit and to conserve moisture.

is dug up and hauled away when trees and evergreens are balled and burlapped. For example, using the figures for root ball sizes from the American Standards for Nursery Stock and average soil bulk densities, it can be determined that digging up an acre of five-year-old evergreens removes 165 tons of soil.

Site Selection.   Soil is one of several factors to consider in selecting a nursery site. The land should be level, so erosion can be controlled. It should be well drained, because wet soils hamper nursery operations. A pH between 5.5 and 6.5 is preferred for most nursery crops.

Soil texture is an important concern for nursery crops. Plants to be harvested bare-root (soil is shaken off the roots) should be grown on sandy or sandy loam soil. Coarse soils most easily shake off plant roots without damaging them. In nurseries where stock is balled and burlapped (soil ball stays on roots), finer-textured soils are better. Soils high in sand do not stick together, so it is hard to dig up an intact soil ball in sandy soils (figure 16-7). Silt loam or clay loam soils are best for balling and burlapping.

Figure 16-6. A nursery soil is always bare and is very exposed to rainfall. Very little organic matter returns to the soil from the crop. It is difficult to preserve soil structure and control erosion in this situation.

Soil Management. Nurseries, perhaps more than any other operation, need to make a serious effort to maintain organic matter levels. Crop rotations of nursery stock with green manure are very typical. Sudangrass, alfalfa, or other vegetation is used to build the soil. Many growers apply animal manures and/or sewage sludge. A few growers plant cover crops between nursery rows or mulch in the rows with chopped hay or wood chips.

A new option could be called a "living mulch." Winter rye is planted in the nursery in early fall, allowed to grow, overwinter, and develop in the spring. When the trees begin putting on growth, kill the rye with an herbicide. The stand of dead rye acts like a mulch, protecting soil and moisture. Even more, chemicals given off by the rye suppress weed growth (allelopathy).

Before planting stock, a soil sample should be taken from the top two feet of soil. Based upon the results of the soil test, the correct amount of potash, phosphate, and lime (or sulfur if the pH is too high) is mixed into the soil. Each year thereafter, high nitrogen fertilizers with ratios of 4-1-2 or 3-1-1 may be

Figure 16-7. It is difficult to dig up an intact soil ball around plant roots in a coarse soil.

Crop	Lb N/1000 Ft^2	Lb N/Acre
Deciduous plants	5	225
Narrowleaf evergreens	4	175
Broadleaf evergreens	3	125

Figure 16-8. Guidelines for fertilizing nursery stock. The best practice is to follow local guidelines.

topdressed in the fall or spring. Figure 16-8 suggests how much nitrogen should be applied.

CONTAINER GROWING

One of the most demanding ways to grow a plant is to grow it in a pot. A containerized plant requires constant attention to watering, fertilizing, and other practices. Despite this, more and more plants are being grown in containers. Not only are greenhouse growers growing mums, poinsettias, and other flowers in pots, over the past few years, more and more nurseries have started to grow shrubs, evergreens, and trees in containers (figure 16-9). There are a number of reasons for this change. One important reason is that the container grower has complete control over soil conditions, making it easier to grow a large, uniform crop of quality plants. More recently, the business of landscaping the *inside* of buildings with potted plants has grown rapidly. Apartment dwellers and even homeowners now garden in containers.

Growing plants in containers differs from growing plants in the ground in one key way: the plant's root system is confined to a small soil volume that must supply all the plant's water and nutrient needs. This means the container grower waters and fertilizes far more often than those who till the ground. This section will look at four major topics: the naturally poor drainage of potted soil, types of potting soil, soluble salts, and soil sterilization.

Container Drainage. A pot of soil is, by definition, poorly drained because of the shallow soil profile. To understand this statement, compare soil in a pot with soil in the ground. Recall from the discussion in chapter four that capillary action "pulls" water into the drier soil below a wetting front. In a deep soil profile, then, lower layers of soil pull water downward.

Figure 16-9. Growing nursery stock in a container presents special challenges. One key to successful growing is the use of highly porous potting mixes.

In a pot, the soil column ends abruptly at the bottom of the pot. The bottom layer of soil has no capillary connection to deeper soil and the last bit of water cannot drain away after watering. Thus, in spite of drainage holes, a layer of soil on the bottom of the pot remains saturated after drainage ceases. This layer is called a "perched water table" (figure 16-10). As a result, potted soil is wetter and has less air after drainage than the same soil in the ground.

The difficulty is the short water column—no "depth" to pull water down, no heavy mass of water pushing down by gravity. Therefore, the taller a pot is, the less severe the problem (figure 16-10). A six inch pot filled with a standard greenhouse mix has a porosity of about 20%. Some greenhouse containers are no more than an inch deep. With the same mix, such a pot has a porosity of perhaps two percent!

Because of poor drainage, a potting mix must be highly porous, with very large pore spaces. The large pore spaces apply less capillary force to hold the water in the pot. There are two approaches to making a porous mix. One is to mix materials into a field soil to make a porous "soil-based mix." The second method is to omit the soil altogether, making a "soil-less" mix out of other materials.

The key to making potting mixes is to use large particles to create large pore spaces. For example, one simple soil-less mix is half sand and half peat. When a fine sand is used to prepare the mix, the percentage of the total soil volume filled with air after watering is about 5%. The same mix with coarse sand has an air capacity of 16%. Thus, the mix with coarse sand holds more than three times as much air as the fine sand mix.

Potting Mixes. Good potting mixes have a high holding capacity for air and water. To accomplish this the mix needs large particles that can absorb water. Picture a pot full of shredded sponge. Each piece of sponge can soak up water, and the empty spaces between the pieces can hold air. Growers may not use

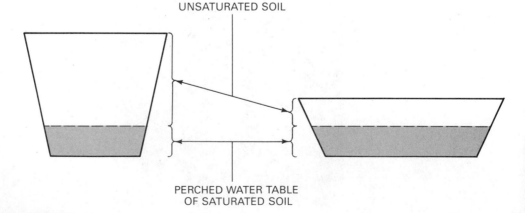

UNSATURATED SOIL

PERCHED WATER TABLE
OF SATURATED SOIL

Figure 16-10. A perched water table in a pot causes poor drainage. Highly porous soils and fairly tall pots help solve the problem.

shredded sponge, but they produce soil-based or soil-less mixes that behave the same way. The mixes contain varying amounts of three materials:

- *Soil* is the main part of soil-based mixes. It cannot be used alone, but must be mixed with the other two materials. Soil helps a mix hold water and nutrients, and helps buffer the soil from rapid changes in moisture, pH, and others. However, the fine particles retard drainage and lower aeration. In shallow pots, soil should be avoided or used only as a small percentage of the mix. In pots large enough to have a deep soil column, soil may be used more freely. The exact amount of soil depends on its texture—the coarser the soil, the more can be used—the depth of the pot, and watering practices. Soils used in potting mixes should be loamy with good structure and be free of pesticide residues.
- *Coarse aggregates* are large, inorganic particles used to create large pores in the mix. Coarse aggregates include coarse sand, perlite (large granules of lightweight expanded volcanic glass) and vermiculite (expanded mica). Shredded plastics and other materials are also used.
- *Organic amendments* hold water and may themselves help porosity. Shredded sphagnum peat, mined from peat bogs, is in most common use (see chapter eight). Many growers shred and compost tree bark or sawdust. Leaf mold, garden compost, and many other materials may be used.

Growers combine these materials into various mixes to suit their needs. Standard soil-based mixes mostly follow the model set by the John Innes mixes developed in England in the 1930s. Generally, these mixes consist of loam, peat, a coarse aggregate, and fertilizers. Later, soil-less mixes based on mixing peat with a coarse aggregate were developed at the University of California (UC mixes) and Cornell University (Peat-lite mixes). Figure 16-11 briefly summarizes these mixes.

	Common Potting Mixes				
	John Innes	UC Mix D	Cornell Peat-Lite A	Cornell Peat-Lite B	Bark Mix
Loam	7	—	—	—	—
Sand	2	1	—	—	1
Peat	3	3	1	1	—
Perlite	—	—	—	1	—
Vermiculite	—	—	1	—	—
Composted bark	—	—	—	—	2

Figure 16-11. For these common potting mixes, the numbers represent parts. For example, Peat-lite A consists of 1 part peat and 1 part vermiculite. These mixes also contain some combination of lime, gypsum, fritted trace elements, superphosphate, and other fertilizers. These mixes are only a sample of the large number of mixes that are in common use.

Many growers now use mixes based on composted hardwood bark chips or pine bark. Bark chip mixes have very high porosity, and they seem to suppress many harmful soil organisms.

Soil Sterilization.   Soil-based mixes contain weed seeds, insects, nematodes, and other parasitic fungi and bacteria. Of special concern are fungi that destroy young seedlings. To kill these organisms, some growers treat the soil with chemicals. The soil must "air out" for several weeks after such a treatment to avoid injuring crops planted in the mix. More commonly, soil is sterilized by heat. Many growers use steam—normally a temperature of 212°F. However, this high temperature may cause three problems:

- High heat kills the bacteria that convert ammonia to nitrates (nitrification) but not the ones that change organic matter to ammonia (ammonification). This break in the nitrogen cycle can cause a buildup of ammonia to harmful levels.
- High heat raises the solubility of manganese, which can result in toxic levels.
- High heat creates a "biological vacuum," destroying organisms that could compete with pathogens if the soil were reinfected.

Several methods avoid these problems. First, lime can be used in a potting mix to maintain the pH between 6.0 and 7.0. At this level, manganese is fairly insoluble, reducing toxicity problems (see chapter ten). Second, potting mixes should be used soon after treatment, before high levels of ammonia can build up. Third, "live steam" (temperature 212°F) can be replaced by "aerated steam" (temperature 160°F to 180°F). Special devices inject air into live steam to lower the temperature. This temperature reduces ammonia and manganese problems, and it also allows some helpful organisms to survive while killing pathogens.

Soluble Salts.   To grow potted plants successfully, a great deal of fertilizer (fed through irrigation water or incorporated into the potting mix) must be poured into a small soil volume. Most irrigation water contains dissolved salts, and most fertilizers are salts. Therefore, one of the most troublesome problems of growing in pots is the buildup of soluble salts. These practices help a grower avoid the problem:

- Use high quality water (figure 6-22). Measuring devices like those shown in figure 16-12 can be used to measure salts in irrigation water. Many soil testing labs can also test water salinity.
- Saline water can be treated with special devices, but the process can be expensive for large amounts of water. A common water softener merely replaces calcium ions with sodium ions and does not lower salinity. Some California nurseries, contending with salty irrigation water, have installed very large—and expensive—systems to remove salts from their water supply.
- Use low-salinity fertilizers (see figure 13-26). For instance, potassium nitrate is a less saline source of potassium than potassium chloride.

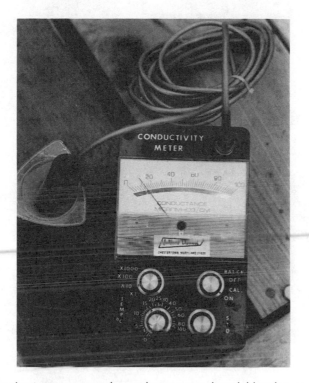

Figure 16-12. A conductivity meter can be used to measure the soluble salt content in greenhouse soils.

- Make sure all pots have good drainage. When watering, use enough water so that some water leaks from the drainage holes. Drainage water leaches out excess salts.

Water Pollution.   The public is concerned about water pollution (see chapters five and fourteen); this concern strongly affects nurseries and greenhouses. For example, container nurseries have pots sitting on the ground, and fertilize through overhead sprinklers. Most of this water—as much as 70%—lands between, rather than in the pots. Since a container field is watered and fertilized daily, the potential for water pollution is great.

One answer, where feasible, is to use trickle irrigation (figure 16-13). Little water or nutrients are wasted this way. Second, greater use of slow-release fertilizers added to the pots could reduce the amounts added to irrigation water. Third, pots should be set on a sealed surface that stops leaching into the soil. Last, the land surface should be carefully graded so water runs off into sealed holding ponds, where the water can be pumped for reuse.

Figure 16-13. Trickle irrigation on boxed stock in a California nursery. Trickle irrigation can reduce water use and fertilizer pollution.

LANDSCAPING

Landscaping makes our surroundings more beautiful and more livable. While most of us cannot live in a setting like the one shown in figure 16-14, even modest landscapes improve our lives. Landscapers should know about soil so that landscape plants will remain healthy, attractive, and easy to care for.

Most transplanted shrubs and trees survive—yet many fail to thrive, and too many actually die. These failures cost money, in replacements and damaged customer relations. Often building contractors are responsible (see title page of this chapter), sometimes homeowners fail to provide proper care, and other times a landscape professional is at fault.

When plants fail to thrive, more often than not, one must look to the roots for answers. That means an understanding of soils is essential. Here are just two examples:

- A few inches of soil are added to the yard to raise the grade. The blanket of new soil cuts off oxygen to tree roots; trees decline for several years than die. Many trees are very sensitive to raised grades.

Figure 16-14. Cheekwood Gardens in Nashville, Tennessee. Knowledge of soil is important to landscaping success, whether the project is an elaborate estate or a backyard garden.

● Many soils are not acid enough for the trees being planted in them. In the author's home city, pin oaks are planted by the thousands, yet most suffer iron chlorosis from unsuitable soil pH.

More than most, landscapers must understand how soils, roots, and water interact. Landscape soils are complex compared to field soils, because they are radically altered by construction and landscaping. Soils of different textures are placed together, and as chapter four points out, that affects water movement. Comparing fine to coarse soils:

1. Because of small pore spaces, water moves through clay more slowly. That is, clay has a lower hydraulic conductivity.
2. Because of small pores, fine soil has a stronger capillary force. That is, it has a higher matric potential. Thus, water will not pass out of clay into sand until the clay is saturated (see figure 4-12 and explanation).
3. Any sharp interface between soils of different textures retards the movement of water and the growth of roots across the boundary.

Here are two examples of how these rules affect landscaping:

Example One: A yard soil is coarse, droughty sand, so the contractor spreads three inches of clay loam on top. Sod is laid above that topdressing. When the sod is watered, the top three inches must saturate before any water drains into the sand (rule one and three). After watering, the topdressing is water-soaked, the sand still mostly dry. When the top three inches dries out, there is little deep reserve. Thus the turf alternates between being water soaked and dry.

Example Two: Age-old practices advise filling in a tree planting hole with a nice soil amended with lots of organic matter. That creates two interfaces—between the root ball and planting soil, and between the planting and surrounding soil. Not only will water movement be altered, but roots may grow into the planting soil, but not into the surrounding native soil. Research conducted at the Oklahoma Agriculture Experiment Station indicates no benefit from amended planting soil.

Now let's see how our knowledge of soil can help us understand proper landscape practices.

Site and Plant Selection.　Too often, landscape designers pick a plant only because it would look pleasing in a certain spot, not because it would grow well in that spot. Mismatches between plants and site, only too common, lead to unhealthy plants and endless maintenance as grounds managers try to keep the plants looking attractive. Landscape designers should know which plants grow well in their area and select plants that match the soil drainage, pH, salinity, and degree of compaction on the landscape site.

A thorough designer will actually examine the soil on a site. Salinity and pH can be checked with a soil test or portable testing device. Compaction can be checked simply by seeing how hard it is to shove a screwdriver into the ground, taking soil type and moisture into account. Soil color provides guidance on drainage, or a very simple, crude percolation test can be done. Dig a one foot hole into the ground, fill with water. After it drains fill again, and see how much drains in a couple of hours. Experience will make these simple tests most valid.

Avoiding Compaction.　Most landscape sites are slightly to severely compacted during construction. Studies on the effects of compaction on tree growth showed interesting results. Compaction caused shallow root systems. Often, roots from deep in the planting hole actually grew upward along the sides of the hole towards the surface, rather than outward into the surrounding soil. Failure to explore the native soil results in drought damage, nutrient shortages, and poor anchorage.

Tillage helps tear up the compaction, but often little can really be done about it. However, landscapers can avoid adding to compaction while working. Landscapers who drive on a site with trucks full or rock mulch only compound the problem. While wheelbarrows require more worker effort, they compact soil much less than dump trucks. When heavy vehicles must be driven, especially over turf, laying heavy planks under the wheels avoids compaction.

Transplanting. The key to successful transplanting is rapid root growth. After all, for plants dug out of a nursery, 98% of the root system is lost. Even containerized stock must grow new roots quickly, because it can't survive long on just the soil mass it was planted with. How do landscapers ensure that the roots of newly planted trees and shrubs will grow quickly? Recent work by the Oklahoma Agriculture Experiment Station questions many standard transplanting practices. Besides the examples listed earlier, they have other suggestions.

Research shows that plastic mulch, commonly used in shrub beds to keep out weeds, restricts root growth by keeping oxygen out of the soil, as pictured in chapter one (figure 1-10). Organic mulches, like wood chips, keep down weeds and conserve moisture without robbing the soil of oxygen. Below the wood chips, one could place landscape fabric rather than plastic sheeting. This woven material is porous, allowing movement of air and moisture but restricting weed growth.

Landscapers have also avoided fertilizing when planting, fearing root injury. Yet farmers have long banded fertilizers within inches of planted seeds. Both nitrogen and phosphate are needed for new root growth. There are two ways to safely fertilize newly planted trees. One is to drop some slow-release fertilizer into the planting hole. Various slow-release products are sold for this purpose. The second way is to spread fertilizer over the soil surface after planting.

A recommended procedure for transplanting is as follows:

1. Check the drainage of the site. If poor, drainage systems may need to be installed. If that is impractical, then create berms for planting. Berms are attractive, rounded, raised beds that will get the plants above the wet soil.
2. If using balled and burlapped stock, try to find stock with soil of texture similar to the soil on the site. If using container stock, make sure the plant is not pot-bound, and that lots of nice white root-tips are visible.
3. Prepare a large planting hole. It should be much larger than the width of the root system, but not deeper. If dug deeper, the plant will settle and end up too deep. This large hole will provide loose, uncompacted soil for the roots to grow into.
4. If the soil is just a bit less well-drained than ideal, plant a little high so the plant is on a low mound. Never plant deeper than the plant was growing in the nursery.
5. Backfill the hole with the same, unamended soil that came out of the hole. Drop some slow-release fertilizer into the hole.
6. Water in the plant carefully.
7. Mulch around the tree with landscape fabric and about three inches of coarse mulch. Avoid fine mulches; they can retard aeration. Fertilize the mulch a bit.

Amending Soil pH. All too often, the native pH of a soil is not proper for all the plants one wants to use. Obviously, this calls for designers and others to check for pH, an easy enough task with a soil test or even a piece of pH paper. Once the pH is known, a designer can make informed plant selections.

Nevertheless, there are times when one needs to change the pH of a soil. Since most landscape plants prefer acid soil, acidification is most common. One common suggestion for acid-loving plants is to dig an especially large planting hole, and then put half by volume peat moss into the backfill. This will work temporarily, but as the peat decays, the pH will return to normal.

Three chemicals are used to acidify soils more permanently: sulfur, aluminum sulfate, and iron sulfate. There are disagreements about which is best, but the author prefers sulfur as being inexpensive, easy to use, and because of concern with aluminum toxicity. Table 10-19 (in chapter ten) suggests amounts to be used. Mix the sulfur into the soil. The pH of calcareous or very alkaline soils may be very difficult to change even by this means.

Landscapers should be aware of the many sources of calcium that can raise pH in a landscape site. Sources even include the lime in a concrete foundation, limestone rock mulches, and calcium-rich irrigation water. Besides avoiding these sources, periodic light treatments can maintain acidity. Annual use of acid-forming fertilizers like ammonium sulfate, or special acid preparations can be used. Sulfur-coated urea, a slow release fertilizer, will also enhance soil acidity.

Trees that do begin to suffer chlorosis can be treated by sulfur, spread over the soil surface. While this changes the pH of only the top inch or so, there are enough roots there to provide adequate iron for the whole tree. Foliar feeding with chelated trace elements can provide a temporary green-up in some cases.

Fertilizing Established Trees. Once landscape plants are growing on the site, fertilizing will continue to improve growth. Most homeowners and landscapers fertilize turf, but few are aware of how important it is to feed trees and shrubs. Young shade trees reach full shade tree size much more quickly, and will resist drought and stress, when watered and fertilized.

Timing strongly affects landscape plant response to fertilization. The best time to fertilize woody plants is when the woody plants are starting to go dormant. Nutrients absorbed in the fall will promote rapid growth the following spring. Early spring is also a good time to fertilize landscape plants. In cold climates, woody plants should not be fertilized from midsummer to early fall because it could spur late growth and keep the plant from hardening off for winter.

Nitrogen is the most important element for trees. To understand how to fertilize trees, one must first know where tree roots are (figure 16-15). Trees may root deeply, but 80% of a tree's feeder roots are in the top foot of soil. These feeder roots reach far from the trunk of the tree. One misconception is that tree roots all grow within the dripline of the tree. In fact, they often extend triple that distance from the tree trunk.

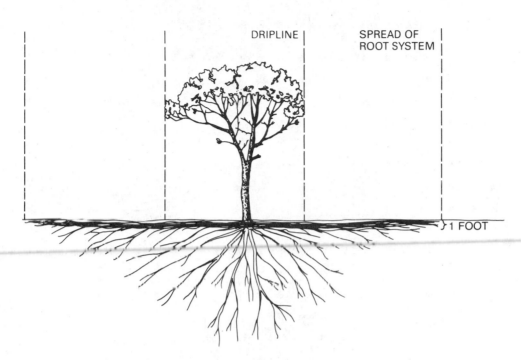

DRIPLINE

SPREAD OF
ROOT SYSTEM

1 FOOT

Figure 16-15. Tree roots cover a very large area and extend far beyond the dripline of the tree. Most feeder roots grow in the top foot of soil.

Shade trees are most simply fertilized by topdressing the soil over the whole root system of the tree. To be most accurate, dig a few holes to find how far the root system reaches. If this is not practical, assume the root system extends 100% beyond the dripline. Then topdress a high nitrogen fertilizer (2-1-1, etc.) over the entire root system at the rate of 3 1/2 pounds of nitrogen per 1,000 square feet. Nitrogen will, of course, leach into the root zone. If the shade tree grows in a lawn, split the fertilizer into two applications to avoid overfeeding the turf. One application should be in the late fall, the other in spring.

Another method of feeding shade trees is called *perforation*. This method is more difficult but it is better if the tree lacks phosphorus or potassium. The amount of nitrogen to be applied can be calculated by the method described previously. Another method, which is simpler but less accurate, is based on the trunk diameter (caliper) measured 12 inches above the ground. If the caliper is less than six inches, apply one-quarter pound of nitrogen per inch of caliper. For trees more than six inches in diameter, one-half pound of nitrogen is added per inch of caliper. A complete fertilizer high in nitrogen, such as a 2-1-1 ratio fertilizer, should be used. This fertilizer is applied by drilling many holes in the soil and dividing the fertilizer by the number of holes. Fertilizer is placed into each of the holes. The holes are drilled 12 to 18 inches deep in a grid pattern with 24-inch centers under the tree. Figure 16-16 shows how this is done.

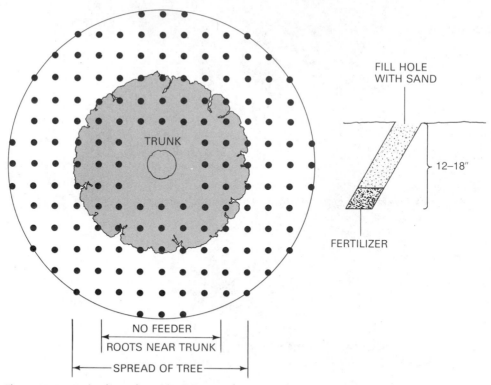

Figure 16-16. To feed trees by perforation, a grid pattern of holes is prepared. The required amount of fertilizer is divided by the number of holes to be filled. After the fertilizer is placed in the holes, sand is added to fill each hole. Perforation is useful for getting phosphate into the root zone, but this technique is not needed for nitrogen.

Some workers feed trees by injecting a solution through a soil needle. In this system, a solution is pumped out of a tank under pressure and forced through a tube inserted into the soil. The pattern for injection would be the same as in figure 16-16, except that it should be on 30-inch centers. Inexpensive versions of "root feeders" are sold in most garden centers.

Turf.   Except in arid regions, turf is the most common element of a landscape. Turf not only looks nice, it protects the ground from erosion. Turf may be planted by seed, sod, or in some cases, sprigs and plugs. Whatever method, good soil preparation is important to success.

Many turf specialists no longer recommend topdressing the existing soil with "black dirt" before planting turf, because the interface is a problem. If any is added, it should be tilled into the existing soil to avoid the sharp interface.

Chemical amendments should be tilled into the soil before planting, based on good soil tests. If recommended, phosphorus should be added now because of its immobility and importance for good root growth and turf spread. Potassium, which promotes good wear tolerance and resistance to disease and cold, also would be best incorporated now. A pH of 6 to 7 is fine; use lime or sulfur to reach that range if needed.

After the soil is tilled to incorporate amendments and to loosen the soil, make sure it is carefully raked or harrowed to make a fine seed bed free of bumps or pits. Plant as recommended locally. Frequent irrigation is needed while the turf is being established.

Turf is fed by topdressing with a high-nitrogen fertilizer. Often a fertilizer with a ratio of 20-1-3 is suggested, but seek out local recommendations. Fertilizers with a good water-insoluble nitrogen (WIN) content are best. In areas where soil salinity is a problem, turf fertilizers should be of low salinity, and care must be taken to avoid overfertilization.

A typical suggestion is about three and a half pounds of nitrogen per thousand square feet applied in three or four applications each year. The largest application, about 1¼ pounds of nitrogen, occurs in the fall. Two smaller applications follow in early and late spring, with a light fertilization in midsummer. One study also indicated that fertilizing turf with iron chelates has a number of helpful effects, and reduced the need for nitrogen.

Core aeration is a most useful turf soil management practice. To aerate, a machine, preferably one with hollow tines, creates small pits in the lawn. This breaks up compaction and increases aeration. Aerate only when the soil is moist; wet soil will compact worse, while turf in dry soil is under stress. Schedule aeration when the turf is growing actively. This means spring and early fall for cool-season grasses and late spring and early summer for warm-season grasses.

For season-long green color, irrigation of turf is usually needed. One expert suggests adding about an inch of water when 30-50% of the lawn shows signs of water stress—a slight rolling of the older leaf blades. Some grasses can be allowed to go dormant in a dry season, but excessive dryness can damage turf.

Xeriscaping. Landscapers—and their customers—of the more arid regions of the United States face the problem of water shortages. The problem has been dramatically compounded by the planting of landscapes adapted to the more moist eastern and northern states. The Fresh Water Society estimates that an average Eastern city dweller uses about 120 gallons of water a day; in Denver 230, in Phoenix 250, and in one small retirement community in Arizona, 750 gallons. The biggest contribution to this consumption is irrigation to support moisture-loving landscapes.

An answer to excess water use is 'xeriscaping' (from the Greek word 'xeros,' or dry). Xeriscaping is landscaping adapted to dry climates. It consists of these strategies:

- Use of low-moisture plants, often native to the area.
- Use of "hard" landscape materials such as paved surfaces, boulders, or pebbles.
- Efficient use of available water.

A wide range of plants are available that will thrive under low-moisture conditions. The most dramatic of these include cacti, succulents, and yuccas (figure 16-17). Being native to desert conditions, they are widely used in the xeric landscape. If purchasing such plants, be sure they have not been "rustled" from the wild.

However, xeriscaping need not mean only a mixture of sand and cacti. A number of shrubs and trees can tolerate dryness. For instance, Mexican redbud *(Cercis reniformis)* and New Mexico privet *(Fostiera neomexicana)* can replace their more moisture-loving relatives, Eastern redbud *(C. canadensis)* and Common privet *(Ligustrum vulgaris).* Examples of non-thirsty flowers include Madagascar Periwinkle *(Catharanthus roseum),* coreopsis *(Coreopsis spp.),* or showy sedum *(Sedum spectabile).* Some short-grass prairie grasses can be grown for lawn with minimal watering, such as blue gramma grass, the state grass of Colorado.

For even lower water usage, some of the plants can be replaced by "hard" features. Turf areas may be replaced by paving, such as brick patios, or by

Figure 16-17. A xeriscape at an old mission church in California. The plantings consist of cactus and succulents. A xeriscape need not be this severe.

mulching with pebbles. Some shrubs can be replaced by boulders, pole tips set upright in the ground, or similar features.

Lastly, the xeric landscape strives to conserve water. Plants can be installed into beds recessed below grade several inches to catch water, then mulched to slow moisture loss. Also, plants tend to be planted less densely, reducing competition for water. Trickle irrigation greatly reduces water consumption.

Arid regions that might practice xeriscaping contend with soil problems other than dryness. These can include alkaline soil pH, and saline or sodic soils. These problems call for selection of tolerant plants, or treatment as described in chapter ten. Incorporation of peat and sulfur into beds is helpful. These soils often suffer from the presence of a lime-cemented hardpan called *caliche.* If severe, the caliche may need to be broken up before planting.

Even in more humid parts of the country, elements of xeriscaping have value. This was forcibly brought to the attention of many in the summer of 1988, when serious drought brought widespread watering bans to the Midwest. Many now look to efficient irrigation and drought-tolerant plants as wise gardening. The use of native prairie plantings is another option.

SUMMARY

This chapter discussed how soils are managed to grow fruits, vegetables, and ornamental plants. An irrigated, coarse soil allows early production of vegetables, while a medium soil gives the best production. Vegetable soils must be very well drained. Many growers make good use of animal and green manures to make up for the small amount of organic matter most vegetable crops produce.

Fruits and field nurseries need a deep, well-drained soil. Clean-cultivated fruits and nurseries need level land and close attention to erosion control. Because the crops stay in the ground for many years, the land must be carefully prepared before planting, including drainage, terracing, irrigation installation, leveling, and plowing lime and nutrients into the soil. In the years after planting, nitrogen is the most important nutrient. Maintenance of soil organic matter is a challenge for these growers.

Soil in pots is poorly drained because the soil column ends at the bottom of the pot. To overcome this, potting mixes are very porous and consist of coarse aggregates such as sand, an organic material like peat moss, and sometimes soil. Soil-based mixes must be sterilized by heat or chemicals to kill weed seeds and soil pathogens.

For landscape plantings to grow well, the soil must be handled properly. Plants are chosen that will thrive in the soil on a landscape site. Landscapers must be aware of the effects that textural interfaces have on the movement of water and the growth of roots. Proper soil preparation for planting turf and good transplant practices are needed for healthy plants. Fertilizing established turf and trees also keeps them growing well. Xeriscaping is landscaping adapted to dry climates.

REVIEW

1. Organic soils are ideal for all vegetable crops (True/False).
2. Good drainage is critical for vegetable crops.
3. Nitrogen is the most important nutrient for established fruit crops.
4. Apples are usually clean cultivated.
5. For digging up shrubs bare root, a clay loam would be most suitable.
6. The deeper a pot is, the better its drainage.
7. Peat-lite type mixes contain about 25% loam.
8. The key to establishment of a new plant in the landscape is rapid root growth.
9. Never fertilize a tree when transplanting.
10. Xeriscapes cannot usually contain any flowers.
11. The best time to fertilize trees and turf is in the _____ .
12. The most important nutrient for trees and turf is _____ .
13. Starter solutions for vegetable transplants are high in _____ .
14. Coarse aggregates improve _____ in potting mixes.
15. A nursery digging trees by balling and burlapping should have a _____ -textured soil.
16. One answer to poor drainage in the landscape is to plant on _____ .
17. Most landscape plants prefer an _____ soil.
18. Strawberry fruit can be kept clean by the use of _____ .
19. Fertilizers in pots raise not only the nutrient level but _____ .
20. One fertilizer recommendation for turf is about _____ pounds of nitrogen per year.
21. A lawn covers 30,000 square feet. How much IBDU (31-0-0) should be used to fertilize this lawn in a year?
22. What soil texture is best for early vegetables?
23. Name a low-salinity nitrogen fertilizer (figure 13-26).
24. List four materials that will lower soil pH.
25. How much 20-10-10 would you use in one year to fertilize a shade tree whose root system extends 30 feet from the trunk? (Remember, the area of a circle equals 3.14 times the radius squared.)
26. Explain why topdressing a yard with good "black dirt" before planting turf may not be a good idea.
27. Why are starter fertilizers for lawns usually high in phosphorus? When should one not use a high phosphorus starter?
28. Why do yards in many parts of the country tend to be too alkaline for many landscape plants?
29. You plant a container shrub, with a very coarse soil root ball, into a fine textured soil. Using the concepts of matric potential and hydraulic conductivity, describe what would happen if you water frequently and heavily. Too seldom? How should you water until the plant is established?
30. Explain xeriscaping.

SUGGESTED ACTIVITIES

1. This little experiment will demonstrate the effect of the depth of a soil column on drainage. Soak a common kitchen sponge until it is saturated. Now hold it up horizontally (but not flat). When it stops dripping, turn to a vertical position. What happens now that the column has become deeper?
2. Evaluate a soil for landscaping, then suggest plants that would be suitable for that soil in your area:
   a. Use a texture ribbon test, a quick percolation test, and figure 3-23 to determine drainage and water holding capacity.
   b. Check for compaction and structure.
   c. Use a soil test kit or pH paper to measure soil reaction.
3. Make a list of flowers, trees, and shrubs for your area that are tolerant of dry soils.

chapter seventeen
Soil Classification and Survey

A soil surveyor at his work, completing the survey of Chisago County, Minnesota. He will gather information important for the proper use of these soils. This is an example of a career activity available to a soil scientist. For those interested in the land, and who enjoy being outside, such a career could be a good choice.

OBJECTIVES

After completing this chapter, you should be able to:

- describe the current USDA soil classification system
- explain how soil surveys are prepared and used
- list soil capability classes

TERMS TO KNOW

diagnostic horizon	phases	soil survey
families	soil association	soil taxonomy
great groups	soil classification	subgroups
land-capability classes	soil order	suborder
land-capability subclasses	soil series	

At the end of the 1800s, public leaders began to realize that land in the United States was being damaged because of poor land policies. This realization led to a series of public efforts to conserve soils—efforts that continue today. A start was made at the turn of the century when the government began to survey and classify the soils of the United States.

SOIL CLASSIFICATION

Soil survey depends on a system of grouping soils of like properties. *Soil classification* helps us to understand, remember, and communicate knowledge about soils.

The Russian soil scientist V. V. Dokuchaiv first suggested a way to classify soils around 1880. He proposed that soils were natural bodies created by soil-forming factors. This proposal formed the basis of a classification system that soil scientists began using to survey U.S. soils.

Over the years, the United States has used three classification systems. The first, simple system was used in the first quarter of the 1900s. It grouped soils based on the soil-forming factors that created them, using terms such as "brown forest soil" or "black prairie soil." This system became more detailed over time and reached its peak form in 1949. The 1949 system remained in use until 1960. The language of the 1949 system may be familiar to some readers, who may have heard such terms as chernozem, chestnuts, and podzols.

The USDA introduced the current soil classification system in 1960 and then adopted an improved form in 1975. The new system differs from earlier efforts in that it places more emphasis on soil traits than on soil history. It resembles the way plants and animals are grouped according to a system known as *taxonomy*. Taxonomy is a grouping of objects at several levels to show how they relate. Figure 17-1 compares the taxonomy of living things with the taxonomy of soils.

As shown in figure 17-1, the new system has six levels of classification. The highest level, the *soil order*, is the broadest group. The system recognizes 10 soil orders, which are described in figure 17-2. These orders are based mainly on the presence or absence of certain key horizons in the soil profile, called *diagnostic horizons,* and on average temperatures and rainfall. An Alfisol, for instance, has a subsurface horizon with a clay accumulation, a medium to high base supply, and moisture at least 90 days of the growing season. Note that the names of all soils orders end in the suffix "ol."

Each order is divided into several *suborders*, the next highest level in the soil taxonomy. The suborder members of an order differ mainly in wetness or temperature. A Boralf, for instance, is a suborder of Alfisols, which is found in cold regions. The name of a suborder includes a Latin or Greek root that provides information about the suborder and ends in several letters that identify the order to which it belongs. These letters come from the first letters of the order name. A Bor*alf*, for instance, is an *Alf*isol. The letters "bor" come from the Greek "boreas," meaning north. A map of the orders and suborders of the United States appears in appendix three.

Suborders are further subdivided into the *great groups, subgroups,* and soil *families*. The naming system for these levels is beyond the scope of this text. The higher levels of soil taxonomy interest national, state, and regional planners and soil scientists. Those who use soils at the local level, like farmers, builders, or county extension agents, are more concerned about the official lowest soil grouping, called the *soil series*.

Soil Classes	Living Classes
Order (10)	Kingdom
Suborder (47)	Phyllum
Great Group (185)	Class
Subgroup (970)	Order
Family (4,500)	Family
Series (10,500)	Genus
(Phases)	Species

Figure 17-1. The USDA soil classification system lists six levels of soil classes (*left*) and gives the number of units of each in the United States. Phases are also listed, but these are not an official part of the system. On the right, the classes of living organisms are listed for comparison.

Soil Orders	%	Description	Use
Alfisols	13.4	Forest soils of cool moist climates, light colored, slightly to moderately acid with illuvial layer high in silicate clays. Medium to high base saturation. Common to northcentral and mountain states. *Typical profile:* O-A-E-Bt-C	Cropland, forest, range
Aridosols	11.5	Arid soils common to southwest United States, often alkaline with salted horizons, thin or no O or A horizon. High base. *Typical profile:* A-Bt-Ck or Ckm, Cy, Cz	Range, irrigated farming
Entisols	7.9	Very young soils in new parent materials or where alluvial deposition or erosion limits profile development (slopes). *Typical profile:* A-C	Range, cropland, forest, wetlands
Histosols	0.5	Organic soils *Typical profile:* O1-O2-O3-C	Wetlands, forest, cropland
Inceptisols	18.2	Young soils, with only those horizons that form quickly. Little elluviation. Weak B horizon visible by color or structure; no illuviation. *Typical profile:* A-Bw-C	Cropland, forest, range

Figure 17-2. This simplified listing of the soil orders of the United States gives the order names, percentage of American land of the order, description of the order, and dominant uses in order of importance.

Soil Orders	%	Description	Use
Mollisols	24.6	Prairie soils of the Great Plains. Dark, thick, good structure, high base A horizon. May have illuvial or calcareous subsoil. Low to moderate rainfall. *Typical profile:* A1-A2-A3-Bw-C	Cropland, range
Oxisols	—	Highly weathered tropical soils. Has subsurface horizon low in weatherable minerals but high in aluminum or sesquioxide clays. In United States, found only in Hawaii and Puerto Rico. *Typical profile:* A-Bo (or Bv)-C	Cropland
Spodosols	5.1	Light colored, acid forest soils of cool humid regions as in northeast United States. Coarse soil, high in silica, has illuvial subsoil layer with humus, sesquioxides, aluminum, or iron. Low base saturation. *Typical profile:* A-E-Bs (or Bhs)-C	Forest, cropland
Ultisols	12.1	Highly weathered soils of warm climates, low base. Subsoil layer with illuviated silicate clays. Common to southeast United States. *Typical profile:* A-E-Bt-C	Forest, cropland
Vertisols	1.0	High in swelling clays. When dry, large deep cracks form that surface soil falls into, mixing the soil. Found in southern states, especially Texas. *Typical profile:* A-AC-C	Cropland, range

Figure 17-2. *(Continued)*

Soil Series.   Soil scientists divide the greater soil groups into smaller units called *soil series*. Each of these units is distinct from other units and is the same as the polypedon described in chapter two. All soils in any one series have very similar soil profiles except for minor changes in the texture of the A horizon. In placing a series, the following properties of the soil profile are studied:

- Order of the horizons in the soil profile.
- Thickness of each horizon.
- Texture, pH, color, and structure of each horizon.
- Parent material.
- Organic matter content.

- Soil depth to bedrock.
- Presence of soil pans.
- Types of clay present.
- Other notable features.

In the United States, each series is given the name of the town, county, or other location near where the series was first identified. Examples include the Zimmerman Series, named after the town of Zimmerman in east central Minnesota or the Saybrook Series, found near the central Illinois town of Saybrook. There are about 10,500 such series in the United States and each is a separate and distinct soil body.

Phases of Soil Series. The series is the lowest official category in the soil taxonomy. However, in practice, a series is subdivided further into *phases*. A phase is a variation in the series based on some factor that affects soil management, such as surface texture. Phases are important subdivisions of a soil series that provide the detail needed for decisions about the use of a specific site.

Textural phases, previously called soil types, reflect small variations in the texture of the surface horizon. These phases are named for the series, followed by the textural class of the surface horizon. Members of the series called Ontario, for instance, include, the Ontario sandy loam, Ontario fine sandy loam, and Ontario loam. There can be no such thing as Ontario clay because the great difference in surface texture would place the soil in a different series.

Other important differences from the series norm can also create a phase, such as slope, degree of erosion, stoniness, or depth to bedrock. Thus, one might have an Ontario loamy sand, eroded phase. The same phase could also be called an Ontario eroded loamy sand.

SOIL SURVEY

The USDA developed the soil classification system for use in soil surveys. Soil surveys classify, locate on a base map, and describe soils as they appear in the field. These soil surveys are performed under the auspices of the National Cooperative Soil Survey Program, a joint effort of the USDA Soil Conservation Service (SCS) and state Agricultural Experiment Stations (see chapter twenty). Most of the actual surveying is done by soil scientists of the SCS.

Field Mapping. For detailed mapping, a soil scientist walks the land to survey it (figure 17-3). Frequently he or she stops to probe the soil. By studying the soil profile, that spot can be placed in the correct series (using the factors listed earlier). The surveyor also notes slope, evidence of erosion, and other interesting features. With this information, the surveyor draws the soil series or phase on a base map. These are the *mapping units*.

The SCS uses aerial photographs as a base map. Figure 17-4 shows a base map being prepared in the field, using an aerial map. Aerial photographs make

Figure 17-3. This soil scientist is conducting a soil survey to determine texture, structure, soil types, erosion, and other soil factors. Then he marks the boundaries of the mapping unit on an aerial photograph. (USDA, Soil Conservation Service)

good base maps because they show landscape features, including ponds, woods, and sand pits. Figure 17-5 shows a map of a small farm as it was drawn by an SCS surveyor.

When the survey is complete, the resulting map is copied neatly. Maps show the boundaries of the mapping units, with each unit identified by a code. The codes vary from state to state. Note the codes shown in figure 17-5. These codes may have one, two, or three parts. The main group of digits refers to the soil series. The mapping unit labeled 179, for instance, is a member of the Braham series. In addition, a unit may be labeled with codes to indicate the slope and erosion. If the latter two codes are absent, one assumes a nearly level relief with no erosion. Figure 17-6 gives the codes and other symbols that indicate different features in the field. The following codes are used on the map in Figure 17-5.

    158   Zimmerman series, no slope or erosion
  158C   Zimmerman series, 6%-12% slope, no erosion
158C1   Zimmerman series, 6%-12% slope, slight erosion

Figure 17-4. A base map being recorded on an aerial photograph during a soil survey

Maps may indicate the series by its intials rather than by a number code. Thus, the Zimmerman soils may be labeled Zi, ZiC, and ZiC1.

Modern technology is becoming an increasingly useful tool for soil surveys. Even satellites are employed for remote sensing of earth forms. Such technologies aid, but do not replace, the activities of a soil surveyor doing field mapping.

Mapping Units. Different mapping units are used in soil surveys, depending on how large an area the map or survey covers. For small areas, the mapping units are detailed phases of the soil series. For larger scale maps, the units may be higher levels of the soil taxonomy, like families or great groups. Suborders are the mapping units on the national soil map in appendix three. For most county maps, phases of soil series are the basic mapping unit.

For some land, different soils are so mixed that many cannot be separated on the scale of the map. For instance, tiny pockets of one soil may be mixed into a larger soil unit. Therefore many mapping units contain more than one series, family, or whatever level is being used. As an example, one of these mapping units is the *association*. An association consists of one or more major soils and one or more minor soils. For instance, Zimmerman soils usually appear beside two other series called the Isanti and Lino. In this area, glaciers carved out a landscape of fairly level outwash soils (Zimmerman), with scattered poorly drained low

Figure 17-5. Soil map of a farm prepared by an SCS surveyor. The numbers are codes for the following soil series:

75 Blufton	179 Braham
123 Dundas	225 Nessel
132 Hayden	540 Seelyville
158 Zimmerman	544 Cathro

(USDA, Soil Conservation Service)

spots (Isanti), and other soils. Because they are so mixed with each other, they appear as one mapping unit on some maps as the Zimmerman-Isanti-Lino association. This association is 45% Zimmerman, 25% Isanti, 15% Lino, plus small amounts of three other series.

Soil Survey Reports. A completed soil map becomes part of a *soil survey report*. A soil survey report has four major parts: 1) a set of soil maps, 2) map legends that explain the map symbols, 3) descriptions of the soils, and 4) use and management reports for each soil. All these parts provide much useful information about the soils, including:

- Taxonomy of the soil—telling the order, suborder, and other classes.
- A brief description of the soil. For instance, for mapping unit 544 in figure 17-5, the Cathro series, the description reads: "The Cathro series consists of very poorly drained soils formed in deposits of herbaceous organic material over loamy sediments in depressions. This soil is black muck 23 inches thick.

			Slope	

Legend	Percentage of Slope	Description
A	0–2	Nearly level
B	2–6	Gently sloping
C	6–12	Sloping
D	12–18	Strongly sloping
E	18–30	Very strongly sloping
F	30–60	Steep
	More than 60	Very steep

Large gully

Deep, caving gully

Drainage ditch

Small potholes

Wet or seep spot

Small steep area

Severely eroded spot

Rock outcrop

Small area of sand

Small area of gravel

Surface stones or boulders

Small area of high lime

	Erosion

Legend	Description
0	No erosion
1 or P	Slight, 0 to 1/3 topsoil gone
2 or R	Moderate, 1/3 to 2/3 topsoil gone
3 or S	Severe, 2/3 or more topsoil to 1/3 subsoil gone
4	Heavy subsoil erosion, deposition of eroded soil

Figure 17-6. Soil mapping symbols give information about slope, erosion, and landscape features.

The substratum is a greyish brown sandy loam. Slopes are less than 2%. Most areas are used for woodland."

- Soil properties of each horizon, including texture, bulk density, permeability, available water, pH, salinity, and other features. Engineering properties are also listed.
- Rating of suitability for engineering projects like landfills, buildings, and roads. Problems are mentioned. For instance, the Cathro is listed as poor for most projects because of ponding.
- Suitability for water management projects like reservoirs, drainage, and irrigation. Problems are mentioned. The Cathro, for instance, is poor for digging aquifer-fed ponds because of a low refill rate.

- Suitability for recreational development like playgrounds and campgrounds. Problems are mentioned. The Cathro is listed as poor because of ponding.
- Potential for cropping, including capability class and projected yields for common crops grown under high management. The Cathro, for instance, cannot be cultivated unless drained. If drained, one can expect 50 bushels of corn per acre, 3.5 tons of grass hay per acre, 6 tons of Reed Canarygrass per acre, and 55 bushels of oats per acre.
- Woodland suitability, including problems and suggested trees to plant. The Cathro is rated as poor for woodlands, but certain trees that are tolerant of wet soil may be planted.
- Information about good plants for windbreaks.
- Potential as a habitat for wildlife. The Cathro is rated as good for wetland plants and animals, but poor for others.

Types of Survey Reports. Several levels of maps are prepared for those with different soil use concerns. The least detailed maps cover the largest areas, like states or regions. Depending on the detail needed and area covered, mapping units for such maps are associations of great groups, suborders, families, or even soil series. Such maps show broad-scale soil associations for region-wide land-use planning.

County soil survey maps employ phases of soil series as the mapping units. These maps are accompanied by complete discussions of land use. These maps give enough detail to allow management decisions about specific sites. County maps are useful to county planners, housing developers, civil engineers, and many others.

Farm maps (figure 17-5) may be prepared by SCS field personnel at the request of individual growers. The mapping units for farm maps are phases of soil series. Farmers can use these maps to plan their own farm operations.

Survey Report Uses. Soil maps are the heart of good land-use planning. Soil maps give the information needed to make good land-use decisions—whether the decision maker is a national planner or a farmer or home builder. At the national level, for instance, the USDA has inventoried soil resources of the United States and kept track of them from soil maps. Regional land planners can use state and county maps to help make reasonable use of the land.

Engineers also need soil maps. For instance, civil engineers planning a new road will study maps to find routes with good soils for roadbeds. Planning commissions searching for new landfill sites will begin with soil maps.

New growers or growers planning to expand find soil maps useful for choosing new land. Instead of driving all over a region searching for the right land, one can target certain prime areas on soil maps.

Growers can use soil maps in many other ways. The information in soil surveys helps in planning irrigation or other engineering projects. For instance, the grower who owns the farm in figure 17-5 dug a pond in the wet Cathro soil in hopes of irrigating out of the pond. Had he read a soil report first, he would have known the pond would refill too slowly to be used for this purpose. Surveys also give farmers guidance as to what yield he or she should be getting and other useful information.

For a grower, an important use of soil maps is to prepare the field map, which is a most useful planning tool.

Field Maps. An SCS farm map makes a good base for a grower's own field map. He or she traces the SCS map on a sheet of clean paper. SCS maps use a standard scale of eight inches to the mile, which may be too small. The grower can redraw it at a larger scale if needed.

With this map, the farmer divides the farm into fields and labels each part. By basing the fields at least partially on soil mapping units (or capability classes, to be covered next), fields will be uniform for cropping and for soil sampling. The grower can now use these blank maps as recordkeeping and planning tools for numerous uses, including:

- Noting crop rotations.
- Tracking manuring or other practices that affect fertilization.
- Recording pesticide applications.
- Making notes of problem spots in the field, like wet areas or large rocks.
- Mapping locations of irrigation and drainage systems.

Computer Applications. Soil mapping information can be assembled in a computer for faster, more knowledgeable natural resource or farm management decisions. At the Pennsylvania Agricultural Experiment Station, data from more than 3,000 different mapping units in the state have been fed into a computer (R. L. Cunningham, G. W. Peterson, C. J. Sacksteder, "Microcomputer Delivery of Soil Survey Information," *Journal of Soil and Water Conservation*, 39 (4): 241-243, 1984). From this data base a variety of soil survey questions can be answered. The system will also generate a variety of specific maps, like a map of the slope classes for a specific area.

At the Agricultural Extension Service of the University of Minnesota, the FARMMAP program is being developed for aiding the decision making of individual growers. Using soil surveys, a grower constructs a soil map of his or her farm on a minicomputer. This computerized map can be used to help compute fertilizer or herbicide recommendations and to make soil conservation or other management decisions. The program can be used in conjunction with a grower's own field map.

LAND CAPABILITY CLASSES

Soil maps provide the basis for placing soils into land capability classes. This system indicates the best long-term use for land to protect it from erosion or other problems. The uses include cropping, pasture, rangeland, woodland (for lumber), recreation, and wildlife. The classes are not designed for all horticultural crops or crops that need very special management.

For example, flat land with deep rich soil can sustain long-term heavy cropping without erosion. It has few limitations and can be used for any of the listed uses. Sloping land, on the other hand, must be managed carefully to avoid destructive erosion and should not be "overfarmed." Sloping land has more serious limitations.

Capability Classes. The United States Soil Conservation Service recognizes eight land capability classes. These are numbered by Roman numerals I to VIII. Class I soils have the fewest limitations and Class VIII soils are so limited as to be totally unsuitable for agriculture. Erosion hazard due to slope is the main criterion, but other criteria are used as well. Figure 17-7 shows sample uses for each class. Note that there are fewer safe uses for each succeeding class.

Class I soils have few limitations. They can be heavily cropped, pastured, or managed for woodlands or wildlife. Crop cultivation is the most profitable use of Class I soils. These soils are well-drained and nearly level (0% to 2% slope). They have good water-holding capacity and are fertile. Ordinary cropping practices, such as liming, fertilizing, and crop rotation, keep these soils productive. In some soil maps, Class I soils are colored light green. Few American soils fall into this class (figure 17-8).

Class II soils are also suitable for all uses, but they have mild limitations that need moderate soil conservation or other measures when cropped. Problems include (1) gentle slopes (2% to 6% slope); (2) moderate erosion hazards; (3) shallow soil; (4) less than ideal tilth; (5) slight alkali or saline conditions; or (6) slightly poor drainage. Class II soils may be colored yellow on a soil map (figure 17-9).

Class III soils can grow the same crops as Class I and II soils. However, serious problems need to be addressed, such as (1) moderately steep slopes (6% to 12%

Use	Class							
	I	II	III	IV	V	VI	VII	VIII
Row crops	X	X	/					
Hay, small grains	X	X	X	/				
Pasture	X	X	X	X	X	X		
Range	X	X	X	X	X	X	/	
Woodland	X	X	X	X	X	X	/	
Recreation, wildlife	X	X	X	X	X	X	X	X

Figure 17-7. Suitable uses for soil capability classes. The higher the class number, the more limited is the number of safe uses. A single slash indicates that very careful management is needed or the soil cannot be used safely for this purpose every year.

Figure 17-8. Class I land consists of level, deep, rich soil that is excellent for cultivated crops. (USDA, Soil Conservation Service)

slopes); (2) high erosion hazards; (3) poor drainage; (4) very shallow soil; (5) droughtiness; (6) low fertility; (7) moderate alkali or saline conditions; or (8) unstable structure. Special conservation methods are needed. Growers should limit the number of row crops grown and favor close-growing crops. This is the lowest soil class that can be used safely for all crops, but only if it is used carefully. Class III soils may be colored red on a soil map.

Class IV soils are marginal for cultivated crops (figure 17-10). Limitations are those listed for Class III but are more severe. Slopes may be 12% to 18 percent. Row crops cannot be grown safely but close-growing crops may be. Crops that cover the soil completely, like hay crops, are best. Careful erosion control measures must be practiced. Class IV soils may be colored blue on a soil map.

Class V soils are not suited to cultivated crops, but may be used for range, pasture, woodlands, and recreation. These soils are level, have little erosion hazard, but are limited by factors such as (1) flooding; (2) short growing season; (3) rockiness; or (4) wet areas that cannot be drained. Class V land may be colored dark green on soil maps.

Class VI soils are unsuitable for cultivated crops, but they may be used for pasture, range, wildlife, and woodland. Problems may include (1) steep slopes

Figure 17-9. Class II land is often gently rolling land. In this case, it is being used as wooded pasture, but it could be more profitably cropped. (USDA, Soil Conservation Service)

(18% to 30% slope); (2) severe erosion hazard; (3) established severe erosion; (4) stoniness; (5) shallowness; or (6) drought. Class VI soils may be colored orange on soil maps.

Class VII soils have the same problems as Class VI but are more severe. It is difficult to maintain high quality pasture, but the land may be used for range, woodlot or forest, recreation, or wildlife if it is carefully managed. Slopes may be greater than 30%. Class VII soils may be colored green on soil maps.

Class VIII soil cannot support any commercial plant production, even timber. They may only be preserved for recreation, wildlife, or for beauty (figure 17-11). Sandy beaches, rock outcroppings, and heavily flooded river bottoms are examples of Class VIII land. Class VIII lands may be colored purple on a soil map.

A soil class may be upgraded if the problem is removed. For instance, a Cathro soil (figure 17-5, mapping unit 544) is so wet as to be placed in Class VII. An artificially drained Cathro, however, may be moved to Class IV. Permanent irrigation, land leveling, and other practices may also upgrade a soil class.

The eight classes can be simplified to soils that can be used for cultivated crops (Classes I-III), marginal land for cropping (Class IV), and lands not suitable for cropping (Classes V-VIII).

Figure 17-10. A Class IV soil in Idaho. This field is too steep for safe culture of row crops. The large circle on the hill is a *slump,* caused when a section of saturated soil slides downhill. (Courtesy of University of Idaho Cooperative Extension Service)

Land Capability Subclasses. All classes except Class I have one or more limitations. The subclasses indicate the factors that limit soil use by means of a single letter code added to the class number. A Class IIe soil, for instance, is slightly limited by erosion hazards; a Class VIe soil is very limited by erosion hazards. The letter codes are as follows:

- e—Runoff and erosion. Land with slopes greater than 2% are those that need some form of water control.
- w—Wetness. These soils may be poorly drained or occasionally flooded (figure 17-12). Soils of Classes IIw and IIIw are mostly wet soils that may be farmed if drained. Many Class IVw and lower soils are classed as wetlands by the United States Fish and Wildlife Service and are best left as wildlife habitat (see discussion in chapter six).
- s—Root zone or tillage problems. These soils are shallow, stony, droughty, infertile, or saline. Wind and water erosion may be problems.
- c—Climatic hazard. Areas of rainfall or temperature extremes make farming difficult. Examples include deserts or the Far North.

Figure 17-11. The North Shore of Lake Superior is typical of Class VIII land. It cannot be farmed or lumbered, but it is a major recreational area.

Soil Use Maps. Land capability classes rate soils for their use in agriculture. One could modify a soil map to replace soil series identifications with use classes, and have a map that will show suitability of the area soils for agriculture. This would be a type of soil use map.

Similarly, a variety of other soil use maps could be derived from soil surveys. For instance, one could draw a map of suitability for home drain fields, or woodlot production. The soil scientist, using the survey, rates each soil for the amount of hazard involved in that soil use. The map is then colored with a coded system. Soils colored green have a low hazard (meaning, "go ahead"), yellow is medium, and red suggests a high hazard. When complete, one would have a colored map showing clearly the best and worst soils in the area for that particular use.

Appendix five shows how one might judge soils for a number of uses. From this appendix, students could judge soils as a lab exercise, or even try to prepare their own land use maps.

Figure 17-12. This Class IIIw land could be drained for farming, but a better use may be to conserve it as wetlands. (USDA, Soil Conservation Service)

Lands of the United States. The United States is fortunate to have a great deal of good farmland, more than any other nation. Figure 17-13 summarizes the capability of U.S. soils. Appoximately 45% of our soil is rated in Classes I to III. This is soil on which nearly any crop can be grown. Most of the rest of the U.S. land is suitable for some form of commercial production like grazing or woodlands.

Good farmland is not evenly distributed over the United States. Figure 17-14 shows the average percentages of land falling into Classes I-III, IV, and V-VIII for regions of the mainland United States. The Corn Belt states have the highest percentage of good farmland, followed by the Northern Plains states and the Delta states. Much of the land of the West is too mountainous to be useful for cultivated crops.

Generally, land use in the United States follows capability. Land in the top three classes is used primarily for cropland. The remaining land is used for pasture, range, or forest. However, the most severe erosion in the nation occurs on marginal land that is cropped by farmers. As Figure 17-15 shows, soil loss increases when farmers use less suitable land. The next chapter will discuss erosion control practices that allow growers to safely use Class II, III, and IV land.

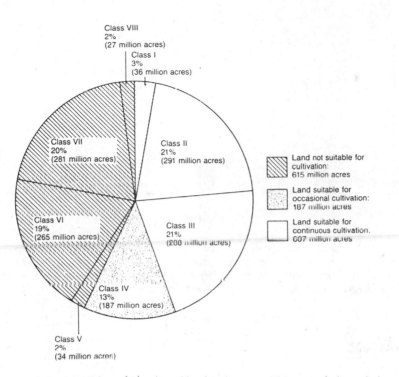

Figure 17-13. Land capability of nonfederal rural land in the United States, excluding Alaska. (USDA Preliminary Report 1982 National Resource Inventory, 1984)

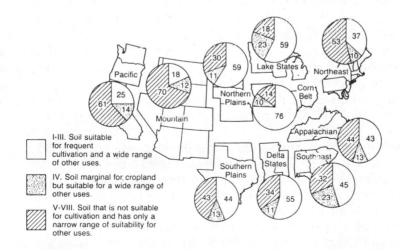

Figure 17-14. Land capability of regions of the United States. (USDA *1980 Appraisal Part I: Soil, Water, and Related Resources in the United States*)

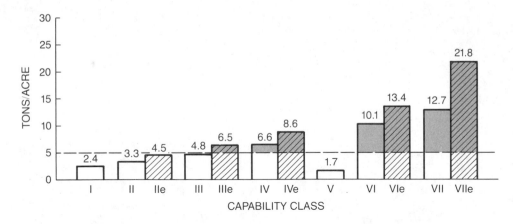

Figure 17-15. Average rate of sheet erosion by capability class in 1982 for cultivated cropland. A loss of about 5 tons/acre is generally the highest acceptable level. Most erosion occurs when farmers cultivate marginal land. Note that even Class IIe land nearly exceeds acceptable rates. (USDA Preliminary Report 1982 National Resource Inventory, 1984)

SUMMARY

Soil survey efforts in the United States began around the turn of the century. At first, a simple soil class system based on soil formation factors was used. This system was refined over time until the current system came into use.

Soil scientists presently classify soils according to their soil properties and profiles. This soil taxonomy has six levels. The top level consists of 10 soil orders in the world. Each order is further divided into suborders, great groups, subgroups, families, and series. The important level to an individual grower is the soil series and its unofficial subdivision, the soil phase.

Soil scientists survey land and prepare a soil map based on this classification system. The surveyor studies the top five feet of the soil profile, and also notes slope, erosion, and other features. A soil survey report includes the map plus printed information about the soils on the map and their suitable uses. These reports are then used by regional planners, engineers, growers, and other individuals.

The information in a soil survey places the land into one of eight capability classes. Classes I, II, and III are suitable for cultivated crops. Class IV is marginally useful for cultivation. Classes V to VIII are restricted to noncultivated uses. A number of factors are used to classify the soil, principally erosion hazard. Other factors include drainage, droughtiness, and extreme climates.

Soil capability classes tell us what soils need treatments such as drainage, irrigation, and desalination. Even more important, soil classes guide us in conservation efforts. The next chapter covers soil conservation.

REVIEW

1. A classification of a soil is based on its age (True/False).
2. There are seven levels to the soil taxonomy.
3. A soil series is named after the person who first described it.
4. A soil phase is a variation of a soil series.
5. Soil maps are drawn on a topographical map.
6. An association contains several soil series.
7. Class V soils are too steep for agriculture.
8. The top division of soil taxonomy is the soil order.
9. An Aquoll is a suborder of Entisol.
10. Soil surveys are conducted from an airplane.
11. A wet Entisol is assigned the suborder _____ . (see appendix three)
12. The order of largely prairie soils is _____ .
13. For county level soil maps, _____ are used as mapping units.
14. The suitability class with the least restrictions is _____ .
15. The hazard for a Class IIe soil is _____ .
16. The names of all soil orders end in the letters _____ .
17. A suborder named an Orthent belongs to the order _____ .
18. A soil order not found in the contiguous 48 states is the _____ .
19. A polypedon is also called a _____ .
20. Most soil surveys are performed by an agency called the _____ .
21. What could you say about a soil identified on a soil map as a "160D2?" See figure 17-6.
22. How is a soil series named?
23. What two classes of land cover the most acreage in the United States? See figure 17-13.
24. What soil classes are suitable for cultivated crops?
25. List the six levels of soil taxonomy.
26. What does a soil surveyor look for?
27. Explain what a soil phase is.
28. Identify the soil on the very southern tip of Texas. What is the moisture and temperature there? See appendix three.
29. Explain what a soil use map is for.
30. Here is the soil profile of the Elloam soil of north-central Montana. It formed under a dry climate on calcareous, clay loam glacial till. Using appendix four, figure 17-2, and your knowledge of soil formation, identify the soil order for the Elloam.

    E (0-4 inches) Light brownish loam, platy, slightly acid.

    B_t (4-8 inches) Brown clay, columnar, mildly alkaline.

    BC_K (8-14 inches) Grayish brown clay loam, prismatic, strongly alkaline and calcareous, many soft masses of lime.

C1$_K$ (14-33 inches) See BCea horizon.

C2$_y$ (33-57 inches) Grayish brown clay loam, blocky, calcareous, few lime masses, high gypsum.

SUGGESTED ACTIVITIES

1. Obtain the most current and detailed soil survey for your area. The teacher can prepare a worksheet asking questions about selected soils.
2. After doing 1 above, visit one of the areas on the map. Compare the soil report to the actual site.
3. Find your locality on the general soil map of the United States. What does it have to say about soils of your area?
4. Judge a sample soil as explained in appendix five.

chapter eighteen
Soil Conservation

Here's the way it was in Oklahoma, 1937. Erosive farm practices and drought caused such dramatic wind erosion that many farmers, such as this one, had to pack up and leave. These folks came to be called "Okies." But erosion didn't cease when the Dust Bowl years were over. Erosion continues to damage the face of America. Read this chapter to understand and prevent soil erosion.

(Courtesy of USDA-SCS)

OBJECTIVES

After completing this chapter, you should be able to:

- list the effects of soil erosion
- describe how soil erosion occurs
- list the types of water and wind erosion
- calculate soil loss from water erosion on a field
- describe ways to prevent erosion

TERMS TO KNOW

ephemeral gully	sheet erosion	Universal Soil Loss
gully erosion	splash erosion	Equation
rill erosion	surface creep	Wind Erosion
saltation	suspension	Equation

EROSION

Each year nearly five *billion* tons of soil wash or blow from the farmlands of the United States. This quantity is equivalent to losing the full plow layer from five million acres of farmland. Most of the loss—3.5 billion tons—results from water erosion. The remaining 1.5 billion tons are lost in wind erosion (figure 18-1).

Soil scientists follow the rule of thumb that one acre of most land can afford to lose between one and five tons of soil each year, because soil formation can balance this loss. The average soil loss to water erosion on cropland is thought to be 4.8 tons per acre per year, close to the limit. Added to this amount, however, is an average soil loss of 3.3 tons per acre per year to wind erosion. As noted in chapter seventeen, most erosion occurs on erosion-prone land. About 44% of

American cropland suffers soil losses greater than the acceptable limit. Figures 18-2 and 18-3 show where water erosion takes place in the United States.

The source of the above data and figure 18-1, the 1982 National Resource Inventory of the USDA, was repeated in 1987. Early data shows that water erosion decreased on the average by a half ton per acre between the years 1982 and 1987, from 4.8 to 4.3 tons per acre per year. This decrease does not reflect erosion controls resulting from the Food Security Act of 1985 (see chapter twenty), which had not yet been put into place. Rather, the improvement was likely due to an increase in voluntary erosion practices, especially conservation tillage.

During the same period, wind erosion rose slightly, from 3.1 to 3.3 tons per acre per year. The increase was probably due to the dry years seen during the 1980s in the Great Plains.

	Sheet and Rill Erosion	Wind Erosion (tons/acre/yr)	Total Erosion
Cultivated Cropland	4.8	3.3	8.1
Pastureland	1.4	0.0	1.4
Rangeland	1.4	1.5	2.9

Figure 18-1. Estimated average annual loss of topsoil (in tons/acre) to erosion in nonfederal lands of the United States. (USDA, Preliminary report of 1982 National Resource Inventory, 1984)

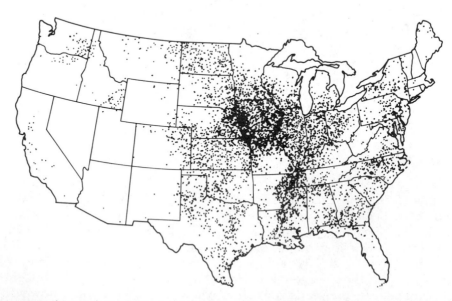

Figure 18-2. Sheet and rill erosion on cropland, in tons/acre, in 1977. One dot equals 250,000 tons of soil eroded annually. (USDA, *1980 Appraisal Part II: Soil, Water, and Related Resources of the United States.*)

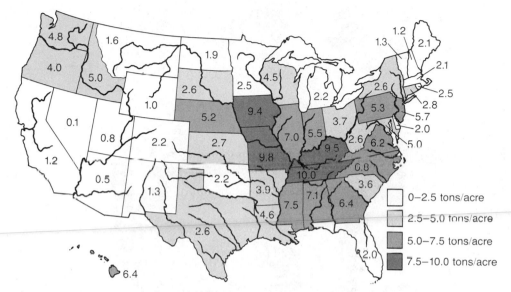

Figure 18-3. Average yearly sheet and rill erosion in 1982 on cropland (tons/acre), by state excluding Alaska. The national average is 4.8 tons per acre. (USDA, Preliminary report, 1982 National Resource Inventory, 1984)

What damage is caused by this amount of erosion?

- Erosion first removes the topsoil. Topsoil affords the best root environment by providing the best structure, the most air, and an active population of living organisms. Once the topsoil is lost, only the less productive subsoil remains.

- The topsoil contains most of the soil's organic matter and plant nutrients. Erosion carries away nitrogen, phosphorus, and any nutrient stored mostly in organic matter.

- As erosion strips away the soil surface, the profile becomes thinner, decreasing the root zone. This is a particular problem on already shallow soils. A major effect of this shrinking root zone is a reduced value of total water-holding capacity.

- Gullies cut up fields into odd-shaped pieces and make it very difficult to operate farm equipment.

- Eroded soil contains nutrients and pesticides that pollute lakes and streams. For instance, large fish kills have occurred in streams fed by runoff water from fields treated with soil insecticides. A Conservation Foundation study of 1985 estimates that 30% of American waters are exposed to enough runoff to chronically affect fish.

- The soil washed away by erosion settles in streams, lakes, harbors, and reservoirs. About 760 million tons of farmland sediment reach surface waters each year. The sediment fills in lakes (figure 18-4) and creates a need for

Figure 18-4. When this lake was filled in by sediments resulting from erosion, there was a loss of recreational value, and a probable decrease in the property values of the homes fronting the lake. (USDA, Soil Conservation Service)

expensive dredging to clear harbors and waterways. It reduces the ability of streams to carry water, resulting in an increase in flooding. Sediment also fills in reservoirs so they hold less water.

Cost of Erosion.   It is only fairly recently that data have become available (such as the 1977 and 1982 National Resource Inventories quoted in several figures) to attach values to the cost of erosion. There are actually two separate sets of costs: (1) the costs to the farmer and consumer of production losses, and (2) the costs to the public of pollution and sedimentation.

Analysis of recently available date (Crosson and Stout, *Productivity Effects of Cropland Erosion in the United States,* Johns Hopkins Press, 1983 and "New Perspectives on Soil Conservation Policy," *J. Soil Water Cons.* 39(4):222-225, 1984) indicates that the cost of production losses to an *average* grower is not as great as previously thought. For instance, forecasts based on present trends and commodity prices indicate that an average corn grower loses about $0.33 per acre per year to erosion. The total economic loss from lost production is about $40 million per year. These losses are, however, cumulative. In the second year, the corn grower will lose $0.66, and so on through the years. In the long term, losses become significant.

Far greater than production losses due to erosion is the cost of off-site effects such as sedimentation or pollution. The costs of sedimentation and pollution add up to about $3.1 billion a year compared with the $40 million for productivity losses and damage is not borne by growers but by society at large. Thus growers have little financial interest in the cost of erosion and place a lower priority on the problem of erosion than on other farming problems.

The nation has been battling erosion since the 1930s, yet it remains a major problem. This data indicates why. The short-range cost to most growers is not great, and the cost of installing conservation measures can be high. Surveys of farmers also show that most underestimate how easily their fields can erode. Further, over the past 40 years, a stream of technological improvements, including fertilizers and improved crop varieties, has masked the effects of erosion. Growers have every reason to assume such advances will continue. However, when the soil becomes thin enough, technology will not save yields.

However, some scientists are suggesting that the direct cost to growers has been set too low. Neglected costs include lost fertilizer, lime, and pesticides, additional equipment needs to work harder subsoils, deposition in drainage and irrigation structures, and many others. To these scientists, direct costs are high enough to justify erosion control expenses for the grower.

Conservation efforts have long been voluntary, with the government helping but not forcing growers to reduce erosion. Recent proposals for controlling erosion have been less voluntary. Increasing awareness of the large public cost of sedimentation and pollution may well lead to growers eventually being required, rather than asked, to control erosion.

HOW WATER EROSION OCCURS

Erosion follows three steps. First, the impact of raindrops shatters surface aggregates and loosens soil particles (figure 18-5). Some of these particles float into soil voids, sealing the soil surface so water cannot readily infiltrate the soil. The scouring action of running water also detaches some soil particles. Second, the detached soil grains move in flowing water and are carried down slopes. Finally, the soil is deposited when the water slows down. These three steps are knows as *detachment, transport,* and *deposition* (figure 18-6).

Erosion is a form of work, and work takes energy. The energy for water erosion comes from the energy of a falling raindrop or running water. The amount of energy in a moving object (or water) is the product of the mass (weight) of the object and its velocity (speed) squared. Expressed as an equation,

$$E = mv^2.$$

The energy of a falling raindrop, for instance, relates to its size and especially to its speed. A two-inch per hour rainfall has the same energy as a one-pound object

Figure 18-5. Raindrop impact on the soil (Courtesy USDA-SCS)

Figure 18-6. Soil erosion on this hill occurred in three steps: raindrops and running water detached soil particles, transported them down the hill, and deposited them at the bottom of the hill. (USDA, Soil Conservation Service)

falling 47 feet onto one square foot of soil. The erosive energy of running water also depends on the volume of water and its speed of flow. Since the velocity factor in the equation is squared, the energy contained in running water relates mostly to its velocity.

With high erosive energy, water can detach and move *larger* soil particles. It can also move *more* soil particles. Thus erosive energy relates directly to the amount of soil carried off a field. Deposition occurs when the energy of running water decreases—such as when it slows down at the foot of a slope. This energy concept will help in the understanding of four erosion factors: (1) soil texture and structure, (2) slope, (3) soil cover, and (4) roughness of soil surface.

Texture and Structure. Texture has two effects on soil erosion. First, texture influences the infiltration rate of water. If rainwater infiltrates the soil quickly, less water runs off. With a lower volume of running water, less soil can be transported. Second, particles of different sizes vary in how easily they can be detached. Silt particles are most easily detached, so silty soils are liable to water erosion.

Structure also influences infiltration—good structural grades like granules reduce runoff. The strength of soil aggregates is important too, since strong peds better resist the impact of raindrops. Because of the importance of organic matter to structure, the organic matter content of a soil has a strong bearing on the soil's erodability. Compaction, loss of organic matter, and destruction of soil peds by tillage all reduce infiltration and increase the volume of water available to transport eroded soil. The combined effects of organic matter content, texture, and structure are called the *erodability* of a soil.

Slope. Slope has two components—length and grade. On a steep slope, water achieves a high runoff velocity, increasing its erosive energy. On a long slope, a greater surface area is collecting water, increasing flow volume. On a longer slope, running water can also pick up speed. Thus, the steepness of a slope and its length contribute to erosive energy. Figure 18-7 shows that long gentle slopes can have the same erosive potential as short, steep slopes. This fact helps explain why many growers underestimate how erodable their land is.

Slope	Slope Length (feet)
4%	1,000
6%	200
8%	100
10%	50
12%	30
14%	20

Figure 18-7. Both slope grade and length affect soil loss. All of the slopes listed have an equivalent soil loss.

Surface Roughness.   A rough soil surface impedes the downhill flow of water, slowing its velocity. If the roughness takes the form of ridges across a slope, water can pond behind the ridges, decreasing the volume of runoff water. However, if enough water collects behind a ridge, it overflows the ridge and wears it away. Thus, roughness can fail to stop erosion during very heavy rains, on long slopes, or if sealing of the soil surface stops infiltration.

Surface roughness depends largely on tillage practices. The seedbed resulting from conventional tillage is smooth, while that from chisel plowing is rough. Tillage across slopes acts to impede downhill flow; tillage up and down the slope promotes downhill flow.

Soil Cover.   Bare soil is fully exposed to the erosive forces of raindrop impact and the scouring of running water. Soil cover reduces the energy available to cause erosion. A mulch or cover of crop residues absorbs the energy of a falling raindrop, lessens detachment, and reduces the sealing of the soil surface. Mulches also slow down runoff water. A complete crop cover like turf or hay has the same effect, plus plant roots hold soil in place (figure 18-8).

Figure 18-8. A complete vegetative cover, such as this crownvetch on a roadside embankment, almost eliminates soil loss. (USDA, Soil Conservation Service)

Crops that are less close growing, such as row crops, have a slightly different effect. As these crops close in between the rows, they form a canopy over the soil. This canopy intercepts rainfall and absorbs most of its impact above the soil surface. When water drips off the plant leaves, it again gains velocity, and thus energy. The energy of these drops depends on the height of the crop canopy, but it is not as great as free-falling raindrops. Thus it is important for erosion control that crops cover the soil surface as quickly and completely as possible. Crop canopies have no effect on runoff speed or volume, unlike mulches, so have a less protective effect.

Types of Water Erosion. A raindrop strikes the soil surface forcefully. The impact shatters soil aggregates and throws soil grains into the air. On a slope, water begins to flow downhill, carrying detached soil grains with it. This water joins other flowing water, increasing in speed, volume, and soil-carrying capacity. This order of events leads to five types of erosion. All five types can occur at the same time on any given slope.

- *Splash erosion* is the direct movement of soil by splashing. A soil grain can be thrown as far as five feet by raindrop splash. These splashed particles fill the voids between other aggregates and seal the soil surface.
- *Sheet erosion* is the removal of a thin layer of soil in a sheet. On gentle slopes, or near the tops of steeper slopes, water moves in tiny streams too small to be noticed. This gives the impression of losing soil in a thin sheet. The eroded knolls shown in figure 18-9 are an example of sheet erosion. Sheet erosion may go unnoticed until the subsoil appears.
- *Rill erosion* is visible as a series of many small channels on a slope. Water tends to collect in channels, picking up energy as it runs down the slope. As a result, running water carves out small but visible channels called *rills*. A rill is small enough to be filled in by tillage. Figure 18-10 shows rill erosion on a roadside embankment.
- *Ephemeral* gullies are large rills. The channel is small enough that tillage equipment can cross it and largely, but not completely, fill it in by tillage. During another heavy rainfall, water will collect in the old channel, and erosion will begin here.
- *Gully erosion* is the most highly visible erosion. Gullies are so large that equipment cannot cross them (figure 18-11). Gullies usually begin to form near the bottom of a slope or on steep slopes, where running water has enough force to carve a deep channel. Gully heads may back up the hill as water running into the gully collapses the sides.

Each type of erosion is important to understand for different reasons. Sheet erosion is a hidden soil loss, since there are no visible signs until the subsoil appears. Rill erosion can also be hidden, because each tillage causes the rills to disappear. The amount of the hidden erosion can be easily underestimated by a grower.

Figure 18-9. The loss of soil by sheet erosion probably went unnoticed by the farmer until the subsoil appeared on these eroded knolls. (USDA, Soil Conservation Service)

Figure 18-10. Erosion on this hillside has formed rills, ephemeral rills, and gullies. Rills can be filled in by tillage, ephemeral rills can be crossed but not completely filled in, and gullies cannot be crossed by farm equipment. (USDA, Soil Conservation Service)

Figure 18-11. Gullies are large channels that usually form where water collects as it runs down a hillside. (USDA, Soil Conservation Service)

The distinction between regular rill and ephemeral erosion is also important. Tillage does not fill in the ephemeral channel, thus it can act as a "seed" for gully formation. Some soil conservation measures must treat these two forms of erosion differently. The Universal Soil Loss Equation, which is the main tool for estimating erosion rates, predicts only sheet and rill erosion, not ephemeral erosion. Thus, the equation can seriously underestimate soil loss on fields with a great deal of ephemeral erosion.

Gullies are the most dramatic image of soil erosion. However, much more soil is lost over most fields by sheet and rill erosion. Gullies chop a field into inconvenient shapes and add to the sediment load of streams.

PREDICTING SOIL LOSS: THE UNIVERSAL SOIL LOSS EQUATION

Using the soil loss factors described, an equation has been developed to predict the average soil loss from sheet and rill erosion on any specific site. The Universal Soil Loss Equation (USLE) was developed over several years from some 10,000 plot-years data from 49 test sites around the country (figure 18-12). A grower can use this equation to decide what conservation practices are needed to keep soil losses within acceptable levels. The USLE also helps growers determine the most economical way to preserve the soil.

Figure 18-12. A rainfall simulator being used by research scientists in South Dakota. The booms imitate rainfall and various treatments are used to model soil loss. (Courtesy of USDA)

The USLE, which is still in the process of being refined, does have weaknesses one should be aware of. For instance, the equation predicts sheet and rill erosion only. If a slope shows signs of ephemeral or gully erosion, the results from the equation will understate the amount of soil loss. The equation predicts the *average* soil loss over time. A specific, highly erosive rainfall may cause far more erosion than is predicted by the equation. Nor does the USLE provide an accurate estimate of soil losses from snowmelt runoff. Also, applications of the equation to rangeland, to farmland in some parts of the country, have not been fully reliable.

Tolerable Soil Loss. At the heart of the USLE is the assumption that a certain soil loss can be tolerated because soil-forming processes will replace some of the lost soil. The equation can then be used to determine if soil loss exceeds this amount. In virgin grasslands, erosion may amount to an inch every 5,000 years. In the Palouse area of the Pacific Northwest, erosion levels sometimes reach 50 to 100 tons per acre per year. This is a loss of an inch of soil every one and a half to two years. Between these two extremes, what soil loss can be tolerated?

Soil scientists have decided that soil losses between one and five tons per acre per year can be tolerated, depending mostly on the quality and depth of a soil. It is assumed that erosion damages a deep soil least. The amount that can be tolerated is symbolized by the letter "T." Each soil series has its own value of T, which is noted on a soil survey report. However, not all soil scientists consider "T" to be a reliable guide to the need for erosion control measures.

The Universal Soil Loss Equation. The USLE is based on a standard test plot, which represents an average eroded site. This plot has a 9% slope 72.6 feet long. The slope is kept in clean-tilled fallow, using conventional tillage up and down the slope. The equation works by comparing a specific spot to this test plot.

The equation reads as follows:

$$A = R\ K\ LS\ C\ P$$

"A" is the tons of soil lost per acre each year. Obviously, "A" should be less than "T." To solve for A, values are inserted for the six *variables* and are multiplied. The variables are

- *R—rainfall and runoff factor.* R is based on the total erosive power of storms during an average year. R depends on local weather conditions. The *isoerodent map* of figure 18-13 shows R values for the United States.
- *K—soil erodability factor.* K depends on texture, structure, and organic matter content. Soil survey reports give the value of K for mapped soils. They may also be calculated.
- *LS—slope factor.* L compares the slope length and S compares the grade with the standard plot. L and S are separate factors, but they can be treated as one variable, "LS." LS values can be determined from the chart in figure 18-14.
- *C—cover and management factor.* C compares cropping practices, residue management, and soil cover to the standard clean-fallow plot. C values are calculated from detailed tables and are valid only within the area for which they are calculated. Many SCS offices prepare simplified tables for use in the field, and some have computerized the computations. Figures 18-15 and 18-16 provide samples of such tables and are included here for use in the solution of USLE problems presented in this text. Check local SCS offices for local charts. If necessary, the values can be calculated (see the reference listed in the suggested activities section of this chapter).
- *P—support practice factor.* P compares the effect of contour tillage, contour strip-cropping, and terracing with the test plot. The LS factor accounts for terraces primarily, but they are included here because they are contoured. The handling of terraces will be covered later. Figures 18-17 and 18-18 give P values.

Sample Solution. For a sample solution of the USLE, use mapping unit 158C in figure 17-5. The slope is 2% to 6% and about 100 feet long. (Ask yourself: How do I know the grade?) According to the soil survey for this area, K = 0.18 and T = 5.0. Assume the grower uses conventional tillage to grow continuous corn

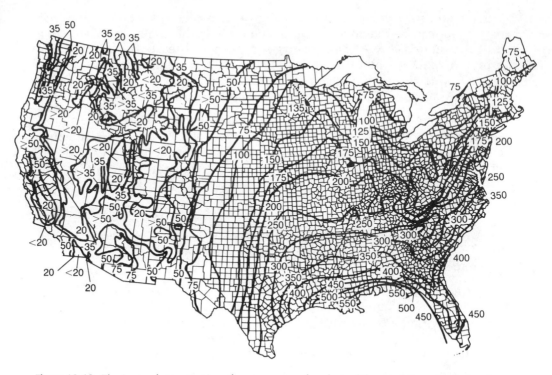

Figure 18-13. The isoerodent map gives the average yearly values of the rainfall erosion index, factor "R." The two darkened counties are used as examples in the text, and the dotted line refers to the sample problem. (USDA, Handbook 537)

and plows up and down the slope in the fall. The farm is in east-central Minnesota—the county is shaded on the isoerodent map.

We begin with the formula:

$$A = R\ K\ LS\ C\ P$$

The value of R is read from the isoerodent map. Since the farm location does not lie on a curve, draw a line between the two curves it lies between and estimate the value of its place on that line (see figure 18-13). In this case, R = 135:

$$A = (135)\ K\ LS\ C\ P$$

The value of K is given in the soil survey report.

$$A = (135)\ (0.18)\ LS\ C\ P$$

The value of LS can be obtained from figure 18-14. The slope is somewhere between 2% and 6%; let's use 5.5%. Find the slope length on the bottom axis

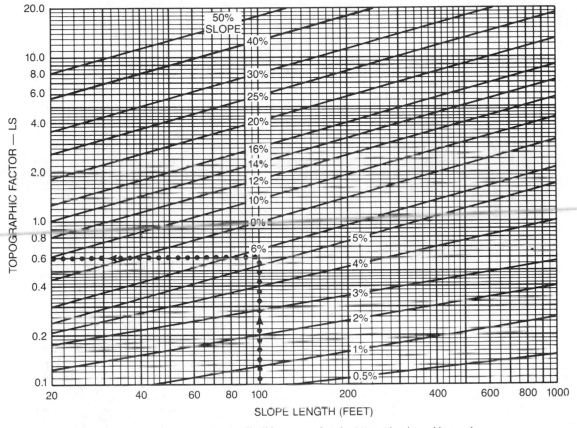

Figure 18-14. The slope effect chart provides the "*LS*" factor used in the USLE. The dotted line refers to the sample problem in the text. (USDA, Handbook 537)

(100) and follow up the chart until it touches the 5% slope line. Now go up to where the 5.5% line would be if there was one. From this point, move to the left to the vertical axis. The *LS* factor is shown as *LS* = 0.6.

$$A = (135)\ (0.18)\ (0.6)\ C\ P$$

The first three factors are fixed for the site—they are not changed by anything the farmer does, short of land leveling, terracing, or incorporating organic matter to lower *K*. Simplify the equation by multiplying the three factors:

$$A = (135)\ (0.18)\ (0.6)\ C\ P = (13.75)\ C\ P$$

Refer to figure 18-15 to find the *C* value under conventional fall plowing, *C* = 0.37

$$A = (13.75)\ (0.37)\ P$$

Crop Sequence	Conventional Plowing		Conservation Tillage		
	Fall	Spring	30%	40%	50%
Continuous Soybeans	0.48	0.45	0.30	0.24	0.20
Corn, soybeans	0.41	0.37	0.24	0.19	0.16
Continuous corn	0.37	0.36	0.19	0.15	0.12
C, SB, SG	0.32	0.29	0.17	0.14	0.11
C, C, SG	0.30	0.27	0.14	0.12	0.09
C, SB, C, O, M, M	0.20	0.18	0.12	0.11	0.09
C, C, C, O, M, M	0.17	0.15	0.10	0.09	0.08
C, C, O, M, M, M	0.11	0.10	0.08	0.07	0.06
Permanent grass	0.003–0.013				
Grazed forest	0.01–0.04				
Ungrazed forest	0.001–0.003				

Figure 18-15. Some crop management factors for southern Minnesota. These numbers apply to the averages of crop rotations using corn (C), soybeans (SB), small grains (SG), oats (O), and meadow (M). The percentages under conservation tillage refer to the percentage of soil covered with residue. (Soil Conservation Service, Minnesota office)

Crop Sequence	Conventional Plowing		Conservation Tillage			
	Fall	Spring	30%	40%	50%	80%
Continuous Soybeans	0.44	0.39	0.32	—	—	0.16
Corn, soybeans	0.37	0.33	0.30	0.24	—	0.10
Continuous Corn	0.34	0.29	0.18	0.15	0.12	0.05
C, SB, SG	0.27	0.25	0.15	0.13	0.11	0.05
C, C, SG	0.28	0.23	0.15	0.13	0.10	0.05
C, SG, SB, SG	0.25	0.21	0.14	0.12	0.10	0.05
C, SG	0.25	0.21	0.14	0.12	0.10	0.05
C, SB, C, O, M, M	0.18	0.15	0.11	0.10	0.09	0.04
C, C, C, O, M, M	0.17	0.14	0.09	0.08	0.07	0.03
C, C, O, M, M, M	0.11	0.11	0.08	0.07	0.06	0.03
Permanent grass	0.003–0.013					
Grazed forest	0.01–0.04					
Ungrazed forest	0.001–0.003					

Figure 18-16. Some crop management factors for central Illinois. The 80% cover column applies to good coverage under a no-till system. Soybean residue in no-till is figured at 40% coverage, the best possible. Soybeans are planted to row widths greater than 20 inches. Conservation tillage is fall chisel plowing with two secondary tillage operations before planting. The chart assumes a high level of management. (The chart is to be used in the review.) The figures on this chart were calculated from Illinois tables by the author to fit the same format as figure 18-13. (Illinois SCS Technical Guide Section I-C, 1982)

Land Slope Percent	P Value	Maximum Slope Length
1 to 2	0.60	400
3 to 5	0.50	300
6 to 8	0.50	200
9 to 12	0.60	120
13 to 16	0.70	80
17 to 20	0.80	60
21 to 25	0.90	50

Figure 18-17. Support practice factor "P" for fields that are contour tilled. The figures are unreliable if the slopes are longer than indicated. (USDA Handbook 537)

Land Slope Percent	P Value			Strip Width	Maximum Slope Length
	A	B	C		
1 to 2	0.30	0.45	0.60	130	800
3 to 5	0.25	0.38	0.50	100	600
6 to 8	0.25	0.38	0.50	100	400
9 to 12	0.30	0.45	0.60	80	240
13 to 16	0.35	0.52	0.70	80	160
17 to 20	0.40	0.60	0.80	60	120
21 to 25	0.45	0.68	0.90	50	100

Figure 18-18. Support practice factors for contour strip-cropping. Column A is for a four-year rotation of row crop, small grain, meadow, meadow. Column B rotation is row crop, row crop, winter grain, meadow. Column C values are for alternate strips of row crops and small grain. (USDA Handbook 537)

The farmer used no support practices, so $P = 1$:

$$A = (13.75) (0.37) (1) = 5 \text{ tons/acre/year}$$

The computed soil loss equals the acceptable level for this soil. However, since the equation, if anything, understates losses, the farmer should try to curtail some of the erosion. Are there any practices that can help lower the erosion level? What happens if the farmer changes to contour plowing rather than up and down plowing? For contour plowing, P changes from 1.0 to 0.5; the other factors remain the same:

$$A = (13.75) (0.37) (0.5) = 2.5 \text{ tons/acre/year}$$

The equation says that simply by changing to contour tillage, the farmer will cut erosion in half. The new level is well below five tons per acre. Changing from moldboard plowing to chisel plowing, leaving a 30% cover, would save even more soil:

$$A = (13.75 (0.19) (0.5) = 1.3 \text{ tons/acre/year}$$

Applications of the USLE. The most obvious way to use the USLE is to predict erosion from a certain field, and to help select the best control measures. The examples showed how this can be done.

The USLE can also be used to identify erodable lands. Land planners could use this information to prepare use maps based on erodability. Or officials could identify lands most in need of assistance.

One proposed scheme defines an "erosion index," or EI, as follows:

$$EI = \frac{RKLS}{T}$$

The formula calculates the "native" erodability of land, such as slope and texture, and excludes management practices. This is divided by T to get a multiple of the acceptable soil loss. This index could be used to identify the soils most in need of good management practices.

CONTROLLING WATER EROSION

All methods of controlling erosion are based on one of the following three actions:

- Reducing raindrop impact to lessen detachment. This can be done by growing vigorous crops that fill in the canopy quickly, by leaving crop residues on the surface, by mulching, or by growing a total vegetative cover.
- Reducing or slowing runoff. This lessens detachment by scouring and reduces the amount of soil that can be transported. Avoiding compaction, maintaining organic matter levels, and subsoiling help water infiltrate the soil. Contour practices and conservation tillage both reduce runoff.
- Carrying excess water off the field safely by use of grass waterways or tile outlets.

Conservation Tillage. Conservation tillage, described in chapter fifteen, sharply reduces sheet and rill erosion. It is the lowest cost conservation method per ton of soil saved and carries other benefits as well, which are described in chapter fifteen. For these reasons, conservation tillage is rapidly becoming the most widely accepted method for controlling soil losses.

The effectiveness of conservation tillage depends on a rough soil surface and the amount of residues left on the soil surface, especially during the critical period before the crop canopy fills in. At least 30% of the soil surface should be covered by mulch at crop planting. Figure 18-19 can be used to convert the values in pounds per acre given in figures 15-1 and 15-2 to percentage of soil coverage. Coverage can be measured directly by stretching a 50-foot cord, marked every six inches, diagonally across several rows. The number of marks touching a piece of crop residue is the percent of coverage (figure 18-20). For instance, if 35 marks touch crop residues, the coverage is 35%. This procedure should be repeated in

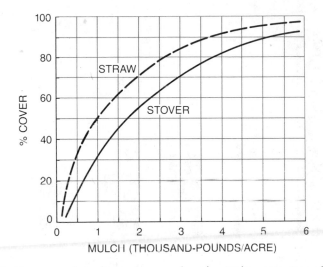

Figure 18-19. The conversion of tons per acre of residue to the percentage of soil surface covered. This chart, when used with figures 15-1 and 15-2, can help a grower plan how to achieve a desired percent of coverage. (USDA Handbook 537)

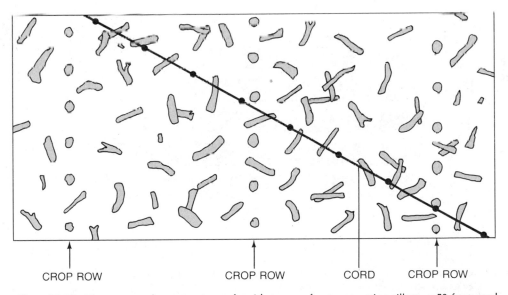

Figure 18-20. To measure the percentage of residue cover for conservation tillage, a 50-foot cord marked every six inches is stretched across crop rows. The number of pieces of residue touched by a mark is the percent of coverage. The soil must be at least 30% covered by crop residues for conservation tillage to be effective.

several parts of the field and the results averaged. Figure 18-21 shows the effect of several residue levels according to the USLE.

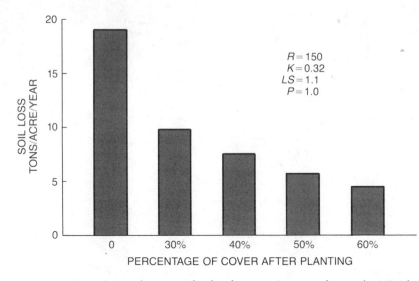

Figure 18-21. Effect of several crop residue levels on erosion according to the USLE for continuous corn. (Soil Conservation Service, Minnesota office)

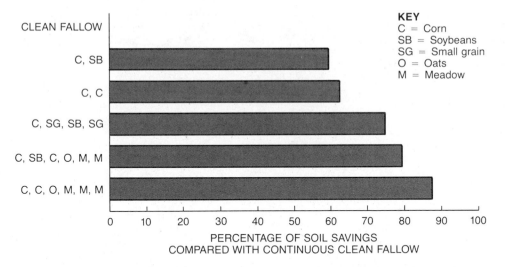

Figure 18-22. Effect of several crop rotations on soil erosion. Other USLE factors remain constant. (Soil Conservation Service, Minnesota office)

Crop Rotation. Crop rotation reduces erosion if a close-growing crop like small grains or forages is included. The effect of these close-growing crops is to reduce the detachment and transport energy of water. Also, it improves the soil's physical properties so that water seeps into the soil better. Figure 18-22 shows the effect of several rotations.

Figure 18-23. This grassed waterway collects runoff water and carries it safely into a farm pond. The contour strips empty into the waterway. (USDA, Soil Conservation Service)

Grassed Waterway. A grassed waterway is a shallow, sodded, wide ditch that runs down a slope (figure 18-23). It is designed to carry excess water off the field safely. Grassed waterways serve several purposes:

- Waterways can be built to prevent gullying where water naturally gathers on a slope.
- Waterways can be used to collect excess water from tillage contours.
- Waterways may also serve as outlets for terraces.

Small waterways may be built with grading tools mounted on farm equipment. Larger ones require the use of construction equipment to grade the shallow ditch. Grassed waterways must be carefully maintained and fertilized. Equipment should not be turned on a waterway, and tillage tools should not be dragged across it. Any damage should be repaired immediately with sod.

Contour Tillage. Contour tillage works best on permeable soils in areas of low intensity rainfall. Generally, moderate slopes suffering from rill erosion benefit most from contour tillage (figure 18-24). Simple cross-tillage is inadequate if the slope has much ephemeral erosion. Cross tillage does not erase ephemeral rills, and the remaining depressions act as channels for water flow. Therefore, land

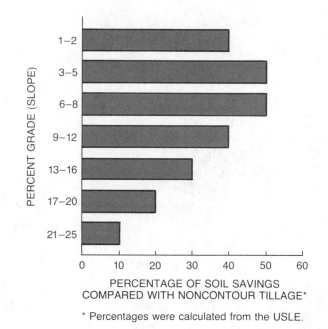

* Percentages were calculated from the USLE.

Figure 18-24. Effect of contour tillage on soil erosion for several grades (slope). Other USLE factors remain constant. Contour tillage works best on moderate slopes of 3% to 8%. On steep or long slopes, water overflows the ridges and wears them away.

with ephemeral erosion may need to be smoothed by more than simple tillage before it can be contoured.

Where runoff is not too great, contour tillage can stand alone. It is often helpful, however, to gently slope contours toward a waterway. In this way, water that may overtop the ridges can flow toward the waterway and be carried off the field.

To establish contours, one surveys a guideline across the slope. The line is either parallel to the slope or dips gently toward a waterway. A new guideline is added when the slope changes. When plowing the field, begin plowing at the guideline and plow parallel to it.

Strip-Cropping. Strip-cropping can be used in all conditions along with contour tillage. The strips of close-growing crops slow down runoff and filter out soil eroded from row-cropped strips. Strip-cropping works best in areas of moderate rainfall, on permeable soils, and on uniform slopes. There are three types of strip-cropping:

- *Buffer strips* correct an eroded area by planting it in grass. Grass buffer strips are much narrower than the cultivated strips between them. Buffer strips are not as effective as the other types.
- *Field strips* are placed straight across a slope, but they may not follow the actual contour. Where slopes are irregular, it is difficult to design strips that

follow the contour exactly. In such cases, field strips may be more practical than contour strips.
- *Contour strips* follow the contour and may empty into a waterway. Where the slope changes often, contour strips are more difficult to establish than field strips, but they are more effective.

As in contour tillage, guidelines are needed to guide plowing for strip-cropping. Strips of equal width are placed across the slope and are planted to alternating row crops, small grains, and meadow.

Improving Organic Matter. Improving, or at least preserving, organic matter can greatly reduce erosion because moisture will seep into the soil more quickly. Growers that make an effort to increase soil humus benefit by reduced erosion.

Terraces. Where strip-cropping cannot halt erosion, terraces may be built. Long or steep slopes on impermeable soil, for example, require terraces. Terraces are costly to install and are used most commonly for valuable crops or where there is a shortage of good land. In general, two kinds of terrace are used:

- Level terraces parallel the slope and do not empty into a waterway. This type of terrace is used where the soil is permeable enough so water can seep in once it is captured in a terrace.
- Graded terraces are needed where the water cannot soak in enough. These may slope gently towards a waterway or be drained by an underground tile outlet.

Several terrace designs are shown in figure 18-25. Of these designs, the broad-based terrace is most common. Terrace construction begins by designing them to fit conservation needs without overly hampering farming. The land is surveyed and the terraces are marked on the slope. Figure 18-26 shows a terrace under construction.

Growers must be careful to maintain terraces. Obviously, tillage follows the terrace. If a grower uses a moldboard plow, he or she must plow correctly. The moldboard throws soil off to the side, so it can change the terrace. Some growers who purchase new, larger equipment fail to maintain terraces properly because the terraces are in the way. Without proper care, terraces cannot be effective.

Terraces are accounted for in the USLE by the *LS* factor. Terraces break up a slope into several shorter slopes. In solving the USLE, the length of a single terrace is taken as the length of the entire hill. Depending on terrace design, the slope factor may or may not change. It is assumed that most of the eroded soil within the cropped part of a terrace is deposited in the terrace channel, and that a smaller amount leaves the field in waterways or outlets. For the *P* factor, use the contour tillage or strip-cropping value.

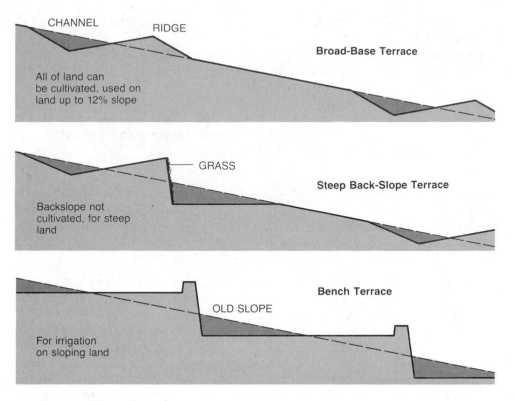

Figure 18-25. Each terrace design serves special purposes. The broad-based design is the most common. The scale is exaggerated in these drawings. The broken lines show the old grade.

Diversions.   Diversions are large-capacity terraces that divert runoff from higher elevations. Diversions are not farmed but are covered with grass. Their uses include

- Protecting fields from runoff flowing from higher elevations (figure 18-27).
- Diverting water away from active gully heads.
- Diverting water from feed lots, farmsteads, or other sensitive areas.

WIND EROSION

Wind erosion accounts for about one-third of the soil loss in the United States, mostly in the Great Plains states. In five states wind erosion far exceeds water erosion. These states are Colorado (average 9.3 tons per acre per year), New Mexico (5.2 tons), Texas (13.2 tons), Montana (8.3 tons), and Nevada (9.2 tons). Other areas with wind erosion problems include the muck and sandy soils of the Great Lakes states and the Gulf and Atlantic seaboards. Figure 18-28 shows amount of wind erosion for each state in average tons per acre per year.

Figure 18-26. Terraces under construction (USDA, Soil Conservation Service)

Dry areas with high winds are most likely to experience wind erosion. At greatest risk is soil kept bare by clean-till summer fallow.

Cause of Wind Erosion. Figure 18-29 shows the effect of wind blowing across a bare soil. A very thin layer of still air covers the soil surface, but many soil particles stick up above the layer. When the wind reaches 10 to 13 miles an hour at a height of one foot above the surface, the soil grains begin to move.

First, wind begins to roll soil grains, which are in the size range of 0.004 to 0.02 inches (0.1 to 0.5 mm). These grains are fine to medium sands. Suddenly a sand grain jumps straight into the air, rising as high as 12 inches. Wind blows the sand grain several feet. The grain strikes the ground, where it may bounce up again or knock loose some other particles. This process is called *saltation* and causes 50% to 75% of all wind erosion. In fact, more than 90% of all movement occurs within one foot of the soil surface.

Very fine silt and clay particles are too small to be picked up by the wind. However, the impact of a sand grain moving by saltation may knock the dust into the air. Once the wind hits it, dust rises high into the air and is carried long distances. This process, which is called *suspension*, accounts for about 3% to 40% of all wind erosion. Silt particles move most easily by erosion.

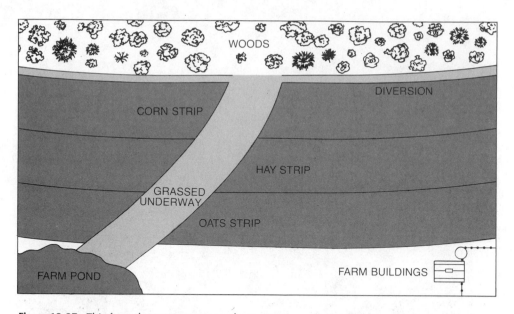

Figure 18-27. This farm demonstrates several conservation practices. A grassed waterway has been planted where water collects and directs the water to a pond. Contour strips empty into the grassed waterway. A diversion protects the field from water flowing from woods above the fields and also empties into the waterway.

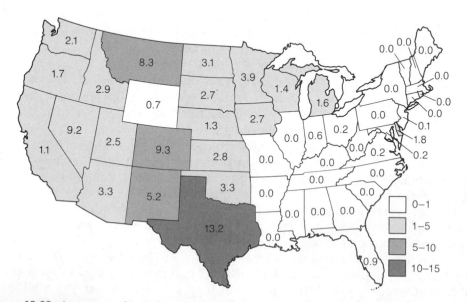

Figure 18-28. Average yearly wind erosion in 1982 on cropland (tons per acre) by state, excluding Alaska and Hawaii. (USDA, Preliminary Report, 1982 National Resource Inventory, 1984)

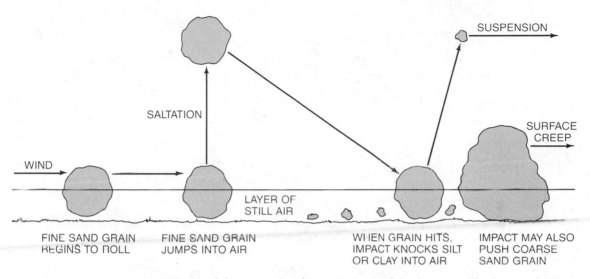

Figure 18-29. It is the saltation of fine sand that triggers wind erosion. Fine sands are large enough to protrude above the layer of still air, but small enough to be picked up by the wind. Surface creep and suspension both depend upon the impact of fine sand in saltation.

Coarse sand particles, ranging in size between 0.02 and 0.04 inches (0.05 to 1.0 mm), are too large to be kicked into the air. However, under the impact of saltating sand grains, they can roll along the ground. This is known as *surface creep.* It accounts for 5% to 75% of wind erosion.

Like water erosion, the detachment and transport of soil particles by wind are functions of energy. The higher the wind velocity, the greater the energy. Blowing soil itself contributes to erosive energy. A soil grain has greater mass than air, thus the impact of a soil grain on soil carries more energy. As a result, wind erosion has an avalanche effect—as wind blows across an open field, more and more soil is picked up and the erosive energy increases, causing even more soil to be picked up.

When the wind velocity dies down, so does its energy. Soil particles in saltation or suspension come to rest when the wind dies down. They may also fall out on the lee (downwind) side of an obstruction, the way snow gathers on the lee side of a snow fence (figure 18-30).

Effects of Wind Erosion.   Like water erosion, wind erosion removes the best soil first—the topsoil. It carries off fine soil particles especially silt and organic matter. The shift toward a coarser soil texture reduces nutrient holding and water-holding capacity. Further, windblown soil particles "sandblast" young plants (figure 18-31), tattering leaves and tearing away plant cells. Young plants can easily lose half their dry weight when exposed to sand-laden winds. Off-site damages can also be severe—and expensive. Blowing soil can fill road

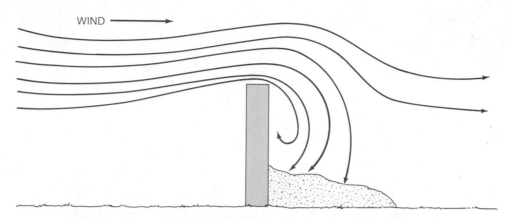

Figure 18-30. Wind-blown soil is deposited on the lee (downwind) side of obstacles. Therefore, it will be caught by windbreaks, snowfences, ridges in the soil, crop strips, and ditches.

Figure 18-31. Windblown soil "sandblasting" young crop plants. This literally tears away living cells. (Courtesy of USDA-SCS)

or drainage ditches (figure 18-32), affect the health of animals and people, increase cleaning and laundry costs, and wear at paint and other surfaces.

Factors in Wind Erosion. The following factors determine the amount of wind erosion.

- *Soil erodability* relates mainly to texture and structure. Soils high in fine sand

Figure 18-32. Windblown soil can fill in roadside or drainage ditches. (USDA, Soil Conservation Service)

are most liable to wind erosion; soils high in clay are least liable. Organic soils are also easily eroded by wind after they have been drained. If soil grains are cemented into larger soil aggregates, they are less likely to be blown away.

- *Soil roughness* makes a larger still air layer at the soil surface. Each clod or ridge also acts like a tiny windbreak to slow the wind and to capture blowing soil.
- *Climatic conditions* that promote wind erosion include low rainfall, low humidity, high temperatures, and high winds. Dry, windy conditions cause faster soil drying, and dry soil is more erodable than moist soil. Dry soil also supports a thinner vegetative cover.
- *Length of field* affects erosion. On the leading edge of a field, there is no wind erosion. As the wind travels across the field, it picks up more and more soil grains, like an avalanche.
- *Vegetative cover* protects the soil, as does a mulch. Bare soil, on the other hand, is fully exposed to the erosive force of wind.

These factors together can be arranged to create a soil loss formula similar to the USLE. It is called the Wind Erosion Equation, or WEE. Since WEE is more complex to apply than the USLE, and generally considered less reliable, it will not be detailed in this text.

Preventing Wind Erosion. Preventing wind erosion means changing the contributing factors listed previously. The following practices help control wind erosion:

- Till at right angles to the wind, leaving the soil surface rough and cloddy (figure 18-33). Lister furrows are very useful because they act as small windbreaks to capture blowing soil.
- Use conservation tillage or subsurface tillage tools to leave crop residues on the soil surface. The residues protect the soil from wind erosion. Various tillage tools can be used, including the standard reduced tillage tools, rod-weeders, and subsurface sweeps.
- Keep the soil covered with vegetation as much as possible. Cover crops of winter grains work very well to protect the soil over winter. Small grains are well suited to places where wind erosion is a problem.
- Plant crops in strips at right angles to the wind. For instance, strips of soil in summer fallow may be partially protected by alternate strips of small grains or row crops.
- Plant windbreaks of trees and large shrubs. Windbreaks shorten the field, reduce wind velocity, and capture blowing soil (figure 18-34). Windbreaks should not be solid; they should block about 50% of the wind.

Figure 18-33. Creating surface roughness on sandy fields in Texas. This will help reduce wind erosion. (Courtesy of USDA-ARS)

Figure 18-34. Windbreaks reduce wind erosion by shortening fields and capturing blowing soil. It is common to see fields in the Sand Plains of the Great Lakes broken into rectangular fields by windbreaks. (USDA, Soil Conservation Service)

- Plant buffer strips as temporary windbreaks. For instance, tall wheatgrass barriers planted north-south, about 50 feet apart, reduce erosion by about 93% (J.K. Aase, et al, "Effectiveness of Grass Barriers for Reducing Wind Erosion," *J. Soil Water Cons.*, 40(4):354-357. 1985). Figure 5-14 shows these strips being used to capture snow.

Critical Periods. Two periods are critical for wind erosion. The first period is when the soil is fallowed. Large clean-fallow fields in the Great Plains are ideal grounds for wind erosion. Subsurface tillage tools that leave crop residues on the surface reduce this problem. Another solution is no-till. Instead of controlling weeds by tillage, herbicides can be used to leave crop stubble and dead weeds standing to protect the soil from wind.

The period between harvest and the following crop cover is the critical period in many areas. A good snow cover protects soil, but if the snow blows off the field, so will soil. Again, conservation tillage protects soil during the critical period. Early harvest crops may also be followed by winter grains to cover soil during the critical period. For instance, after harvesting a pea crop one might plant winter rye. This can be harvested the following year or plowed into the soil before a different crop is planted.

SUMMARY

About two-thirds of the soil lost to erosion in the United States washes away in running water. Water erosion strips the topsoil, reduces yields, and deposits sediments in streams, lakes, and reservoirs. About one-third of erosion in the United States results from wind. Wind also strips the topsoil, blows away the smallest soil particles, and buries ditches and other structures.

Falling raindrops and running water detach soil particles from the soil surface and carry them away. Depending on the slope, erosion removes soil as a sheet or creates rills and gullies. Water erosion is promoted by bare and erodable soil, long or steep slopes, and the lack of conservation practices.

Soil scientists use the Universal Soil Loss Equation to compute soil loss. The USLE accounts only for losses from sheet and rill erosion and will understate soil loss where there is ephemeral or gully erosion. Using the USLE, a specialist can suggest practices to keep a farm productive.

Growing vigorous crops, maintaining organic matter, and avoiding overtillage and compaction help to control erosion. Both conservation tillage and crop rotation sharply curb erosion. Contour tillage, contour strip-cropping, and terraces are effective ways to slow runoff. Where these are not enough to stop runoff, they may be combined with grassed waterways or outlets to carry the excess off the field without erosion.

Wind blows soil off fields by saltation, suspension, and surface creep. Wind erosion is most likely to occur on (1) soils high in fine sand or on organic soils, (2) in hot, dry, windy climates where (3) soil is kept bare, especially for summer fallow. Control practices include breaking the wind, keeping the soil rough, and planting at right angles to the wind. The most effective method is keeping the soil covered by vegetation or crop residues.

REVIEW

1. Nationally, more soil is lost by wind than water erosion (True/False).
2. Growers are the only people to feel the effects of erosion.
3. The faster water moves, the more energy it has.
4. Only steep slopes will erode.
5. Sheet erosion is hard to detect.
6. The USLE predicts only sheet and rill erosion.
7. Silt is the soil particle most likely to be carried long distances by wind.
8. Saltation carries soil particles high into the air.
9. Moist soil is less likely to blow away than dry soil.
10. Eastern states suffer the most from wind erosion.
11. The USLE does not predict erosion from _____ and _____ water erosion.
12. The standard for acceptable soil loss lies between _____ and _____ tons/acre/year.
13. One can protect a possible gully site by constructing a _____.

14. For conservation tillage to effectively prevent erosion, residue coverage must be at least _____ % .
15. Plowing across a hillside is called _____ .
16. _____ particles are moved by saltation.
17. Wind erosion tends to make soil texture more _____ .
18. Large fields subject to wind erosion can be shortened by the use of _____ .
19. _____ and _____ soils are most likely to suffer wind erosion.
20. If the wind comes from the west, plant rows in the direction _____ .
21. List three effects of water erosion.
22. How could you move water off a sloping field safely?
23. Say a topsoil erodes away, exposing the subsoil. What factor of the USLE would change?
24. Describe the three steps of water erosion.
25. You operate a farm in Maclean County, Illinois (indicated in figure 18-13). The soil is a Saybrook, and according to the soils survey report, $K = 0.32$ and $T = 5.0$. One field has a slope of 4%, and is 300 feet long. Calculate average annual soil loss for these two practices, and determine if that rate is acceptable:
    a. fall-plow up and down the slope
    b. contour chisel, leaving 40% cover
26. List three effects of wind erosion.
27. Name the three forms of wind erosion and identify the main soil particles associated with each.
28. Why would a vegetable grower on sandy soil plant winter rye after harvesting?
29. Explain why large expanses of bare ground are most likely to erode by wind.
30. What parts of the country suffer the most water erosion? Wind erosion?

SUGGESTED ACTIVITIES

1. Study USDA Handbook 537, *Predicting Rainfall Erosion Losses*, 1978, by W. H. Wischmeier and D. D. Smith. This is the published guide for using the USLE, and it gives far more detail than is possible here.
2. Visit farms that use different conservation methods.
3. Using local soil survey reports and C-value tables, practice applying the USLE to farms in your area. If no C-value tables are available, study the USDA Handbook 537 cited to learn how to calculate values.
4. Use the technique of figure 18-20 to measure residue cover in several fields.

chapter nineteen

Urban Soil

Timber retaining walls are often constructed in yards with steep slopes. Aside from whatever ornamental values they may have, retaining walls act as an urban form of terrace. Urban soils have special problems that are experienced by those that use them. This chapter will describe some of those problems.

OBJECTIVES

After completing this chapter, you should be able to:

- list five characteristics of many urban soils
- describe ways of dealing with urban soils
- describe erosion control on urban sites

TERMS TO KNOW

debris basins	hydroseeding	rip-rap

Urban soils are those found within a city, town, or metropolitan area. The USDA reports that in 1982 there were 46,627,900 acres of urban land. Roadsides and other built-up land in rural areas may also be considered "urban" in that they share the problems of urban soils. This land accounts for another 26,932,400 acres, for a total of about 5% of the nonfederal land in the United States.

Urban soils present certain difficulties to those who use them, such as landscapers, homeowners, gardeners, or building contractors. This chapter will look first at problems of soils in established urban areas and then at erosion problems in new urban sites.

PROBLEMS OF URBAN SOILS

Compared with rural soils, urban soils have been greatly altered by construction and other activities. Earth has been moved from site to site, grades and drainage patterns changed, foundations dug, and foreign soil brought in. As a result, the problems faced by people using urban soils are different from those encountered by rural growers. Often the problems are more severe.

Figure 19-1 shows an extreme example of the problems of urban soil. A soil map of this city indicates a sandy loam texture in the area where this photograph was taken. A nursing home was being built on this site, which had been a parking lot. The photograph shows the "soil profile" in the parking lot: three inches of tar on top of four inches of concrete, over three feet of very coarse sand. At the spot where this photo was taken, the pavement was removed and a lawn was later planted.

In general, there are extreme variations in soil conditions across a city landscape. These differences are caused by the massive soil moving that goes into the building of a city. Such variation complicates soil mapping and land-use planning. In fact, it is only fairly recently that any detailed mapping of city soils has occurred. Soil moving also causes abrupt changes in the soil profile. For instance, the black soil spread before planting grasses may be of a very different texture than the native soil.

Soil moving often causes very poor soil quality. A less desirable subsoil may be brought to the surface by excavation. At other times, a sandy, stony fill may be

Figure 19-1. "Soil profile" of a city parking lot shows pavement over sand. After the pavement was removed, three inches of black dirt were spread over the sand and sod was planted.

brought to a site. Contractors will usually lay three or four inches of topsoil over this. The topsoil may be helpful, or may interfere with water movement and rooting of the turf, as described in chapter sixteen. In any event, it is far from a return to desirable soil conditions. Urban soils are best improved by the addition of organic matter, such as leaves, grass clippings, compost, or bagged manures. Some cities compost city leaves and offer the compost to city gardeners (figure 19-2).

Aside from extreme soil variations and poor quality soil, three other traits of urban soil can be identified: buried debris, compaction, and lead contamination.

Buried Debris. Urban soil usually has a lot of debris buried in it. During construction of a building, contractors often bury wood or masonry scraps on the site, rather than haul them away. Buried masonry, which contains lime, can raise pH to unacceptable levels. The debris is also a constant source of frustration to gardeners and landscapers trying to work the soil. A new suburb near the author's

Figure 19-2. This city composts leaves gathered in the fall, then offers them to people in the neighborhoods. This arrangement saves landfill space for the city and provides a good organic amendment for home gardens.

home, recognizing this problem, recently passed a law requiring contractors to remove their debris.

In older neighborhoods, the soil contains artifacts from years of habitation. Some very old sites have even been excavated to uncover old bottles, tools, toys, and other relics of the city's past.

Compaction. Urban soils are usually moderately to severely compacted. The compaction results from the use of heavy equipment on the soil during construction. Countless footsteps on yards and parks also cause compaction (figure 19-3). Compaction from construction may kill sensitive trees like oaks, cause greater erosion, and make it more difficult to establish a landscape. Compaction from footpaths promotes the growth of compaction-tolerant weeds (like knotweed) and makes it difficult to establish good turf.

Compaction can be measured by bulk density. Chapter one mentioned that an "ideal soil" is about 50% solid particles and 50% porous space. The bulk density of such a soil is about 1.3 grams per cubic centimeter. As bulk density rises above 1.4, root growth begins to suffer from lack of air and direct physical resistance. At a bulk density of 1.7, roots cannot penetrate the soil.

Figure 19-3. This footpath in a city park illustrates a common problem of compaction resulting from maintenance equipment and countless footsteps.

The bulk density of an average farm soil ranges from 1.2 to 1.6. The people in charge of landscaping around the nation's Capitol measured some bulk densities on the Capitol Mall ranging between 1.8 and 2.2. Recent work done at the University of Minnesota compared soils at some building sites with nearby undisturbed soils. On construction sites, bulk density averaged 1.6 and nearby soils averaged 1.0. The researchers then studied the effect of compaction on the growth of the shrub forsythia. They found that increasing the bulk density 30% caused a loss in root growth of 45% and top growth of 40%.

Planning can avoid some compaction. For instance, on new construction sites, it may be possible in some cases to limit the area driven on by construction equipment. Landscape architects and designers can also help control pedestrian compaction by remembering that people will take the shortest route between two points—they will cut corners, walk across lawns, and generally ignore paths if they are the long way around. Knowing this, designers should carefully analyze expected foot traffic patterns on a new site, and install sidewalks or mulched paths where people can be expected to walk.

Although it is difficult, it is possible to break up compaction if no plants are in the way. Deep tillage will break up compaction. The soil can then be heavily amended to stop further compaction. Digging large, solid particles like fly ash into the soil helps by creating a "skeleton" that resists compaction. Large amounts of organic matter like wood chips or leaf mold also help. Where heavy foot traffic is expected, a deep layer of wood chips cushion the soil.

Another method of dealing with foot traffic is to pave the soil with brick pavers that have large holes built into them. Grass can grow in the spaces, giving the impression of turf. Water and air can also move through the spaces.

If trees or turf already occupy a site, compaction is more difficult to repair without hurting roots. Machines called aerators remove vertical cores from the soil. This process helps break up compaction. For turf, vertical coring to 6 inches breaks up the soil, and makes passages for air and water movement. Machines that *remove* a core should be used, and the cores should be left on the surface as a topdressing. Aerators that *punch* a hole without removing a core squeeze the soil between the holes, increasing compaction.

For trees suffering from the effects of compaction, coring to 18 inches is needed. Unfortunately, there are no machines designed to dig to this depth. Thus, it would have to be done by hand.

Lead Contamination. Lead contamination is a problem unique to urban soils. Lead is a toxic metal that once was added to paints and remains in use as an additive in leaded gasoline. It primarily affects children, causing permanent brain damage, behavioral problems, other health problems, and even death. In 1984 the Environmental Protection Agency (EPA) estimated that 1.5 million children in the United States had enough lead in their blood to lower intelligence by at least three IQ points. Lead contamination also has been shown to contribute to high blood pressure in adults.

Urban children acquire lead by ingesting contaminated soil or paint chips or by breathing air-borne lead. In old, especially run-down neighborhoods, chips of lead-based paint may be eaten by infants and young children. Paint chips may also contaminate soil around a building. The main source of soil lead is automobile exhaust from cars using leaded gas. In spite of measures to promote the use of unleaded gas, in 1983 45% of all gas sold was still leaded. Some of this was used in old cars that were designed for leaded gas, some in modern cars designed for unleaded gas. The EPA estimates that a city the size of Minneapolis, Minnesota is dusted annually with 2,000 tons of lead from automobile exhaust.

The following suggestions may be useful in controlling health problems from lead:

- Use unleaded gasoline. Many feel that even older cars can be operated with unleaded gasoline and suffer no damage to the engine. Certainly modern cars are damaged by leaded gas, increasing maintenance costs. At the time this text was being written, the EPA had just accelerated its timetable for removal of all lead from gas. Many critics, however, want lead banned immediately.
- Keep yards covered by a good stand of turf to prevent children from playing in contaminated soil. Sandbox sand can be changed yearly. Children can ingest lead by eating soil, by sticking dirty fingers in their mouths, or eating with dirty hands.
- Cleaning children's hands, keeping the house free of dirt, and other cleanliness measures lower the amount of contaminated soil ingested or breathed by children as dust in the air.
- Remove chipping lead-based paint from old homes, clean up thoroughly, and repaint with new paints.
- Have a child's blood tested for lead levels in high-risk neighborhoods. These include areas with old, run-down housing and areas near heavy traffic.

Many home gardeners are concerned about lead in their garden soil (figure 19-4). The biggest risk is the direct ingestion of soil by children. The other, much lesser risk, is absorption of lead by garden crops, which are then eaten by the family. This is also a problem posed by heavy metals in sewage sludge (see chapter fourteen). In some cities, it may be possible to have garden soil tested for lead.

If a gardener suspects a lead problem, a number of measures may be taken. One would be to dig out the old soil to a depth of six inches, then build an elevated bed with railroad ties or by other means. If the bed were raised six inches above the grade, this would give a total of 12 inches fresh soil. This depth would contain most plant roots. A second measure is to keep the soil pH near neutral, since lead is much more soluble in acid soils. It is also possible that high levels of organic matter from compost might form a complex with lead, tying it up.

Crop choice may be affected by the amount of soil lead. Heavy metals gather in the leafy parts of plants rather than in fruits or seeds. Thus peas, beans,

Figure 19-4. Inner city gardens near major roadways may contain lead from automobile exhausts.

tomatoes, and other seed or fruit crops would be safer than lettuce, spinach, or other leafy crops. Root crops present a danger mainly because it is difficult to clean them enough to remove all the lead that might cling to the outside of the root. Homeowners may prefer flower gardens to vegetable gardens where lead is a serious problem.

URBAN EROSION

During each year of the last decade, more than three million acres of land changed from rural to urban uses. These urban uses include roads, highways, housing developments, commercial developments, and parking lots. Once these urban sites are well established, with turf and trees growing, little erosion occurs.

During the construction phases of urban or road development, erosion can be 10 to 100 times greater than on similar farmland (figure 19-5). Large developments involve large-scale earth moving, leaving the soil bare for long periods of time. Topsoil removal usually exposes the subsoil, which is often more erodable than the topsoil. Areas that normally absorb water, like woods, are often graded. Once stripped, these areas absorb less moisture, causing more runoff. Since roads and parking lots allow no infiltration, even more runoff is produced. When this extra runoff flows over bare soil produced during construction, massive erosion results.

The impact of construction site erosion differs slightly from farmland erosion. Generally, loss of plant nutrients and growth media is a secondary

Figure 19-5. Severe rilling on a roadside embankment caused the ditch along the road to be almost filled with sediment. Much of the lost soil reached a nearby lake. (USDA, Soil Conservation Service)

concern. A developer is concerned with erosion damage—like rilled and gullied land that must be repaired before a site can be sold. In extreme cases, erosion can undercut and collapse roads and foundations. And, of course, urban erosion contributes to off-site problems. About 5% of the sediment load annually reaching surface waters in the United States comes from urban sources.

Controlling Erosion.   Controlling erosion on construction sites requires careful planning, using soil surveys, topographic maps, and other tools. The plans must include provisions for controlling erosion and sedimentation. Methods for controlling erosion may be used by both building contractors and landscapers. There are five general principles for controlling runoff, erosion, and sedimentation:

1. Keep disturbed areas small.
2. Protect disturbed areas.
3. Keep runoff velocities low.
4. Divert runoff away from disturbed areas.
5. Retain sediment on-site.

Keep Disturbed Areas Small. During the planning stage, identify critically erodable areas and plan to avoid disturbing them. Such areas can be left as "green" areas. Disturb only that land being actively built on, rather than stripping the soil over a large area. Where possible, retain natural soil cover.

Protect Disturbed Areas. As soon as possible, vegetation should be planted on disturbed sites. Such vegetation could include a temporary cover of annual grasses, permanent turf, shrubs, or ground covers. On very steep slopes, a layer of rock, like the *rip-rap* in figure 19-6, can control erosion. Retaining walls made of concrete, railroad ties, or stone can be used to terrace slopes (see chapter title page).

Keep Runoff Velocities Low. When grading land, keep slopes short and gradients as low as possible. Keep vegetative cover intact to slow runoff where feasible. Various permanent or temporary control structures may be used. For instance, highway construction crews often make temporary dams of hay bales across ditches on long sloping roadsides to keep running water from picking up speed downhill.

Figure 19-6. A rip-rap cover in a long, sloping roadside ditch. (USDA, Soil Conservation Service)

Divert Runoff Away from Disturbed Areas.   Land or pavement at a high elevation can act as a small watershed, collecting rainwater which can run onto bare soil at a lower elevation. These bare areas should be protected by diversions, and the water led into waterways or outlets. Waterways can be covered by turf (grassed waterways), rip-rap, or even concrete, depending on the amount of waterflow they must handle.

Retain Sediment On-Site.   It is impossible to stop all erosion during construction, but sediments can be captured and retained on the site. This avoids pollution and sediment damage in other areas. Grass strips filter soil out of runoff water, as it does in strip-cropping. Most modern developments include *debris basins* to capture sediments. These are small ponds (figure 19-7) built in the path of drainage patterns. Sediment-laden water flows into the basin, and soil settles out. Later, these ponds can provide an attractive amenity to local housing.

Establishing Vegetation.   Growing vegetation on denuded areas is essential. Permanent turf is the best cover, but mowing can present a problem on steep banks. Grass may also not be adapted to some soils. Herbaceous groundcovers like the legume crownvetch or the attractive iceplant in California are good alternatives. Very closely spaced shrubs, especially those that spread by suckering, may also replace turf. Using all three—turf, groundcover, and shrubs—properly can bring beauty and a wildlife habitat as an added benefit.

Figure 19-7. Erosion control at a housing development consists of a debris basin (background) to capture runoff and trap sediment. In the foreground, the soil has been stabilized by seedling grass and using a mulch to protect the soil until the turf is established. (USDA, Soil Conservation Service)

It is often difficult to establish plants on critical erosion areas, because the soil may be poor and both seeds and soil may wash away. Before planting, make sure all runoff is diverted from the site. The soil should be free of rills and gullies, compaction broken up, and lime and fertilizer added as needed. If groundcovers or shrubs are being planted, wood chip mulches can be used to control erosion.

When an erodable slope is seeded to grasses or legumes, a light mulch is needed to keep soil and seeds from washing away. About 80 pounds of clean hay or straw per 1,000 square feet (1 1/2 tons per acre) provides a satisfactory cover. This mulch should be held in place by netting (such as jute netting), spraying with special asphalt materials, or being punched into the soil with a modified disc. Sometimes, heavy jute netting (a fiber mesh) replaces the mulch. In very difficult areas, seed and mulch may be applied at the same time by *hydroseeding*. In this method, a mixture of water, seed, and chopped hay is blown on a slope from the side of the road.

Sodding provides a quick cover of critical spots like steep slopes. Sodding, if properly done, stops erosion quickly, but it is a costly solution. Because a steep slope is a problem site, careful attention must be paid to the proper sodding techniques. The sod should be staked.

Calculating Soil Loss. The Universal Soil Loss Equation (USLE) can be used to estimate soil losses from construction sites and to identify critical erosion sites. In this application, the main concern is not topsoil loss per se, but movement of sediment. The results are not precise, but they do offer some guidance. Two shortcomings keep the USLE solution inaccurate. The equation predicts soil loss only from small rills and sheet erosion, not large (ephemeral) rills and gullies. These forms of erosion are common on construction sites and can carry a lot of sediment. In addition, some sediments actually leave the site, but some are deposited *on* the site. The equation does not predict where the sediments will be deposited.

The USLE is written as:

$$A = R\ K\ LS\ C\ P$$

"*A*" is the soil loss in tons per acre. Refer to chapter eighteen for instructions in how to solve the equation. The following values can be used to solve the USLE for construction sites:

- *R*. The rainfall factor can be obtained from figure 18-13.

- *K*. The erodability factor can be obtained from soil survey data or calculated from USDA Handbook 537 (listed in the activities section of chapter eighteen). If the topsoil has been removed, then the *K* value for the exposed subsoil must be used.

- *LS.* The slope factor can be obtained from figure 18-14. Slope factors are reliable only up to a 20% slope, and so are not reliable for very steep roadside embankments.

- *C.* The cover and management factor for bare, stripped soil is the same as a clean-fallow plot, so C equals 1.0. If a mulch covers the soil, C is much lower. Figure 19-8 gives some C values for construction sites.

- *P.* The support practice factor is usually 1.0, because few of the support practices are applicable to construction sites. Terraces are included in the *LS* factor, as described in chapter eighteen.

Type of Mulch	Mulch Rate	Land Slope	Factor C	Length Limit
	Tons per acre	*Percent*		*Feet*
None	0	all	1.0	—
Straw or hay,	1.0	1–5	0.20	200
tied down by	1.0	6–10	0.20	100
anchoring and				
tacking equipment	1.5	1–5	0.12	300
	1.5	6–10	0.12	150
	2.0	1–5	0.06	400
	2.0	6–10	0.06	200
	2.0	11–15	0.07	150
	2.0	16–20	0.11	100
	2.0	21–25	0.14	75
	2.0	26–33	0.17	50
	2.0	34–50	0.20	35
Crushed stone,	125	<16	0.05	200
1/4 to 1 1/2 in.	135	16–20	0.05	150
	135	21–33	0.05	100
	135	34–50	0.05	75
	240	<21	0.02	300
	240	21–33	0.02	200
	240	34–50	0.02	150
Wood chips	7.0	<16	0.08	75
	7.0	16–20	0.08	50
	12	<16	0.05	150
	12	16–20	0.05	100
	12	21–33	0.05	75
	25	<16	0.02	200
	25	16–20	0.02	150
	25	21–33	0.02	100
	25	34–50	0.02	75

Figure 19-8. Cover and management factor (C) and length limits for construction slopes. If the slope length is exceeded, higher mulch application rates or some means of shortening the slope is needed. (USDA Handbook 537, Table 9)

SUMMARY

Urban soils are characterized by great variation, poor soil, buried debris, compaction, and lead contamination. Poor soil, buried debris, and compaction increase the difficulty of growing an attractive landscape around homes and in parks. Lead presents a health risk to children. Urban growers can use compost to improve the soil. They should keep the soil pH near neutral to tie up lead.

During the construction of roads or housing, soil is fully exposed to the forces of water erosion. Keeping disturbed areas small, establishing vegetation on the soil as quickly as possible, and diverting runoff from bare soil can help prevent erosion. Sediments can be captured by grass filter strips or catch basins.

REVIEW

1. After a housing development is built, the county soil map for that site should still be accurate (True/False).
2. Construction debris usually lowers soil pH.
3. In many landscape sites, proper designing can help prevent pedestrian compaction.
4. Aeration can help relieve surface compaction.
5. Neighborhoods with new cars suffer the most lead contamination.
6. The greatest danger of lead in a garden soil is that young children may ingest the soil.
7. The "K" value of a soil would often be higher after the topsoil is stripped off.
8. Erosion on a construction site is seldom as severe as on farmland.
9. Parking lots can contribute to urban erosion.
10. Special efforts are required when seeding steep slopes.
11. Small ponds built to catch sediments from construction projects are called _____ .
12. Blowing a mixture of water, straw, and seed into soil is called _____ .
13. Buried masonry _____ soil pH.
14. Two sources of lead contamination include _____ and _____ .
15. A layer of rocks set on the soil to inhibit erosion is called _____ .
16. Suggest three kinds of plants that can replace turf on hard-to-mow slopes.
17. Explain three measures to prevent lead in vegetables grown in an urban garden.
18. List several strategies for controlling the amount of sediment that leaves a construction site.
19. What are the health effects of lead?
20. Assume that behind a new home is a slope that is 50 feet long and 40 feet wide with a gradient of 10%. The home is in central Illinois,

marked on the map in figure 18-13. The subsoil, exposed after grading, has a K value of 0.20. How much total soil will be carried off this slope in an "average" year if left bare? How much if mulched with 1.5 tons/ acre hay? One acre equals 43,560 square feet.

SUGGESTED ACTIVITIES

1. If any large-scale housing projects are being built nearby, visit them and identify erosion and sediment control practices.
2. Talk to a landscape architect about erosion control in the landscape.
3 Find out how soils can be tested for lead in your state. If you know of any community gardens, have their soil tested for lead.

chapter twenty

Government Agencies and Programs

This farmstead in Minnesota has windbreaks to protect its buildings, and the grower uses strip cropping to protect the soil. Government agencies can help farms develop such helpful practices. And not just farmers— the author was also aided by photos supplied by such groups. This chapter will describe some of the agencies and their programs.

(Courtesy of USDA-SCS)

OBJECTIVES

After completing this chapter, you should be able to:

- list federal agencies that assist growers
- describe some of the soil programs of these agencies
- give examples of state and local programs
- describe trends in programs to promote soil conservation

TERMS TO KNOW

Agricultural Experiment Stations	Agricultural Stabilization and Conservation	Soil and Water Conservation
Agricultural Research Service	Service	Districts
Agricultural Cooperative Extension Service	Conservation Reserve Program	Soil Conservation Service

Research yearly comes up with new methods that can help growers, but growers cannot spend much of their time in school learning new methods. Market forces pressure growers to change, not always in ways that are best for the soil. As businesspeople, growers must often put their money where the financial return is best—which is not always for long-term benefit. A number of programs support growers with technical assistance and have information to answer questions such as: How do growers keep up with change? How can they afford soil improvements?

Figure 20-1. Education is an important service. Here the Soil Conservation Service and other organizations helped a community college buy land as a study area. (USDA, Soil Conservation Service)

A network of laws and agencies helps the grower in several ways.

- *Education* (figure 20-1) provides information on new and old methods in the form of publications, workshops, and advice.
- *Technical assistance* (figure 20-2) advises growers on how to complete specific projects like irrigation or grassed waterways.
- *Financial assistance* helps growers pay for farm projects.
- *Research* helps by creating new and better ways to farm.

Government programs operate through a complex web of federal, state, and local agencies. In addition to the government, there are private sources of information and help. We will concentrate here on government programs, especially those in the area of conservation.

ASSISTANCE TO THE FARMER

Help comes to farmers from many sources. One major source is the United States Department of Agriculture (USDA), established in 1862. The USDA includes several agencies that help farmers. Many USDA programs work through state or local groups (figure 20-3). We will look at USDA, state, and private programs.

Figure 20-2. Erosion in a streambed (*left*) caused damage to the cornfield. Growers can obtain help from various agencies to solve this type of problem. (USDA, Soil Conservation Service)

Research and Education.   The main research arm of the USDA is the *Agricultural Research Service* (ARS). It was established in 1953 to do basic and applied research in agriculture. The ARS, of course, does a variety of work, some of which deals with soils. For instance, the ARS is heavily involved in the development of the Universal Soil Loss Equation. They publish an interesting nontechnical magazine that reports on current research, titled *Agricultural Research.*

The *State Agricultural Experiment Stations* were established by Federal law in 1862. They are, of course, state organizations, mostly attached to universities, but are partly funded by the USDA. Their two main goals are research and education. Research is partially funded by a USDA agency, the Cooperative State Research Service.

The information obtained by the ARS and experiment stations is channeled to farmers largely through the efforts of the *Agricultural Extension Service* which is attached to the experiment stations. They receive some of their funding from a USDA agency, the Cooperative Extension service. Cooperative Extension services publish information bulletins, run workshops, and provide expert help to farmers. Most counties in every state have a resident extension agent to give advice to farmers. In addition, most experiment stations have specialists on their staff to provide help in specific areas. For example, there are specialists in

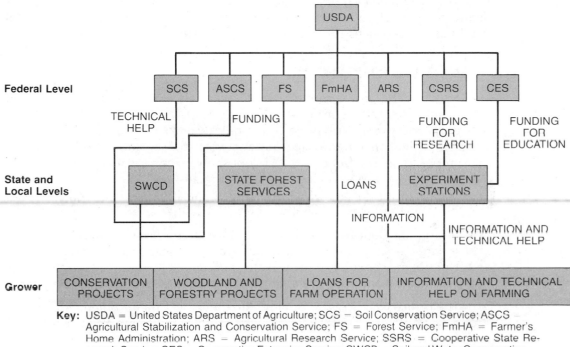

Key: USDA = United States Department of Agriculture; SCS – Soil Conservation Service; ASCS – Agricultural Stabilization and Conservation Service; FS = Forest Service; FmHA = Farmer's Home Administration; ARS – Agricultural Research Service; SSRS = Cooperative State Research Service; CES – Cooperative Extension Service; SWCD – Soil and Water Conservation District

Figure 20-3. Help for growers is available at several levels. U.S. Department of Agriculture agencies provide aid, much of it through local Soil and Water Conservation Districts or Agricultural Experiment Stations.

greenhouse crops, irrigation systems, and many other areas. Cooperative Extension supplies both general education and technical advice.

Business also gives helpful information to farmers. Many cooperatives have agronomists on their staff to help co-op members. Equipment dealers and representatives of chemical companies can also be valuable sources of information. However, one must remember that their job is to sell products.

Financial Assistance. Two programs run by the USDA provide financial help to farmers. The *Farmer's Home Administration* (FmHA) supplies loans to farmers for a variety of needs. Many of these loans help with general farming expenses or with farm ownership. In addition, the FmHA can make loans for special soil purposes, like irrigation and drainage, conservation, reforestation, or grazing land improvement.

The *Agricultural Stabilization and Conservation Service* (ASCS), another USDA agency, supplies money to farmers and others for conservation efforts. The ASCS is involved in a number of programs, as will be discussed later in the chapter.

The Production Credit Association (PCA), not a USDA agency, also supplies loans for general production expenses. Some of these may be soil-related. Local banks also make loans to cover production expenses.

CONSERVATION EFFORTS

Many of the programs mentioned so far help with short-term efforts that have a very direct effect on farm profit. Such efforts include fertilization and seed selection. Some long-term projects also quickly make money for the farmer, for instance, irrigation and drainage. Soil and water conservation efforts, however, have more long-term benefits that do not show up immediately on the farmer's income statement. We will look at these programs next.

A major thrust of the USDA efforts since the 1930s has been soil and water conservation programs. At this time, the USDA defines several priority problems:

- A large percentage of American land has excessive erosion.
- In large areas of the West, groundwater for irrigation is being depleted.
- Flooding is common in many watersheds.
- Sediment, plant nutrients, and pesticides continue to pollute our waters.

The primary agencies for dealing with these and other problems are the Soil Conservation Service (SCS) and the Agricultural Stabilization and Conservation Service (ASCS).

Soil Conservation Service.   Congress set up the *Soil Conservation Service* in 1935 to carry out a national program of soil and water conservation. Originally the only goal of the SCS was to control erosion, but within a few years the agency ran numerous programs. Some of these programs include

- Primary responsibility for soil surveys in a joint effort with Agricultural Experiment Stations. As time allows, SCS surveyors can do farm surveys for growers.
- Helping to carry out flood prevention programs.
- Helping city and county officials with land-use planning (figure 20-4).
- Developing useful plants for conservation.
- Measuring the winter snowpack on mountains in the West to forecast the amount of water available the following year.
- Gathering important data about soil and water resources for federal planning. These data-gathering projects, called the National Resource Inventories, are the sources of some of the data used in this text.
- Provide technical help for soil and water conservation efforts by farmers (figure 20-5).

The last role is the core program of the SCS. Conservation efforts are organized through several programs, including the Agricultural Conservation Program, the Great Plains Conservation Program and others. The SCS maintains a network of soil scientists in counties of the United States to serve these programs.

Figure 20-4. The Soil Conservation Service helps to ensure that land is used properly when it is being developed for urban uses. Here, a soil scientist surveys a development site. (USDA, Soil Conservation Service)

Many SCS activities are channeled through local agencies called Soil and Water Conservation Districts.

Soil and Water Conservation Districts. Most of the present conservation programs were authorized in the late 1930s. It was felt that these federal programs should operate through some local authority. In 1937 President Franklin Roosevelt proposed to all state governors a model for creating *Soil and Water Conservation Districts* (SWCD).

All states but one now have such districts. Generally the boundaries of a SWCD follow county boundaries (figure 20-6). Each district is governed by locally elected or appointed boards. Members of these boards are usually district growers. The districts plan for and carry out programs that they feel have a priority in their district. These priorities vary. In the eastern United States, water erosion control is often a high priority. In the West, irrigation, salination, or sometimes wind erosion are important. The actual role and duties of the SWCD vary from state to state, since they are state and not federal agencies.

All districts have formal agreements with the U.S Secretary of Agriculture to set up programs. The secretary, in return, agrees to help the districts. Most

Figure 20-5. A newly planted grassed waterway installed with the help of the Soil Conservation Service.

districts have SCS soil scientists assigned to them for technical aid. In fact, it is through the SWCD that the scientists operate. A prime example of the cooperation of the SCS and the SWCD is the Agricultural Conservation Program.

Agricultural Conservation Program. The *Agricultural Conservation Program* (ACP) was authorized by Congress in 1936. The ACP is concerned with (1) controlling erosion on farmland, (2) controlling pollution and sediments from farmland, (3) improving water quality in rural America, and (4) helping growers meet national and state guidelines. Two USDA agencies, the SCS and ASCS, run the program through local Soil and Water Conservation Districts.

The program is run by the ASCS. It provides money on a cost-sharing basis for conservation programs. Cost sharing means that growers pay part of the cost, and the ASCS pays part. They can also enter into long-term plans (3 to 10 years) approved by the district. The SCS provides technical help, except for the forestry projects. These are aided by the National Forest Service.

Figure 20-6. Most of the United States is divided into Soil and Water Conservation districts, usually along county lines. Local district committees set conservation priorities. It is through these committees that many programs operate. (USDA, Soil Conservation Service)

Specialized Programs. Several special programs like the Agricultural Conservation Program have been set up to help certain regions or for certain uses. The *Great Plains Conservation Program*, for instance, was set up in 1956 to meet the special needs of farmers and ranchers of the Great Plains. This program is run much the same as the one described above. It is interesting to note that during dry periods in 1976 and 1980 in South Dakota, twice as many nonparticipants in the program had severe drought-related problems as had participants.

A number of other programs operate in a similar fashion:

● *Appalachian Land Stabilization and Conservation Program* is designed to prevent erosion in the Appalachian states. This program has gotten little funding in recent years.
● *Water Bank Program* is designed to preserve wetlands in farming areas.
● *Rural Clean Water Program* is designed to help farmers control water pollution from their farms.
● *Rural Conservation and Development* is designed to help local natural resource programs (figure 20-7). The FmHA also makes loans under this program.

Figure 20-7. The Rural Conservation and Development program funded this project to promote commercial wild rice production in one area. (USDA, Soil Conservation Service)

STATE AND LOCAL EFFORTS

In addition to federal programs, states and localities also have laws and programs. Foremost of these, of course, are the Soil and Water Conservation Districts. Obviously, this text cannot list all the state and local efforts. A few examples are worth noting, however.

About 15 states have passed erosion and sediment control laws, with more laws being passed yearly. About six states have cost-sharing programs for conservation efforts like ASCS programs. Maryland requires all counties to adopt erosion control laws, with the conservation districts being heavily involved. Minnesota enjoys a program called *Reinvest in Minnesota* that sets aside some land for wildlife cover.

Many local and state laws involve controlling land use. Many of these are zoning laws. For instance, many outer suburban areas limit how far land can be subdivided. To save farmland, many states have "green acre" laws, which give tax breaks to land in developing areas that is kept as farmland. In Illinois, anyone wishing to subdivide certain farmland must make a proposal to the local SWCD. The SWCD then submits an opinion to the local governing body that grants permission for such subdivisions. Colorado has set up land-use controls involving state and county commissions. Hawaii has zoned the state into use categories based on soil surveys. A few counties in the country have even restricted Class I land to agricultural use.

TRENDS

Many people are looking closely at farmland losses to erosion and urbanization, sedimentation, and pollution from farms. Critics point out that after nearly 50 years of federal conservation efforts, erosion continues to be a major problem. During the 1970s, it even increased because more land was used for row crops. Meanwhile, housing, highways, and other urban uses continue to eat up agricultural land (figure 20-8).

In the United States, control over land use falls almost entirely to the landowner. Therefore, most programs have been voluntary. For instance, most SCS offices wait for farmers to come to them for help. Progress in preventing erosion has been slow because of the costs of conservation efforts, because a few farmers are not educated to the need for erosion control, and because conservation runs up against market forces that squeeze the farmer.

Figure 20-8. A shopping center is being planned to replace this farmland. However, in the future, we will probably see greater efforts to slow the loss of farmland to nonfarm uses. (USDA, Soil Conservation Service)

As world food demand rises and the United States agricultural base shrinks, we are likely to see a shift away from totally voluntary efforts to more demanding programs. These could fall into two areas: soil conservation and land-use control.

Conservation. In the 1980s, federally funded conservation programs have become more aggressive. The prime act is the 1985 farm act, or the Food Security Act of 1985, which included a number of erosion control and related provisions. These provisions include:

- Conservation Reserve Program (CRP) buys ten-year conservation easements from growers. The CRP targets highly erodable land. In exchange for payments, the grower plants the land to permanent cover such as grasses or trees. The program allows for some cost sharing between the grower and the government.
- Conservation compliance provisions require growers to develop conservation plans to remain eligible for other USDA program benefits.
- "Sodbuster" provisions allow the denial of federal price supports to farmers who plow out highly erodable lands that are not currently being cultivated.
- "Swampbuster" sections of the act can deny eligibility for other USDA programs to farmers that drain and farm certain wetlands.

A goal of the CRP is to set aside about 40–45 million acres of the most erodable lands by 1990. At the time this was being written, good progress was being made. Texas had the highest sign-up rate, and the SCS estimates that about 125 million tons of soil per year were being saved there.

One fear of the soil conservation community is that when easements expire growers will plow out the land and resume cropping. Partly for this reason, the SCS favors tree plantings, which are less likely to be destroyed. It is to be hoped that conservation progress will not be lost.

Land-Use Controls. Increasingly, states are passing laws to slow the loss of prime farmland to other uses. At this point, this is mainly done by taxation. In this method, prime farmland being changed to other uses is taxed at a higher rate to make the change less profitable. A few areas allow local governments to buy land-use rights for farmland. That is, the owner owns the land and can farm it. However, the government can refuse to allow nonfarming use of the land because they bought the use rights.

It is possible that the future will see strict mandatory land-use controls as are practiced in some places. For instance, laws might appear that allow Class I and Class II land to be used only for farming.

SUMMARY

Several groups assist growers with education, technical help, financial aid, and research. Many of these groups are part of the United States Department of

Agriculture. USDA research is carried on by the Agricultural Research Service. The Farmer's Home Administration makes loans for farming expenses, while the Agricultural Stabilization and Conservation Service helps with the cost of conservation efforts. The Soil Conservation Service provides technical help for soil and water conservation projects, among its many duties.

Almost every county in the United States is also a Soil and Water Conservation District. In these state-run districts, local growers set conservation priorities for their area. Attached to these districts is an SCS soil scientist.

Most counties also have a county Cooperative Extension agent. These agents represent the state Cooperative Extension service and are there to advise growers. Part of their role is to spread to farmers results of research from the ARS and state Agricultural Experiment Stations. The Extension service also has specialists on staff at the Experiment Stations and runs educational workshops and publishes bulletins.

All the programs mentioned above (except the SWCD) are related in some way to the USDA. In addition, many states have laws and programs that promote soil conservation. Many states also try to control the conversion of prime farmland to other uses. The 1985 farm bill included a number of provisions, such as *CRP* and *sodbuster* to control erosion and protect wetlands.

REVIEW

1. The Soil Conservation Service provides money for conservation projects (True/False).
2. Agriculture Experiment Stations are funded entirely by the USDA.
3. Much of the soil conservation programming operates through the Soil and Water Conservation Districts.
4. The CRP aims to reduce erosion by preventing growers from farming any erodable land.
5. Several of the provisions of the 1985 Food Security Act involve eligibility for other USDA programs.
6. The board members of a _____ are usually elected or appointed from local growers.
7. If a farmer wanted his farm surveyed, he would contact the _____ .
8. The SCS provides technical aid, and the _____ financial aid for conservation projects.
9. For general farm advice, one can visit the county _____ .
10. Conservation easements of the CRP last _____ years.
11. Who conducts county soil surveys?
12. Agriculture research is sponsored by what agencies?
13. Explain how soil surveys and capability classifications could be used for setting land use policies.

14. If a grower wanted help installing a terrace system, what agencies would he/she go to and what would they offer?
15. What would you see as conservation priorities for your area?

SUGGESTED ACTIVITIES

1. Visit your local SCS office and SWCD office. Find out what local projects they have been involved in.
2. Visit your county extension agent or have her visit your class.
3. Find out what extension specialists are in your state.

appendix one
Some Basic Chemistry

ATOMS AND ELEMENTS

Elements are the building blocks of all matter. Examples of the 105 known elements include oxygen, carbon, and iron. Each element is assigned a symbol made of a letter or letters from the English or Latin word for the element. For example, oxygen is "O," carbon is "C," and iron is "Fe" from the Latin "ferrous."

The smallest unit of an element is the *atom*. While modern models of the atom are more complex, a simple picture of the atom, called the Bohr model, helps us understand how chemical processes occur in the soil. Atoms are made of three particles: a negatively charged *electron*, a positively charged *proton*, and a neutral *neutron*. According to the Bohr model, protons and neutrons inhabit a *nucleus*, and electrons circle the nucleus like planets orbiting the sun, as in figure A1.

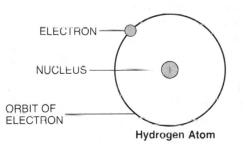

ELECTRON

NUCLEUS

ORBIT OF
ELECTRON

Hydrogen Atom

Figure A-1. Hydrogen atom

The simplest element is hydrogen, which is composed of a single proton and electron. Elements get successively heavier and more complex as more protons, neutrons, electrons, and electron orbits are added. Oxygen, for instance, has eight of each particle and electrons occupy two "orbits." The total weight of all the protons, neutrons, and electrons in an atom of an element is its *atomic weight*. Electrons are very light and contribute little to atomic weight. Thus the atomic weight of oxygen, with eight protons and eight neutrons, is about 16.

COMPOUNDS

Atoms combine to form *molecules*. A molecule is symbolized by writing the atomic symbols of each element in the molecule, with a number in the form of a

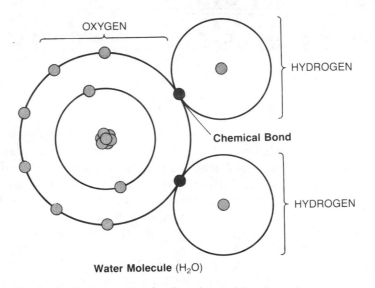

Water Molecule (H_2O)

Figure A-2. Water molecule (H_2O). In some molecules, chemical bonds result from sharing of electrons.

subscript to tell how many atoms of each. Thus water, or H_2O, is a molecule with two atoms of hydrogen combined with one atom of oxygen. One way for molecules to form is for atoms to join by sharing electrons, as shown for the water molecule in figure A-2. This is one way to form a *chemical bond* between two atoms.

A collection of like molecules is a *compound,* as in the compound water. Silicon dioxide, or SiO_2, is a compound containing two oxygen atoms attached to a silicon atom. Silicon dioxide is the mineral quartz, the major component of sand and one of earth's most common minerals. Pure solid compounds of the earth's crust, like quartz, are called *minerals. Rocks* are mixtures of minerals. Granite, for instance, consists of the minerals feldspar, quartz, and others.

ORGANIC COMPOUNDS

The common minerals are *inorganic,* distinguishing them from a special class of compounds that are labeled *organic.* Organic compounds all contain carbon and hydrogen, and most have oxygen, sulfur, nitrogen, or other elements. Organic compounds are the stuff of life; all life is made of them. When this text refers to organic matter, it refers to organic compounds in the soil, all of which come from living creatures and plants.

IONS

A normal atom or molecule has an equal number of electrons and protons; their charges balance and the net charge is zero. Sometimes an imbalance occurs, and the resulting atoms or molecules are called *ions*. For instance, when salt (NaCl) dissolves in water, the molecule breaks apart into sodium ions that are short one electron and chlorine ions with an extra electron. Thus, each sodium ion carries a positive charge, while chlorine ions have a negative charge:

$$NaCl \rightarrow Na^+ + Cl^-$$
$$salt \rightarrow sodium\ cations + chlorine\ anions$$
$$solid\ salt \rightarrow salt\ in\ solution$$

Positively charged ions are called *cations*, negatively charged ones *anions*. A *solution* results when compounds disassociate in water with ions dissolved in it.

CHEMICAL AND PHYSICAL REACTIONS

Two types of reactions are important in soils: *chemical reactions* and *physical reactions*. Chemical reactions involve the actual rearrangement of atoms to form new molecules and compounds, as in the reaction of hydrogen and oxygen to form water, shown in equation (a). The reaction, as written, states that two hydrogen molecules (in the natural environment, hydrogen and oxygen come as two-atom molecules) combine with one oxygen molecule to form two molecules of water.

(a) $$2H_2 + O_2 \rightarrow 2H_2O$$
(b) $$H_2O\ (liquid) \rightarrow H_2O\ (solid\ ice)$$

In physical reactions, the physical form but not the chemical form changes, as in the freezing of water in reaction (b) or the disintegration of rock by physical forces. Both physical and chemical weathering contribute to soil formation.

Reaction (a) suggests that a chemical reaction is a one-way process. In actuality, all reactions go in both directions at once but will favor one side of the reaction or the other. One might compare a reaction to a room full of people. Assume that 10 people are in a room, and they are asked to go to one side of the room or the other. People that go to the left side are "*A*" people, those that go to the right are "*B*" people. In this "reaction," assume seven people go the left, and three go to the right. We could write this "reaction" as:

$$A \rightleftharpoons B$$
$$(7\ people) \qquad (3\ people)$$

In a real reaction, the people don't stop; a few continue to walk back and forth across the room, but the numbers of people will continue to balance out to 7:3 (*A* in figure A-3). This balance is called the *equilibrium* of the reaction and is constant for a given reaction at a given condition.

If the conditions of a reaction change, the equilibrium can shift to try to achieve a new balance, Say that in the room full of people two of the three people on the right side of the room leave altogether. To achieve a new equilibrium, the reaction will shift to the right—that is, more people will walk to the right side of the room than back to the left until a new balance is achieved (see *B* in figure A-3). On the other hand, if a bunch of people enter the room on the left, the room is "unbalanced" and the "reaction" again shifts to the right to achieve a new equilibrium (see *C* in figure A-3). This process of shifting equilibrium is important, because it governs chemical reactions in the soil that relate to farming activities such as liming and fertilizing with ammonia.

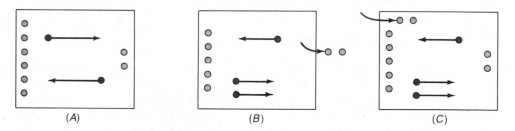

Figure A-3. Chemical reaction equilibria as represented by movement of people in a room. *(A)* Reaction in equilibrium; *(B)* reaction in which equilibrium is disturbed by people leaving room to right. Reaction shifts to right to compensate for change. *(C)* Reaction in which equilibrium is disturbed by people entering room from left. Reaction shifts to right to compensate for change.

OXIDATION-REDUCTION REACTIONS

A very important type of reaction in the soil is the *oxidation-reduction reaction*. Technically, an oxidation occurs when an element loses an electron in a reaction (oxidation), and some other element gains that electron (reduction). The electron donor (the "loser") is said to be oxidized and the electron acceptor (the "gainer") is said to be reduced. Most commonly, it is oxygen that accepts an electron or oxidizes another element in the formation of a molecule. This text will use oxidation to mean the reaction of an element with oxygen. A well-known oxidation is the reaction of iron (Fe) with oxygen (O) to form rust:

$$4Fe + 3O_2 \rightarrow 2Fe_2O_3$$

Rust is an oxidized form of iron, while metallic iron is the reduced form. Very important oxidized forms of elements in the soil incude carbon (CO_2, carbon dioxide or CO_3^-, carbonate ion), nitrogen NO_3^-, nitrate ion), sulfur (SO_4^-, sulfate ion), and iron (Fe_2O_3, ferric oxide).

ENERGY

Both chemical and physical reactions in the soil depend on *energy*. Energy is simply the capacity to do work. The greater the energy, the more work that can be done. An understanding of energy is important because everything that happens in the soil—the "work" of soil and plants—is fueled by energy. Energy comes in many forms, such as heat, electricity, light, and motion. Here are forms important to soil:

- *Light energy* is energy contained in sunlight. Plants use that energy for photosynthesis, which creates organic matter for the soil (figure 1-3).
- *Chemical energy* is energy contained in a chemical bond. During photosynthesis, plants convert energy in sunlight to energy in chemical bonds. This chemical energy is "stored" for later use by the plant or animals that eat it.
- *Potential energy* is the energy stored in an object that can fall. The energy is released when the object does fall, like rocks falling off a cliff, water falling over a dam, or rain falling from the sky.
- *Kinetic energy* is the energy of a moving object. Soil erosion by wind and water is a result of motion energy. The amount of energy in a moving object is a function of its mass (m) times velocity (v) squared:

$$Energy = mv^2$$

Thus, rapidly moving water can do more work than slowly moving water. The equation says, for instance, that if the velocity of running water doubles, it can do four times as much work (two squared) or cause four times as much erosion.

To understand chemical and physical reactions in the soil, it is important to know two rules. First, *energy can change forms.* For instance, sunlight energy can change to heat energy when it strikes pavement—a fact one can become painfully aware of when walking to the beach barefoot on a sunny day. Second, *matter tries to achieve the lowest possible energy state.* An electrical-generating dam is a good example. Water at high elevation behind the dam is in a high state of energy—it has a lot of potential energy. By flowing downhill through or over the dam, that potential energy is released to energy of motion. When the water reaches the lower level, it is now at a lower elevation, and in a lower energy state. In the process, the "lost" potential energy was converted to motion. That motion energy can, in turn, be changed to electrical energy when it does the work of turning the turbines in the generating plant.

These two rules of energy control all the physical and chemical reactions in the soil. They are especially important to the behavior of water in the soil (chapter four) and erosion (chapter eighteen).

Gradients. One result of the tendency for lower energy states is the concept of *gradient*, which controls the movement of materials in the soil. A gradient is a change of something over distance. The simplest gradient to understand is a slope, or hill, which is an elevation gradient. A cyclist on top of a hill can coast down the hill without pedaling, because he or she is moving to a lower energy state. To go up the hill, however, is to go to a higher energy state. To do that, the cyclist must feed in energy—which is to say, pedal the bike.

The example above presents a rule about gradients—that *movement down a gradient occurs without effort, while movement up a gradient requires an input of energy.*.

Many gradients exist in nature. A spoon in a hot cup of coffee creates a temperature gradient. Heat will flow naturally from the hot end in the cup to the cool end where you hold it. Water will move in the soil from where it is moist to where it is dry (moisture gradient). Vapors from an opened perfume bottle will scent an entire room as they move from where they are concentrated near the bottle to where they are less concentrated (this is called diffusion along a concentration gradient). A similar diffusion through soil water moves soil nutrients towards plant roots.

Remember that normal movement is down a gradient—from where there is more of something, like heat or water, to where there is less. The opposite movement requires an input of energy—which is the same as saying, it takes work. Imagine trying to get perfume vapors back into the bottle.

appendix two
Sedimentation Test of Soil Texture

DESCRIPTION

The sedimentation test is an easy way to measure the percent sand, silt, and clay in a soil sample. It is based on the fact that large, heavy particles will settle most rapidly in water, while small light particles will settle most slowly. The Calgon laundry powder is used to "dissolve" the soil aggregates and keep the individual particles separated.

MATERIALS

- Soil sample
- One quart fruit jar with lid
- 8% Calgon solution—mix 6 tablespoons of Calgon (a laundry powder available in stores) per quart of water
- Metric ruler
- Measuring cup
- Tablespoon

PROCEDURE

1. Place about 1/2 cup of soil in the jar. Add 3 1/2 cups of water and 5 tablespoons of the Calgon solution.

2. Cap the jar and shake for 5 minutes. Leave the jar on the desk and let settle for 24 hours.

3. After 24 hours, measure the depth of settled soil. All soil particles have settled, so this is the TOTAL DEPTH. Write it down and label it.

4. Shake for another 5 minutes. Let stand 40 seconds. This allows sand to settle out. Measure the depth of the settled soil and record as SAND DEPTH.

5. Do not shake again. Let the jar stand for another 30 minutes. Measure the depth, and subtract the sand depth to get the SILT DEPTH.

6. The remaining unsettled particles are clay. Calculate clay by subtracting silt and sand depth from total depth to get CLAY DEPTH.

7. Now calculate the percentage of each soil separate using these formulas:

$$\% \text{ sand} = \frac{\text{sand depth}}{\text{total depth}} \times 100$$

$$\% \text{ silt} = \frac{\text{silt depth}}{\text{total depth}} \times 100$$

$$\% \text{ clay} = \frac{\text{clay depth}}{\text{total depth}} \times 100$$

appendix three
Soil Orders of the United States

Soil Order/ Suborder	(1) Mapping Unit	(2) Moisture	(3) Temp	(4) Fertility
Alfisols				
Aqualfs	A1a	Wet	Moderate	High
Boralfs	A2a	Moist	Cold	High
	A2s	Moist	Very cold	High
Udalfs	A3a	Moist	Moderate	High
Ustalfs	A4a	Dry	Moderate	High
Xeralfs	A5S1	Xeric	Moderate	High
	A5S2		to warm	High
Aridisols				
Argids	D1a, D1S	Very dry	Varied	High
Orthids	D2a, D2s	Very dry	Moderate	High
			to warm	
Entisols				
Aquents	E1a	Wet	Very warm	Moderate to low
Orthents	E2a	Very dry	Cold	Moderate to low
	E2b	Very dry	Warm	Moderate to low
	E2c	Xeric	Warm	Moderate to low
	E2S1	Very dry	Moderate	Moderate to low
	E2S2	Xeric	Warm	Moderate to low
	E2S3	Xeric	Very cold	Moderate to low
Psamments	E3a	Varied	Moderate to	Low
	E3b, E3c		very warm	
Histosols				
	H1a	Wet	Cold	Moderate to high
	H2a	Wet	Warm	Moderate to high
	H3a	Wet	V. warm	Moderate to high
Inceptisols				
Andepts	I1a	Xeric	Very cold	Moderate to high
	IsS1	Xeric	Very cold	Moderate to high
	I2S2	Moist	Moderate	Moderate to high
Aquepts	I2a, I2P	Wet	Varied	Moderate

Soil Order/ Suborder	(1) Mapping Unit	(2) Moisture	(3) Temp	(4) Fertility
Ochrepts	I3a	Moist	Very cold	Low
	I3b	Moist	Warm	Moderate to high
	I3c	Moist	Moderate	Moderate to low
	I3d	Moist	Moderate	Low
	I3S	Moist	Moderate to warm	Low
Umbrepts	I4a, I4S	Moist	Moderate	Low
Mollisols				
Aquolls	M1a	Wet	Varied	High
Borolls	M2a, 2b, 2c	Moist	Cold	High
	M2S	Dry	Cold	High
Udolls	M3a	Moist	Moderate to warm	High
Ustolls	M4a, 4b, 4c M4S	Dry	Moderate to warm	High
Xerolls	M5a, M5S	Xeric	Moderate	High
Oxosolls	—	—	—	—
Spodosols				
Aquods	S1a	Wet	Cold to v. cold	Low
Orthods	S2S1 S2S2 S2S3	Moist	Moderate to v. cold	Low
Ultisols				
Aquults	U1a	Wet	Moderate to warm	Moderate to low
Humults	U2S	Moist	Moderate to v. warm	Moderate to low
Udults	U3a, U3S	Moist	Moderate to warm	Moderate to low
Vertisols				
Uderts	V1a	Moist	Warm	High
Usterts	V2a	Dry	Warm	High
No Soil				
Salt flats	X1	Dry	Warm	—
Rock land	X2	—	—	—

DEFINITIONS

1. The mapping unit codes are indicated on the map (figure A-4).
2. Soil moisture describes the moisture level of the soil during the growing season. Xeric soils are moist during the early part of the growing season but dry during the latter part.

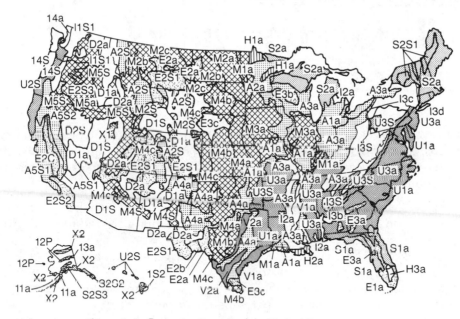

Figure A-4. General soils map of the United States.

3. Soil temperature is the average temperature of the soil, as follows:

Very cold: average annual soil T 32 to 47 degrees Fahrenheit.
Cold: same as very cold except that the summer temperatures are slightly warmer.
Moderate: average annual T 47 to 59 degrees Fahrenheit.
Warm: average annual T 59 to 72 degrees Fahrenheit.
Very warm: average annual T more than 72 degrees Fahrenheit.

4. Soil fertility is a broad comparison based on soil minerals, nutrient availability, and soil moisture.

appendix four
Soil Horizon Symbol Suffixes

These lowercase letters are used as suffixes to label certain types of master horizons. One or more suffixes may follow a master horizon designation—such as Bky—which indicates a B horizon that has accumulated both carbonates (k) and gypsum (y). The symbols and their meanings are as follows:

a Highly decomposed organic material. This suffix is used with the O horizon.

b Buried horizon. Such a soil layer is an old horizon buried by sedimentation or other processes.

c Concretions or hard nodules. A nodule or concretion is a hard "pocket" of a substance like gypsum in the soil.

e Moderately decomposed organic material. Used with the O horizon.

f Frozen soil. The soil contains permanent ice (permafrost).

g Strong gleying. Such a horizon is gray and mottled, the color of reduced (nonoxidized) iron, resulting from saturated conditions.

h Illuvial accumulation of organic matter. The symbol is used with the B horizon to show that complexes of humus and sesquioxides have washed into the horizon. Includes only small quantities of sesquioxides. May show dark staining.

i Slightly decomposed organic matter. Used with the O horizon.

k Accumulation of carbonates (CO_3^-). Indicates accumulation of calcium carbonate (lime) or other carbonates.

m Cementation. The symbol indicates a soil horizon that has been cemented hard by carbonates, gypsum, or other material. A second suffix indicates the cementing agent, such as "k" for carbonates. This is a hardpan horizon; roots penetrate only through cracks.

n Accumulation of sodium. Indicates a high accumulation of exchangeable sodium, as in a sodic soil.

o Accumulation of sesquioxide clays.

p Plowing or other human disturbance. Horizon was heavily disturbed by plowing, cultivation, pasturing, or other activity. Applies to O and A horizons.

q Accumulation of silica (SiO_2).

r Weathered or soft bedrock. Used with C horizon to indicate bedrock that can be dug with spade that roots can enter through cracks.

s Illuvial accumulation of both sesquioxides and organic matter. Both the organic matter and sesquioxide components of humus-sesquioxide complexes are important.

t Accumulation of silicate clays. Clay may have formed in horizon or moved into it by illuviation.

v Plinthite. Previously known as laterite, an iron-rich material common to tropical soils that hardens when exposed to air.

w Development of color or structure. The symbol indicates that a horizon has developed enough to show some color or structure but not enough to show illuvial accumulation of material.

x Fragipan or other noncemented natural hardpans. These are horizons that are firm, brittle, or have high bulk densities from natural processes.

y Accumulation of gypsum ($CaSO_4$).

z Accumulation of salts more soluble than gypsum.

appendix five

Land Evaluation

Land may be evaluated for a number of uses, such as suitability for row crops, home landscapes, or building sites. Such evaluations might be done by soil scientists, land planners, landscape designers, or even students engaged in a soil judging contest. While land evaluation purposes and methods vary between regions of the country, this appendix suggests some simple and general soil features to use when evaluating a soil.

DEFINITION OF TERMS AND CRITERIA

Suitability.   Land can said to be good, fair, or poor for any use. The same terms can also be applied to a particular feature of the land, such as slope or soil texture. Looking at the chart that follows, for each land use there is a series of criteria listed on the same line to the right. Each can be rated, and the "total" used to rate the land:

> *Good* land is well suited to the use being considered. There are no serious limitations or hazards. Ratings for each criteria are good, or perhaps a couple are fair.

> *Fair* land is suitable for the use if a few limitations are corrected. Several of the features are rated as fair.

> *Poor* is unsuitable for the use, or corrections are too expensive to be practical. Even one or two severe limitations may be enough to gain this rating, or a large number of "fair" descriptions.

Slope.   Slope is expressed as a percent. For instance, a slope of 2% suggests a two foot change of elevation over a hundred foot horizontal distance. Figure 17-6 gives some slope ranges that may help interpret these numbers. Changing slope would involve expensive earth moving.

Soil Texture.   Chapter three describes how to determine a soil texture. The textural class groupings used in this chart are grouped below:

Group	Textural Classes
fine	sandy clay, silty clay, clay
moderately fine	sandy clay loam, silty clay loam, clay loam
medium	silt, silt loam, loam
moderately coarse	sandy loam
coarse	sand, loamy sand

Except for small areas, like a garden, texture cannot be changed practically.

Flooding. Flooding concerns how often or how long the land is actually covered with water during the season. Such flooding could be due to stream flooding, heavy rains, or snowmelt. Flooding may be able to be corrected, in some small areas, by earth moving to change drainage patterns. In many cases, only large scale projects such as the construction of levees can solve these problems.

Internal Drainage. Internal drainage can be determined from figure 3-23. Many cases of poor drainage can be corrected by installing drainage systems, and excessive drainage may be improved by irrigation.

Depth to Restrictive Layer. Restrictive layers can interfere with rooting depth and plant anchorage, roadbeds, foundations, and home drain fields. Such layers could be bedrock, soil pans, water tables, or others. Some can be corrected, like ripping a soil pan, while bedrock cannot be altered easily.

Available Water for Plants. The available water is here rated as the number of inches of water held in the top five feet of soil. Since most soils will contain layers, one will have to add the capacity of each layer down to five feet. Figure 6-20 rates the water retention of soil textures. Here, a capacity greater than nine inches for five feet is good, six to nine is fair, and less than six is poor. Poor capacity can be alleviated by irrigation, where practical and acceptable.

Erodability. Erodability is related to slope, texture, and other factors discussed in chapter eighteen. Here, classes are as follows:

slight means that under average conditions there is little chance of excessive erosion. Slopes are gentle and short, internal drainage is good. Wind erosion is unlikely.

moderate could occur on medium or fine texture soil on gentle slopes over 300 feet long, or shorter moderate slopes. Wind erosion is possible due to texture or lack of cover.

severe erodability occurs on slopes over 12%, or areas which experience teaches are quite erodable.

The Erosion Index described in chapter eighteen could be used to infer erodability.

Other Features. Other features may also limit a soil for certain uses. Examples include stoniness, soil pH, soil fertility, or salinity. Some of these can be improved easily, like liming agricultural soil to raise pH. On the other hand, improving soil fertility is more difficult for forest uses.

SITE INSPECTION

Much of the information needed for this evaluation can be obtained from soil surveys. As a lab exercise for students, a soil pit should be dug to allow examination of the soil profile.

Select an area of uniform character, preferably a single mapping unit on a soil map. Dig a pit measuring 3′x3′ at the surface, and about 3½ feet deep. Orient the pit so that the sun shines on the side to be observed. That side should be vertical; the other side can be sloping to save digging.

Some of the information needed can be discovered from examining the pit. Other can be obtained from surroundings, like slope. The instructor will need to provide such information as flooding.

As the student examines each feature, he or she should circle the best description on the chart included in this appendix. Then consider how important the feature is, or how easily it can be corrected. Finally, come up with an overall rating, as described earlier in this appendix.

This appendix is adapted from the University of Minnesota Ag. Extension Service Environmental Education Activity Sheet #1, "Selecting Suitable Uses for Land", by Clifton F. Halsey.

SELECTING SUITABLE USES FOR LAND

USE	SUITABILITY	% SLOPE	SOIL TEXTURE (Surface soil only unless specified otherwise.)	FEATURES OF THE LAND AND SOIL					OTHER IMPORTANT FEATURES NOT EASILY IMPROVED
				FLOODING DURING GROWING SEASON or use (frequency-duration)	INTERNAL DRAINAGE	DEPTH TO RESTRICTIVE LAYER (inches)	AVAILABLE WATER FOR PLANTS (in.) Capacity to 5 deep	EROD-IBILITY	
Home lawns, shrubs and gardens	Good-no important limitations	0-6	medium, moderately coarse	none	well drained, moderately well drained	more than 36	more than 9	slight	
	Fair-1 or 2 important limitations	6-12	moderately fine, fine, coarse	once a year, not over 3 days	somewhat poor, poor	20 to 35	6 to 9	moderate	stony
	Poor-severe limitations	more than 12	organic	more frequent or more than 3 days	excessive	less than 20	less than 6	severe	moderate to strongly alkaline, pH more than 8.0
Cultivated crops	Good	0-6	medium, moderately coarse	none	well or moderately well drained	more than 36	more than 9	slight	
	Fair	6-12	moderately fine, fine, organic	once a year, less than 48 hours	somewhat poor (needs drainage)	20 to 36	6 to 9	moderate	stony
	Poor	more than 12	coarse	more frequent, longer	poor or excessive	less than 20	less than 6	severe	moderately to strongly alkaline pH more than 8.0
Permanent grass pasture	Good	0-18	medium or moderately fine, fine	up to 4 times annually, less than 48 hours	well or moderately well drained	more than 36	more than 9	slight	
	Fair	18-25	moderately coarse	more than 48 hours	somewhat poor, poor	20 to 36	6 to 9	moderate	
	Poor	more than 25	coarse, organic	5 times or more annually	excessive	less than 20	less than 6	severe	moderately to strongly alkaline, pH more than 8.0
Forests—conifers, for wood products, x-mas trees	Good	0-18	(Entire Root Zone) medium, moderately coarse	never	well drained, moderately well drained	more than 20	more than 6		soil fertility: medium to high pH: 5 to 6
	Fair	18-45	moderately fine, coarse	never	somewhat poor, excessive		3 to 6		soil fertility: low
	Poor	more than 45	fine, organic	any flooding	poor	less than 20	less than 3		
Forest—deciduous, for wood products	Good	0-18	(Entire Root Zone) medium, moderately fine	once in 10 years	well drained, moderately well drained	more than 20	more than 9		soil fertility: medium to high pH: 6 to 8
	Fair	18-45	moderately coarse	annual spring flooding	somewhat poor, excessive		6 to 9		soil fertility: low
	Poor	more than 45	coarse, fine, organic	more than once annually	poor	less than 20	less than 6		pH: less than 5, more than 8

SELECTING SUITABLE USES FOR LAND

USE	SUITABILITY	% SLOPE	SOIL TEXTURE (Surface soil only unless specified otherwise.)	FLOODING DURING GROWING SEASON or use (frequency-duration)	INTERNAL DRAINAGE	DEPTH TO RESTRICTIVE LAYER (inches)	AVAILABLE WATER FOR PLANTS (in.) Capacity to 5' deep	ERODIBILITY	OTHER IMPORTANT FEATURES NOT EASILY IMPROVED
					FEATURES OF THE LAND AND SOIL				
Forest—wildlife habitat	Good	0-25	(Entire Root Zone) moderately coarse thru moderately fine	once in 10 years, less than 24 hours	well or moderately well drained	more than 20	more than 9		
	Fair	25-50	coarse or fine, organic	up to 4 times annually, 24 to 48 hours	somewhat poor, excessive		6 to 9		
	Poor	more than 50		more frequent, more than 48 hours	poor	less than 20	less than 6		pH: less than 5, more than 8
Water fowl habitat	Good		Can have surface water continuously, open water all summer, and be more than 3 feet deep.		poor drainage				
	Fair		Can have surface water and open water during spring only, less than 3 feet deep.		somewhat poorly drained				
	Poor		Can have short periods of open water, less than 1 acre		moderately well to excessively drained				
A. Athletic, play and picnic grounds / B. Public camp-grounds, primitive campsites	A. Good / B. Good	0-2 / 0-6	medium, moderately coarse	none during use	well and moderately well drained	more than 36	more than 9	slight	
	A. Fair / B. Fair	2-6 / 6-12	coarse, moderately fine	once-less than 24 hours	somewhat poor	20 to 36	6 to 9	moderate	
	A. Poor / B. Poor	more than 6 / more than 12	fine, organic	2-3 times during season	poor, excessive	less than 20	less than 6	severe	
Nursery	good	0-3	moderately coarse, medium (BR), medium, medium fine (BB)	none	well drained	more than 48	more than 9	slight	
	fair	3-6	coarse, medium fine (BR), moderately coarse, fine (BB)	once a year, not over 1 day	moderately well drained	36-48	6-9		stony
	poor	>6	fine (BR) coarse (BB)	more frequent or longer	excessive, some-what poor, poor	less than 36	less than 6	moderate severe	large stones
Land exposed to erosion during construction or use	Good	0-2	coarse, moderately coarse					none to slight	
	Fair	2-6	medium to fine					moderate	
	Poor	more than 6						severe	

SELECTING SUITABLE USES FOR LAND

USE	SUITABILITY	% SLOPE	SOIL TEXTURE (Surface soil only unless specified otherwise.)	FLOODING DURING GROWING SEASON or use (frequency-duration)	INTERNAL DRAINAGE	DEPTH TO RESTRICTIVE LAYER (inches)	AVAILABLE WATER FOR PLANTS (in.) Capacity to 5' deep	EROD-IBILITY	OTHER IMPORTANT FEATURES NOT EASILY IMPROVED
					FEATURES OF THE LAND AND SOIL				
Houses and other low buildings	Good	0-6	**To 5 feet deep.** coarse, moderately coarse, medium		excessive, well drained, moderately well drained, none	**To bedrock,** water more than 36			
	Fair	6-12	moderately fine	none	somewhat poor	20 to 35			
	Poor	more than 12	fine, organic	any	poor	less than 20			
Basements and utility excavations	Good	0-6	**To 5 feet deep.** medium. moderately coarse, coarse	none	well drained, excessive	**To bedrock,** water more than 5 feet			
	Fair	6-12	moderately fine	none	moderately well drained	3½ to 5 feet			
	Poor	more than 12	fine, organic	any	somewhat poor, poor	less than 3½ feet			
Home sewage absorption fields	Good	0-6	**To 5 feet deep.** medium, moderately coarse	none	well drained	6 feet			
	Fair	6-12	moderately fine	none	moderately well drained	6 feet			
	Poor	more than 12	coarse, fine, organic	any	somewhat poor poor, excessive	less than 6 feet			
Local streets	Good	0-6	**To 5 feet deep.** moderately coarse, coarse	none	excessive, well drained, moderately well drained	**To bedrock,** more than 36			
	Fair	6-12	moderately fine, medium	once a year, not over 3 days	somewhat poor	20 to 36			
	Poor	more than 12	fine, organic	more frequent, more than 3 days	poor	less than 20			

Glossary

Acid soil. A soil that contains more hydrogen ions than hydroxyl ions; soil pH is less than 7.0.

Actinomycete. An order of microbes common to soil; related to bacteria but resembling fungi in having a mycelium. Important decomposers and sources of medical antibiotics.

Adhesion. Force of attraction between two different substances. In soil, used to define the force that attracts water to soil particles.

Adhesion water. Inner layer of water molecules in a water film around a soil particle, held tightly by adhesion so it cannot move or be absorbed by plants.

Adsorption. Bonding of an ion or compound to a solid surface, usually temporarily. In soil, cations are adsorbed on clay and humus particles.

Aeration, soil. Process by which air in the soil is replaced by air from the atmosphere. Related to number, size, and continuity of soil pores and to internal drainage.

Aerobic. An adjective applied to orgaisms that grow, or processes that occur, in the presence of oxygen.

Aggregate, soil. A mass of fine soil particles glued together by clay, organic matter, or microbial gums. Aggregates are part of soil structure.

Air-dry soil. Soil allowed to dry out in open air without heating. Still contains hygroscopic water.

Alfisol. One of ten soil orders. A mineral soil, usually formed under forest, common to northern and midwestern states. Has leached E (A1) horizon with accumulation of bases and clays in Bt horizon.

Algae. Simple chlorophyll-containing plants. Single-celled algae add organic matter to soil by photosynthesis.

Alkaline soil. A soil that contains more hydroxyl ions than hydrogen ions; pH greater than 7.0.

Allelopathy. Suppression of the growth of plants by substances exuded by the roots of other plants. *See antagonism.*

Alluvial fan. A fan-shaped alluvial deposit formed where flowing water slows down and spreads out at the base of a slope.

Alluvial soil. A soil developed from mud deposited by running water.

Amendment, soil. A substance mixed into the soil to improve its properties. Usually applied to materials used to improve physical conditions.

Ammonification. Process by which certain soil microbes convert organic nitrogen to ammonia.

Anaerobic. An adjective applied to organisms that grow, or processes that occur, in the absence of oxygen.

Anchorage. Function of soil to hold plant firmly in place.

Anion. An ion with a negative or minus charge.

Anion exchange capacity. Total sum of the number of exchangeable anions a soil can adsorb, expressed as milliequivalents per 100 grams of soil.

Antagonism. The suppression of the growth of one organism by another organism by the production of toxic or growth inhibiting (antibiotic) substances. Allelopathy is an example of antagonism.

Antibiotic. A substance produced by one species of organism that will kill or inhibit growth of some other organisms. Streptomycin, an antibiotic used in human medicine and produced by actinomycetes, is an example.

Aquifer. An underground formation that holds water. It is porous enough that water can flow through it to a well, so can be a source of groundwater.

Arid region. An area with too little rainfall to produce crops without irrigation. See semi-arid region.

Aridosol. One of ten soil orders, common to arid regions. Soil is dry and low in organic matter.

Arthropods. A phylum of animals with no backbone, jointed body and legs, and usually a hard shell. Includes such soil animals as insects, spiders, sowbugs, and others.

Association, soil. A soil mapping unit in which two or more taxonomic soil units that occur together are combined.

Atom. Smallest particle of an element that can exist alone or in combination. If an atom of an element is divided further, it ceases to be that element.

Autotrophic. Capable of producing one's own food from carbon dioxide and/or carbonates by photosynthesis or oxidation of inorganic compounds.

Available nutrient. An essential element, or nutrient, in the soil in a form that plants can absorb into roots.

Available water. That part of the soil water that can be taken up by plant roots. Mostly cohesion water, or defined as lying between the field capacity and the wilting point (–1/3 to –15 bars, or 3.3 to 1.5 MPa soil matrix potential).

Banding. Method of fertilization by placement of fertilizer, at planting, near the seed. May also be applied to surface or subsurface placement of fertilizer in strips.

Base saturation percentage. Percentage of the cation exchange capacity occupied by cations other than hydrogen or aluminum.

Bedrock. Solid, or consolidated, rock lying under the soil. It may be, but is usually not, the parent material of the soil lying above it.

BOD. Biological oxygen demand. Oxygen used up by the decay of organic materials in water.

Border strip irrigation. Surface irrigation in which water is run into strips bounded by low earthen borders.

Broadcast. Application of fertilizer by spreading on the soil surface, usually before planting and incorporated by tillage.

Buffering. The ability of a solution, like the soil solution, to resist changes in pH when acid or alkaline substances are added. Often used when speaking of soil to describe its resistance to pH changes when limed or acidified.

Bulk density. Mass of oven-dry soil per unit volume, expressed as pounds per cubic foot or grams per cubic centimeter.

Calcareous soil. Soil high in calcium carbonate (lime), usually derived from limestone-rich parent materials. Will bubble or "fizz" if treated with cold, dilute (0.1N) hydrochloric acid.

Caliche. A zone of soil, near the surface, that is cemented by lime (magnesium or calcium carbonates). Common to arid soils.

Capillary. A very thin tube that water will flow into because of adhesive and cohesive forces. In the soil, small soil pores can act as capillaries.

Capillary fringe. A zone of soil just above a water table that is nearly saturated because of capillary rise.

Capillary rise. Movement of water upward in the soil through soil capillaries. Occurs as soil surface dries, drawing moisture from below.

Capillary water. Old term for water held loosely in capillary pores in the soil, capable of moving in the soil and being absorbed by roots.

Carbon-nitrogen ratio. Ratio of the weight of carbon to weight of nitrogen in an organic material. Obtained by dividing the percent carbon by the percent nitrogen.

Catena. A group of neighboring soils formed of similar parent materials under similar conditions at about the same time. The soils differ because of variations in relief and drainage. Synonym: toposequence.

Cation. An ion with a positive charge.

Cation exchange. Exchange between a cation in solution and one adsorbed on a soil colloid.

Cation exchange capacity. Total number of exchangeable cations a soil can adsorb. A measure of the soil's ability to hold nutrients that are cations in the soil. Expressed as milliequivalents per 100 grams of soil.

Cemented. Hardened because soil particles are "glued" together by substances such as lime or iron oxides.

Chelate. A molecule with a metal atom surrounded by a complex organic molecule. Some humus-metal chelates are insoluble. Other chelates protect metals from fixation in the soil. Artificial chelates are used as trace-element fertilizers.

Chemical weathering. Breakdown of rocks and minerals by chemical reactions, mostly with water.

Chiseling. (1) Primary tillage with a chisel plow, which pulls long curved teeth through the soil to loosen it without turning it over. (2) Using a subsoiling chisel plow to break up deep compacted soil layers. (3) Using chisels to inject fluid fertilizers or ammonia into the soil.

Chlorite. A 2:1 silicate clay in which a fourth sheet holds together the 2:1 layers.

Chlorosis. Common sign of nutrient shortage, showing as a loss of normal green color in a plant. Color loss means a failure to make enough chlorophyll and may result from shortage of nutrients involved in chlorophyll formation (nitrogen, sulfur, iron, and others).

Chroma. One of the three variables in the Munsell color system. Chroma is the purity, or strength of a color; its opposite is the amount of grayness.

Clay. (1) The class of smallest soil particles, smaller than 0.002 millimeter in diameter. (2) The textural class highest in clay.

Claypan. A dense subsoil layer with a higher clay content than the soil above it. *See pan.*

Clod. A user-made soil aggregate, produced by tillage when a soil is too wet or dry. Clods may vary in size from a quarter inch to ten inches.

Coarse texture. A soil texture whose traits are largely set by the presence of sand. Includes sands, loamy sands, and most sandy loams.

Cohesion. The force attracting similar substances. In the soil, applied to attraction of water for itself.

Cohesion water. Outer film of water around soil particles, loosely held in place by cohesion to the inner film of adhesion water. Capable of movement in soil and of being taken up by plants.

Colloid. A very tiny particle capable of being suspended in water without settling out rapidly. Soil colloids, mostly humus and clay, have a charged surface that attracts cations.

Colluvium. A deposit of rock and soil resulting from materials sliding down a slope under the force of gravity.

Compaction, soil. The squeezing together of soil particles by the weight of farm and construction equipment, vehicles, and animal and foot traffic. Compaction reduces average pore size and total air space in the soil.

Complete fertilizer. A fertilizer containing all three of the primary macronutrients—nitrogen, phosphorus, and potassium.

Composite sample. The soil sample sent to a soil-testing laboratory, resulting from mixing together many individual samples. It should represent the average soil in a field.

Composting. Piling organic materials under conditions that cause rapid decay. Reduces the carbon-nitrogen ratio and destroys many weed seeds and disease organisms.

Conservation tillage. A tillage practice that leaves crop residues on a rough soil surface to reduce erosion.

Consistence. Characteristics of a soil in its response or resistance to pressure, as described at various soil moisture contents. Such characteristics include stickiness, plasticity, hardness, or friability.

Consumptive use. Amount of water transpired from plants, incorporated into plant tissue, and evaporated from soil.

Contour. An imaginary line across a slope that stays at the same elevation.

Contour tillage. Tillage following the contours of a slope, rather than up and down a slope. Helps prevent erosion and runoff.

Cover crop. A crop planted to prevent erosion on a soil. Cover crops can be planted on soils not currently being farmed or between rows of trees in orchards or nurseries.

Crop rotation. Planting a repeating sequence of different crops on the same piece of land.

Cultivation. Tillage to control weeds and loosen soil.

Cyanobacteria. Certain nitrogen-fixing microbes, previously called blue-green algae, now classified in Monera.

Decomposers. Microbes that obtain their food from dead organic materials, causing decay and decomposition.

Delta. A usually fan-shaped alluvial deposit created where a stream or river enters a body of quiet water like an ocean or a lake.

Denitrification. Chemical reaction caused by certain microbes in the soil that change nitrate nitrogen to gaseous nitrogen or gaseous nitrogen oxides.

Deposit. Loose material left in a new place after being carried by wind, water, ice, gravity, or man.

Desertification. Conversion of land to desert, often caused by overgrazing, deforestation, or other disturbance.

Diagnostic horizon. Any of a series of specific types of soil horizons used to assign a soil to its proper soil order.

Diffusion. Flow of matter through a liquid or gas by the random movement of molecules. In soil science, applied to movement of nutrient ions through soil solution. The movement is caused by a concentration gradient, with the ions moving from more to less concentrated.

Diversion. Changing the direction of movement. In soil conservation, a special terrace built to divert the flow of running water.

Drainage, soil. (1) The speed and amount of water removal from soil by runoff or downward flow through soil. (2) Amount of time when soil is free of saturation.

Drought. A period of soil dryness that seriously harms plant growth.

Dryland farming. Methods of producing crops in low-rainfall areas without irrigation.

Duripan. A soil layer hardened and cemented by silica.

Elluviation. Removal of a material, such as clay or nutrients, from a layer of soil by percolating water.

Eolian soil material. Wind-deposited soil material, mostly silt and fine sand.

Ephemeral erosion. Erosion in which running water makes large rills that are not filled in when the ground is tilled, leaving a channel for further erosion. Soil carried off fields by ephemeral erosion is not predicted by the Universal Soil Loss Equation.

Erodable. Soil that is easily eroded, due to a variety of factors.

Erosion index (EI) or erosion potential. Measurement of the inherent erodability of a soil used without preventive measures. Using the Universal Soil Loss equation, EI = RKLS/T.

Essential element. An element needed by plants for proper growth and reproduction.

Eutrophication. The rapid increase in growth of aquatic plants and algae caused by pollution of water by phosphates and, to a lesser extent, nitrates.

Evapotranspiration. The sum of water lost from soil by evaporation and transpiration.

Exchangeable base. A cation, excluding hydrogen and aluminum, held on cation exchange sites that can be easily replaced by another cation. Considered to be available for plant growth.

Fallow. Soil left idle to accumulate water and/or mineral nutrients.

Family, soil. Level of soil taxonomy just above the soil series.

Fertigation. Fertilizing with soluble fertilizers through an irrigation system, usually sprinkler or trickle.

Fertilizer. A material added to the soil to supply essential elements. State laws may set minimum requirements for materials sold as fertilizers.

Fertilizer analysis. Composition of fertilizer measured by chemical tests. On a bag of fertilizer, would appear as a listing of the percent of each of the nutrients contained in the bag, including primary, secondary, or trace elements.

Fertilizer burn. Damage to plant tissue resulting from overapplication of fertilizer, a form of soluble salt damage.

Fertilizer carrier. A compound mixed into a fertilizer to supply a nutrient.

Fertilizer filler. A non-nutrient material added to fertilizer, for instance, clay, sand, or corncob granules.

Fertilizer grade. The guaranteed minimum analysis in whole numbers of nitrogen, available phosphate, and water-soluble potash, listed as "$N-P_2O_3-K_2O$."

Fertilizer, inorganic. Fertilizer which contains no carbon. For the purposes of this text, fertilizers which are unaltered minerals are considered separately. Urea, while chemically organic, is often classified inorganic because of rapid hydrolysis in the soil to ammonium ions.

Fertilizer, organic. Fertilizer which contains nutrients plus carbon and hydrogen. Often excludes urea. *See above, fertilizer, inorganic.*

Fertilizer ratio. Proportion of the primary nutrients in a fertilizer. Obtained by dividing the grade by the lowest common denominator.

Fertilizer, starter. A small amount of fertilizer placed near seeds or transplants to promote early growth.

Field capacity. The percentage of water remaining in the soil after drainage has just stopped.

Fine texture. Soil with a large amount of clay. Usually includes clay, sandy clay, clay loam, silty clay, and silty clay loam.

Fixation. (1) A process that changes chemicals from soluble or available forms to insoluble or unavailable forms in the soil. (2) Conversion of gaseous nitrogen to ionic forms.

Flood plain. Land near a stream that is commonly flooded when the stream is high. Soil is built from sediments deposited during flooding.

Foliar fertilization. Fertilizing plants by spraying leaves with fluid fertilizers.

Forest soil. Soils developed under forest vegetation.

Fluid fertilizer. Fertilizer used in liquid form, either a solution or a suspension.

Fragipan. Naturally occuring hard, brittle subsoil layer high in clay.

Friable. A consistency term, expressing how easily a moist soil can be crumbled.

Fritted trace element. A slow-release trace element fertilizer in which fertilizer carriers are mixed into glass powder.

Frost wedging. Breakage of rocks caused by pressure created by water freezing in cracks in the rock.

Fungi. Important soil organisms, especially as decomposers. Considered to be either primitive plants or as one of five kingdoms of living organisms.

Furrow diking. Tillage with special tool that creates ridges and furrows with series of small dams and basins in the furrow. The technique retains water in the furrows.

Glacial drift. General term for debris deposited by glaciers. Common in northern tier of states.

Glacial outwash. Glacial drift deposited in water flowing away from a melting glacier. Outwash is sorted by the running water.

Glacial till. Glacial drift that deposits in place as glacier melts, unsorted.

Gley. Soil layer that develops under poor soil drainage conditions, has gray color and mottles.

Gravitational water. Water that moves through the soil under the influence of gravity.

Great group. A taxonomic level of the current soil classification system.

Green manure. A crop grown to be turned under while still green to improve the soil.

Groundwater. Water stored underground in a saturated zone of rock, sand, gravel, or other material.

Gully. A large channel in the soil, caused by erosion, that is deep and wide enough that it cannot be crossed by tillage equipment.

Hardpan. A hard subsoil layer caused by cementation by carbonates or other chemicals; limits root growth and the infiltration of water.

Heterotroph. Organism not capable of producing own food. Obtains energy for life processes by oxidation of organic compounds.

Histosol. One of ten soil orders. Histosols are organic soils.

Horizon, soil. A horizontal layer of soil, created by soil-forming processes, that differs in physical or chemical properties from adjacent layers.

Hue. One of the three color variables in the Munsell system. Refers to the actual color of soil. *See also chroma and value.*

Humus. Decay-resistant residue of organic matter decomposition. Humus is dark-colored and highly colloidal.

Hydraulic conductivity. A trait of soil relating to the ease of water movement in that soil. For example, the finer the soil texture, the lower its hydraulic conductivity.

Hydrologic cycle. The circular route of water from the atmosphere back to the atmosphere after it has undergone precipitation, runoff, percolation, storage, or evapotranspiration.

Hydrolysis. The reaction of a compound with water to create a new compound.

Hydroponics. The culture of plants in nutrient solutions rather than soil.

Hygroscopic water. Water held tightly by adhesion to soil particles. Cannot be used by plants and remains in soil after air-drying. Can be driven off by heating.

Hyphae. Individual strands of the vegetative body of fungi. Hyphae can grow into organic matter to cause decay and can surround a mass of soil particles to make soil aggregates.

Igneous rock. Rock formed from the cooling of molten rock from deep in the earth.

Illuviation. Deposition in a soil layer of materials transported from a higher soil layer by percolating water.

Immature soil. A young soil that is still changing relatively rapidly in response to its surroundings. Usually has little horizon development.

Immobilization. Absorption of an available nutrient by a soil organism or plant, changing it to an unavailable organic form.

Inceptisol. One of ten soil orders, usually young soils with weak horizons.

Infiltration. Downward entry of water into the soil. *See percolation.*

Inoculation. Adding microbes to soil, seed, or culture medium. For instance, to treat legume seeds with bacteria from the *Rhizobium* genus.

Inorganic. Any chemical that does not contain both carbon and hydrogen. Inorganic nitrogen, for instance, includes nitrates (NO_3^-) but not urea (NH_2CONH).

Ions. Atoms or molecules that are electrically charged because of gaining or losing electrons.

Irrigation. Artificial application of water to the soil to improve crop growth, or sometimes for other purposes like activating herbicides.

Isomorphous substitution. Replacement of one atom by another of similar size in a crystal lattice. May result in a charge if the atom has a different charge than the one replaced.

Lacustrine. Mineral sediments deposited in fresh water.

Land capability classes. Eight soil classes ranked for their suitability for agriculture according to risk of erosion and other factors.

Land capability subclass. A subclass of a capability class that designates one of four problems: erodability (e), wetness (w), climate (c) or other soil factors (s).

Land capability unit. A group of soils nearly alike in suitability for plant growth that can be managed the same.

Landscape. (1) The natural features of the earth's surface like hills, trees, and water. Usually the piece of land that can be seen by the eye in a single view. (2) The designed landscape around buildings or other structures of the built environment.

Land-use planning. Developing plans for using land that will best serve the long-term public interest.

Leaching. Removal of soluble material in solution from the soil by percolating water.

Legume. A member of the legume family of plants, such as soybeans, peas, clover, alfalfa, locust trees, and many other economically important plants. Legumes host the *Rhizobia* bacteria that fix nitrogen.

Levee. Alluvial deposit of a shallow ridge along a river, resulting from coarse deposits during flooding.

Lignins. Complex, decay-resistant organic chemicals that glue together cellulose fibers in a plant to make it rigid.

Lime. Materials used to neutralize acidity, containing calcium or magnesium carbonate, oxide, or hydrate.

Load-bearing capacity. Ability of a soil to carry a load like a roadbed or building without shifting.

Loam. A medium soil texture class, in which sand, silt, and clay contribute almost equally to soil properties.

Lodging. Breaking of plant stems because of stem weakness, often caused by excess nitrogen and low potassium.

Loess. Wind-deposited silt.

Luxury consumption. Absorption by plants of more nutrients, especially potassium, than they need at the time. The nutrients may, however, be used for later stages of growth and can be used by animals feeding on the plant.

Macronutrients. An essential element used in large amounts by plants, including nitrogen, phosphorus, potassium, calcium, magnesium, and sulfur.

Mapping unit, soil. Basis for setting boundaries in a soil map. May be phases of series, families, or other taxonomic units, or may be associations of such units.

Marl. A soft limey material deposited in fresh water that can be used as an agricultural lime.

Mass flow. Movement of nutrients by movement of soil water.

Massive soil. A structureless soil in which each soil particle sticks to neighboring particles. Common in C horizons or puddled soils.

Mature soil. A soil in equilibrium with its surroundings, usually with well-developed horizons.

Medium-textured soil. Soils intermediate between fine- and coarse-textured soils. Includes loam, fine sandy loams, silt loam, and silt.

Metamorphic rock. Rock that has been changed by heat or pressure in the earth.

Micelle. An individual particle of silicate clay.

Micronutrients. Essential elements used in small quantities by plants.

Microorganism. An organism too small to be seen without the aid of a microscope.

Mineral. A pure inorganic compound in the earth's crust. Most rocks are mixtures of minerals. Often used in soil science to mean the inorganic solid particles of soil.

Mineral soil. A soil whose traits are determined mainly by its mineral content; mineral soils contain less than 20% organic matter.

Mineralization. Conversion of elements in organic forms to inorganic forms by decay. *See immobilization.*

Minimum tillage. Tillage methods that involve fewer tillage operations than conventional tillage. *See conservation tillage.*

Mixed fertilizer. A fertilizer containing more than one primary nutrient.

Mollisol. One of ten soil orders. A soil with a high organic-matter topsoil and high base saturation; usually formed under prairie vegetation.

Mottling. Spots of different colors in a soil, usually indicating poor drainage. *See gley.*

Muck. An organic soil in which the organic matter is mostly decomposed. *See peat.*

Mulch. A material spread on the soil surface, like straw, leaves, plastic, or stones to protect soil from freezing, raindrop impact, evaporation, and heaving.

Munsell color system. A system used to identify soil color by means of three variables: hue (color), value (intensity), and chroma (purity). Identification is done by comparing soil to a set of standard color chips.

Mycorrhizae. Fungi that form a symbiotic relationship with plant roots. The fungi help the plant absorb water and nutrients, while they receive food from the plant.

Nematode. Small unsegmented worm; many are parasitic on plant roots.

Nitrification. Microbial conversion of ammonium nitrogen to nitrate nitrogen.

Nitrogen cycle. Series of changes of nitrogen from atmospheric nitrogen, fixation in soil, a series of changes in the soil, and returning to the atmosphere.

Nitrogen depression period. Period of time during decay of organic matter in which nitrogen is being used up by microbes faster than being released by decay (immobilization exceeds mineralization). Resulting temporary nitrogen tie-up can cause nitrogen shortage in crops.

Nitrogen fixation. Microbial conversion of gaseous nitrogen to organic nitrogen in the soil.

Nonexchangeable. A term applied to an ion so strongly adsorbed to a soil colloid that it is not normally available to be exchanged with ions in the soil solution.

No-tillage. Method of growing crops that involves no tillage. Seeds are planted in slits in soil and chemicals are used to control weeds.

Order, soil. Highest taxonomic level in the soil classification system. There are ten recognized soil orders.

Organic. A material containing both carbon and hydrogen, and often oxygen, nitrogen, or sulfur. *See inorganic.*

Organic amendment. A soil amendment that is mostly organic matter.

Organic farming. Farming without using inorganic fertilizers or artificial pesticides.

Organic matter. Material of plant or animal origin that decays in the soil to form humus.

Organic soil. Soil containing more than 20% organic matter. Soil properties are dominated by the organic matter.

Oven-dry soil. Soil dried at 105 degrees Celsius until it reaches a constant weight.

Oxidation. A chemical reaction in which an element loses electrons to another participant in the reaction, often oxygen. For instance, iron is oxidized by oxygen to form rust. In this text, oxidation is applied to reactions with oxygen.

Oxisol. One of ten soil orders; highly weathered and leached tropical soils, limited to Hawaii and Puerto Rico in the United States.

Pan. A dense, hard, or compacted layer in soil that slows water percolation and movement of air and obstructs root growth. Pans may be caused by compaction, clay, or chemical cementation.

Parasite. A organism that lives off another organism (the host). The host is injured by the parasite. *See symbiosis.*

Parent material. The unconsolidated mineral or organic matter from which the solum (A, E, B horizons) has developed.

Particle density. The mass per unit volume of soil particles, excluding pore space. Most mineral soils have a particle density of about 2.65 grams per cubic centimeter.

Peat. Undecayed or slightly decayed organic soil, formed underwater where low oxygen conditions inhibit decay.

Ped. A natural soil aggregate, such as a crumb or prism.

Pedon. The smallest soil body. A section of soil that extends to the root depth and has about 10 to 90 square feet of surface area.

Perched water table. A zone of saturated soil that because of obstructions or other reasons is maintained above the normal water table. Also applied to the layer of saturated soil in the bottom of a pot of soil, since excess water cannot drain to the normal water table in the soil.

Percolation. Downward movement of water through the soil profile.

Percolation test. A test for soil drainage that involves timing the speed that water will drain out of a hole dug in the ground.

Permanent wilting point. Water content of soil when a plant wilts and does not recover when placed in a humid chamber. Also called the permanent wilting percentage.

Permeability. Ease with which gases, liquids, and plant roots pass through a specific mass of soil.

pH, soil. Measure of the acidity or alkalinity of a soil. Technically, reciprocal of the logarithm of the hydrogen ion concentration in the soil solution.

Phase, soil. A subdivision of the soil series, designating a difference from the series norm. Includes slope, degree of erosion, stoniness, or surface texture.

Photosynthesis. The reaction, in the presence of chlorophyll, of carbon dioxide and water to form carbohydrates, using light energy.

Physical properties. Traits of a soil caused by physical forces that can be described by physical terms or equations. Examples include soil texture, structure, and bulk density.

Physical weathering. Breakdown of rock particles by physical forces like frost action or wind abrasion.

Plasticity. A consistency term describing how easily a mass of soil can be shaped and molded when wet.

Platy structure. Soil structure in which platelike soil peds lie horizontally in the soil. Common in E horizons or compacted A horizons.

Plinthite. A mixture of various chemicals in the soil that when exposed to cycles of wetting and drying permanently hardens to a bricklike state. Common to tropical soils.

Plow pan. A tillage pan formed just under the plow layer.

Polypedon. A group of similar neighboring pedons that makes up a soil series.

Pore space. Portion of soil not occupied by solid material but which is filled with air or water.

Porosity. Percentage of soil volume not occupied by solid material.

Prairie soil. Soil formed under grassland vegetation.

Precipitation. A form of water falling to earth from the atmosphere; may be rain, snow, mist, or hail.

Predator. An organism that hunts and eats prey.

Primary consumer. Organism that feeds on plants, the second level of the food chain.

Primary nutrients. The three macronutrients (nitrogen, phosphorus, potassium) needing to be added to the soil in the greatest amounts for good crop growth.

Primary producer. Lowest level in the food chain, organisms that produce their own food. Mostly chlorophyllous plants and microorganisms.

Primary tillage. The first step in seedbed preparation in conventional and mulch tillage. Breaks up and loosens soil and buries some or all crop residue.

Prismatic soil structure. Soil structure with large, prism-shaped peds arranged vertically in the soil, usually in the B horizon.

Profile, soil. The vertical section of a soil through all its horizons, ending in the parent material.

Protein. Important nitrogen containing compounds in living matter, from which comes most of the nitrogen in organic matter.

Puddling. Dispersal of soil aggregates, caused by working soil when wet, creating a massive surface layer.

Rangeland. Land in arid or semiarid areas with permanent plant cover used for grazing.

Reaction, soil. The degree of acidity or alkalinity of a soil, expressed as pH.

Reduced tillage. *See minimum tillage.*

Relief. Variations in elevation in the landscape.

Residual soil. Soil formed in place from bedrock, rather than from transported parent materials.

Respiration. Biological reaction in which carbohydrates are broken down to carbon dioxide and water with the release of energy. Opposite of photosynthesis.

Rhizobium. Genus of bacteria that live symbiotically on legume roots and can fix atmospheric nitrogen.

Rhizosphere. A zone of soil immediately surrounding plant roots that supports a high population of microorganisms feeding on organic materials released by the root.

Rill. A channel in the soil caused by water erosion that is small enough to be erased by tillage.

Rip-rap. Large stones laid on top of the soil to protect it from erosion.

Runoff. Water that falls on the soil but fails to be absorbed; flows on the soil surface.

Salination. The accumulation of soluble salts in the soil, usually from irrigation.

Saline seep. Small area of saline soil resulting from summer fallow.

Saline soil. A soil high in soluble salts but without too much exchangeable sodium. The pH is between 7.0 and 8.5; electrical conductivity is greater than 4.0 millimhos per centimeter; exchangeable sodium percentage is below 15.0.

Saline-sodic soil. Soil has both high soluble salt and sodium levels. The pH is between 7.0 and 8.5; electrical conductivity is greater than 4.0 mmhos/cm; exchangeable sodium percentage is greater than 15.0.

Saltation. Movement of soil particles in which wind causes fine sand to hop along soil surface.

Sand. (1) Largest of the soil separates, between 0.05 and 2.00 millimeters in diameter. (2) Coarsest textural class.

Saprophyte. Microorganism that feeds on dead organic matter.

Saturation. (1) All or most soil pores filled with water. (2) The amount of the cation exchange capacity filled with a certain cation.

Secondary consumer. An organism that feeds on a primary consumer; the third level of a food chain.

Secondary nutrients. Those macronutrients used less often as fertilizers than the primary elements. Includes calcium, magnesium, and sulfur.

Secondary tillage. Tillage operation following primary tillage that smooths out the soil for a fine seedbed.

Sediment. A layer of loose material deposited by wind or water. Term often applied to eroded soil deposited off the field, such as in lakes or streams.

Sedimentary rock. Rock made of sediments hardened over time by chemicals or pressure.

Semi-arid climate. Area where rainfall is low enough to limit crop production, but where dryland farming can be used. Evapotranspiration slightly exceeds precipitation.

Series, soil. The basic unit of soil classification, consisting of soils alike in all profile traits except for slight differences in surface texture, slope, erosion, or stoniness. *See phase.*

Sesquioxide clays. Finely divided particles of iron and aluminum oxides, most common in soils of warm, humid regions.

Sewage sludge. Semisolid wastes, collected by sedimentation from sewage, that, when properly treated, can be used as an organic fertilizer.

Sheet erosion. Form of erosion in which soil washes off the land in a thin, uniform layer.

Shifting cultivation. Cropping system common to tropical forest soils. A plot of forest is cut and burned, crops are grown for a few years, then the land is allowed to return to forest to fallow for several years. Also called slash-and-burn agriculture.

Shrink-swell potential. How much a mass of soil swells when wet and shrinks when dry; a function of the amount of swelling clays. An important engineering property of soil.

Silt. Medium-sized soil separate, particles between 0.05 and 0.002 millimeter in diameter.

Single-grained soil. A structureless soil in which each soil grain is loose in the soil. Usually in sand.

Slick spots. Small areas of a field that are slick when wet because of high sodium content.

Slow release. Term applied to fertilizers that, by various means, release their nutrients slowly. Such fertilizers may be only slowly soluble in water or be coated with substances that hinder solution.

SMP buffer test. Method of measuring lime requirement for an acid soil.

Sodic soil. Soil high in sodium and low in soluble salts. The pH is between 8.5 and 10.0; electrical conductivity is below 4.0 mmhos/cm; exchangeable sodium percentage is above 15.0.

Soil. Loose mineral and organic material on the earth's surface that serves as a medium for the growth of land plants.

Soil air. Gas phase of soil; space of soil not filled with solid or liquid.

Soil amendment. Any material added to soil, including organic materials such as compost or inorganic materials such as gypsum, to improve plant growth. While many amendments include nutrients, the term excludes those materials added primarily as fertilizers.

Soil classification. The arrangement of soils into classes of several levels.

Soil conservation. Protection of soil from erosion.

Soil fertility. Ability of a soil to supply the elements needed for plant growth.

Soil loss tolerance. The average maximum yearly loss of soil in tons per acre that will not lead to loss of productivity over time. This amount, usually between 1 and 5, is characteristic of a soil series.

Soil pitting. Creating many small pits on sloped rangeland to retain water, reduce erosion and improve forage growth.

Soil reaction. Degree of acidity and alkalinity of a soil, expressed as a pH value.

Soil sampling. The systematic gathering of soil samples for use in soil testing.

Soil separate. Classes of mineral particles less than 2.0 millimeters in diameter, includes clay, silt, and several sizes of sand.

Soil solution. The liquid phase of soil, consisting of water and dissolved ions.

Soil survey. The examination, description, and mapping of soils of an area according to the soil classification system.

Soil testing. Using various tests to measure properties that affect how well soil will support plant growth.

Soil-water potential. The amount of work a small amount of water can do in moving from the soil to a pool of pure, free water at the same location. Includes matric, gravitational, pressure, and salt potential. Normally, matric potential is the main component, but salt potential is significant in salted soils.

Soluble salts. Salts in the soil that are more soluble than gypsum.

Solum. The upper, weathered part of the soil profile; the O, A, E, and B horizons.

Solution. Water with ions dissolved in it. Applied to soil, the soil solution is the water in the soil with nutrients and other materials dissolved in it.

Splash erosion. The movement of soil particles by splashing from the impact of a drop of water.

Spodosol. One of ten soil orders. Mainly acid, coarse-textured forest soils of cool, humid regions. Has subsoil layer with illuvial accumulation of humus, aluminum, and sesquioxides.

Stomata. Small openings in plant leaves through which oxygen, carbon dioxide, and water vapor are exchanged.

Strip-cropping. Planting different types of crops in alternate strips to prevent wind or water erosion. Strips are usually planted on a slope contour or across the direction of the prevailing wind.

Structure, soil. The arrangement of soil particles into aggregates, or peds.

Structureless soil. Soil with no visible aggregates. See *massive soil and single-grain soil.*

Stubble-mulching. Practice of leaving crop stubble stand between the time of harvest of one crop and beginning stages of growth of the following crop.

Subirrigation. Method of irrigation in which capillary rise from a saturated zone in the soil carries water to plant roots.

Subsidence. The lowering of a surface. Organic soils subside because organic particles are lost to decay or wind erosion.

Subsoil. Soil below the plow layer, generally the B horizon.

Subsoiling. Breaking up compact subsoils or pans by the use of a chisel or other tool.

Subsurface drainage. A method of artificially draining wet soils by burying a system of tiles or perforated pipes that carry excess subsurface water off the field.

Subsurface tillage. Tillage with special tools that are drawn beneath the soil surface to kill weeds and loosen soil without mixing crop residues into the soil for protection against wind erosion.

Surface creep. Movement of sand particles along soil surface by being rolled in the wind.

Surface drainage. A method of artificially draining wet soils by digging a system of ditches that collect water and carry it off the field.

Surface soil. The top few inches of soil, usually mixed during tillage. Often the same as the plow layer.

Surface water. Water in natural or man-made bodies of water on the earth's surface, such as lakes or reservoirs.

Suspension. (1) A system of tiny particles hanging in a liquid or gas. (2) Movement of clay or silt particles in the wind by being suspended in the air.

Symbiosis. Two organisms living in close association to the other's benefit. *Compare with parasitism.*

Symbiotic nitrogen fixation. Fixation of nitrogen by microbes living in symbiosis with plant roots, usually by *Rhizobium* bacteria and several species of actinomycetes.

Talus. Deposits of dry rock and soil that have slid to the base of a slope under the force of gravity.

Taxonomy, soil. System of classifying soils to show their relationships.

Temporary wilting point. Point of plant water content at which plant wilts but at which plant will recover if watered, cooled, or placed in humid air.

Terracing. Construction of a raised or level strip of earth on a slope to control runoff and erosion.

Texture. The relative proportion of the soil separates in a soil.

Tillage. Mechanically working the soil to change soil conditions for crop growth or to kill weeds.

Tilth. Physical condition of the soil in terms of how easily it can be tilled, how good a seedbed can be made, and how easily seedling shoots and roots can penetrate.

Topography. *See relief.*

Topsoil. The A horizon; or soil stirred during tillage.

Trace elements. See micronutrients.

Trickle irrigation. Irrigation method in which water drips out of a specially designed trickler into the soil surface under a plant.

Type, soil. An old term, meaning the same as a textural phase of a soil series.

Universal Soil Loss Equation. Equation developed to predict average soil losses from a soil due to sheet and rill erosion.

Ultisol. One of ten soil orders. Leached soils of warm climates.

Urban land. Areas altered by urban activities and structures, so that the soil is difficult to identify.

Value. One of three variables of the Munsell color system. Refers to the lightness of intensity of color.

Vertisol. One of ten soil orders. Soils high in swelling clays that have deep wide cracks when dry. Surface soil falls into the cracks, causing soil mixing.

Volatilization. Evaporation. Also applies to loss of nitrogen in ammonium fertilizers and urea as soil reactions convert it to ammonia gas which is lost to the atmosphere.

Virgin soil. Soil that has not been disturbed by humans.

Water table. Upper surface of a layer of saturated material in the soil.

Water table, perched. The water table of a layer of saturated soil that is separated from the main water table lower in the ground. Held in place by a layer of rock, compacted soil, or, in potted soils, by the bottom of a pot.

Waterlogged soil. Soil whose pores are filled with water and so are low in oxygen. Caused by high water tables, poor drainage, or excess moisture from rain, irrigation, or flooding.

Watershed. The total land area in which runoff water flows into the same stream.

Water-use efficiency. Crop production per unit of water reaching the land the crop occupies.

Weathering. Natural process that breaks down rock into parent materials.

Index